DATE DUE		
DEC 04 2001			
OCT _ 3 2002			
MAR 26 2007			

PRAISE FOR *REGENERATING AGRICULTURE*

I spent happy and reflective moments reading this book on sustainable agriculture
Ambassador Robert Blake (Committee on Agricultural Sustainability for Developing Countries, Washington)

Informative, comprehensive and full of new empirical evidence on the benefits of sustainable agriculture
Dr Michael Cernea (The World Bank, Washington)

This is the book we have so badly needed... It should be read and acted on by all concerned with agricultural policy, research and extension
Dr Robert Chambers (Institute of Development Studies)

This book is not only about agriculture; it is about people – farmers, extension workers and researchers, and their successes so far in creating a sustainable agriculture for the 21st century
Professor Gordon Conway (Vice-Chancellor, University of Sussex)

An excellent book with clear analysis of the components of sustainable agriculture, illustrated by relevant example
Dr Oliver Doubleday (farmer, UK and Brazil)

Congratulations on what will be an important and needed contribution
Sam Fujisaka (CIAT, Colombia)

Fascinating... an exciting book
Dr Koyu Furusawa (Mejiro Gakuen Schools, Tokyo, Japan)

This book makes a substantial contribution to the essential dialogue on the urgent need for more sustainable natural resource management systems
Dr David Gibbon (University of East Anglia, UK)

Wonderful, extremely well-written. It blends analysis, argument and example very well. This book will challenge you
Ed Mayo (New Economics Foundation, London)

I will recommend it to my students
Dr John Mumford (Imperial College of Science and Technology, London)

The next agricultural revolution will be born out of hope and not despair – and that is what this book is all about. Through its focus on what makes regenerative agricultures work and spread, this book will inform the practices and struggles for sustainability and self-reliance everywhere. An exciting contribution to the debate on sustainable agriculture
Dr Michel Pimbert (WWF International, Geneva)

Authoritative, interesting, well-researched and amazingly well-referenced
Hugh Raven (SAFE Alliance, London)

This book makes a major contribution towards an emerging new development paradigm, as it mounts a severe challenge to the way development theorists and practitioners currently think, behave and act
Chris Roche (Oxfam, UK)

Highly readable, pragmatic and convincing
Professor Niels Röling (Wageningen Agricultural University, The Netherlands)

Compelling evidence about the impact participatory approaches and institutions can have on regenerating agriculture. A must for both practitioners and policy makers
Parmesh Shah (New Delhi and Sussex)

Regenerating Agriculture offers a wide range of readers an opportunity to see where practice and experience are often ahead of theory. It challenges academics to accelerate this process of discovering – and rediscovering ways to feed our present and future generations without degrading the resource base
Professor Norman Uphoff (Director CIFAD, Cornell University, USA)

A very suitable book for a variety of university courses. This is yet another magnificent effort to fill a major gap in the literature that will be important to both academics and development professionals
Professor Mike Warren (Iowa State University, Iowa)

REGENERATING AGRICULTURE

REGENERATING AGRICULTURE

Policies and Practice for
Sustainability and Self-Reliance

Jules N. Pretty

JOSEPH HENRY PRESS
Washington, D.C. 1995

JOSEPH HENRY PRESS
2101 Constitution Avenue, N.W. • Washington, D.C. 20418

The Joseph Henry Press is an imprint of the National Academy Press, publisher for the National Academy of Sciences, the National Academy of Engineering, the Institute of Medicine, and the National Research Council. Any opinions, findings, conclusions, or recommendations expressed in this volume are those of the author and do not necessarily reflect the views of the National Academy of Sciences or its associated institutions.

Library of Congress Cataloging-in-Publication Data

Pretty, Jules N.
Regenerating agriculture : policies and practice for sustainability and self-reliance / Jules N. Pretty.
p. cm.
Includes bibliographical references (p.) and index.
ISBN 0-309-05248-3. — ISBN 0-309-05246-7 (pbk.)
1. Sustainable agriculture. 2. Sustainable agriculture—Government policy. I. Title.
S494.5.S86P74 1995 95-5259
338.1—dc20 CIP

Printed in Great Britain

For

Gill
Freya and Theo

CONTENTS

ABBREVIATIONS

AKRSP Aga Khan Rural Support Programme
BPH brown planthopper
Bt *Bacillus thuringiensis*
CAP Common Agricultural Policy (of the EC)
CIAT Centro Internacional de Agricultura Tropical
CIMMYT Centro Internacional de Mejorimento de Maíz y Trigo
CIP Centro Internacional de la Papa
CGIAR Consultative Group on International Agricultural Research
CMB cassava mealybug
CPRs common property resources
CSE consumer subsidy equivalent
EC European Commission (the policy-making institution of the European Union along with the European Parliament)
EVs extension volunteers
EPAGRI Empresa de Pesquisa Agropecuária e Difusâo de Technologia de Santa Catarina, Brazil
EU European Union (formerly the European Community)
FAO Food and Agriculture Organization of the United Nations
FARMI Farm and Resource Management Institute, Philippines
FFS farmer field school
GATT General Agreement on Tariffs and Trade
GDP gross domestic product
GNP gross national product
GPS global positioning system
ICLARM International Center for Living Aquatic Resource Management
ICRISAT International Crops Research Institute for the Semi-Arid Tropics
IFAD International Fund for Agricultural Development
IFAP International Federation for Agricultural Producers
IIED International Institute for Environment and Development
IITA International Institute for Tropical Agriculture
IPM integrated pest management
IRRI International Rice Research Institute
MAFF Ministry of Agriculture, Fisheries and Food, Britain
MALDM Ministry of Agriculture, Livestock Development and Marketing, Kenya
MVs modern varieties

NARSs	national agricultural research systems
NCS	national conservation strategy
NGO	non-government organisation
NIA	National Irrigation Administration, Philippines
NPK	nitrogen, phosphorus, potassium fertilizer
NRC	National Research Council
NRSP	National Rural Support Programme, Pakistan
NSDS	national sustainable development strategy
OECD	Organization for Economic Cooperation and Development
PAF	Project Agro-Forestier
PATECORE	Projet d'Aménagement de Terroirs et Conservation de Resources
PIDOW	Participative and Integrated Development of Watersheds
PRA	participatory rural appraisal
PSE	producer subsidy equivalent
PTD	participatory technology development
SALT	sloping agricultural land technology
SIDA	Swedish International Development Authority
SCS	Soil Conservation Service (of the US Government)
SPEECH	Society for People's Education and Economic Change
SWCB	Soil and Water Conservation Branch, Kenya
TAC	Technical Assistance Committee (for the CGIAR)
T and V	training and visit (system of extension)
TNAU	Tamil Nadu Agricultural University
UCIRI	Union of Indian Communities in the Isthmus Region, Mexico
UNDP	United Nations Development Programme
UNEP	United Nations Environment Programme
UoN	University of Nairobi
USAID	US Agency for International Development
USDA	US Department of Agriculture
VEs	village extensionists
WHO	World Health Organization of the United Nations
WV	World Vision

A NOTE ON TERMINOLOGY

There is no satisfactory set of terms for describing the broad differences between countries. Where the distinctions were once clearer, so now the boundaries are blurred. In this book, the poorer countries of the world are described as 'Third World countries' or as 'countries of the South'. These are sometimes called 'developing countries' elsewhere. These terms encompass a great variety of economies, ranging from some with GNP per capita of below US$250 to those with over US$2000. The richer countries are called 'industrialized countries', which are broadly defined as those belonging to the OECD.

ACKNOWLEDGEMENTS

I am very grateful to many colleagues and friends who gave critical comment and constructive suggestions on sections of this book or earlier related papers. They are Charles Antholt, Richard Bawden, Josh Bishop, Bob Blake, Roland Bunch, Andrew Campbell, Michael Cernea, Robert Chambers, Gordon Conway, John Devavaram, Barbara Dinham, Oliver Doubleday, Paul Ellis, Sam Fujisaka, Koyu Furusawa, David Gibbon, Irene Guijt, Arturo Gómez-Pompa, Rupert Howes, Meera Kaul Shah, John Kerr, J.K. Kiara, Sam Joseph, Charles Lane, Ed Mayo, Catrin Meir, Neela Mukherjee, John Mumford, Michel Pimbert, Hugh Raven, Niels Röling, Chris Roche, Ian Scoones, Parmesh Shah, Devika Tamang, John Thompson, Koy Thomson, Lori Ann Thrupp, Norman Uphoff, Mike Warren, Yunita Winarto and Jim Woodhill.

Many others too numerous to name have contributed at one time or another and I am grateful for their insights. My colleagues in the Sustainable Agriculture Programme at IIED, Irene Guijt, Ian Scoones and John Thompson, together with Richard Sandbrook, deserve a special mention, as many of the concepts, values and methods have been developed together over many years. I am also grateful for administrative support from Deviani Vyas, Ginni Tym and Fiona Hinchcliffe.

I would also like to thank close members of my family for valuable support and forbearance: Gill, Freya and Theo; Pat and Bryan; and my mother and father, Susan and John.

The views and opinions expressed in this book, together with any errors or omissions, are naturally my own responsibility.

This book has evolved out of work at the International Institute for Environment and Development commissioned and funded by the Swedish International Development Authority on policies for sustainable agriculture. It has also benefited from support to IIED from WWF International for analysis of people and participation issues in protected areas; and from SIDA, the Swiss Development Cooperation (SDC), the German Technical Agency (GTZ), the Ford Foundation, and the British Overseas Development Authority for support for a collaborative research project investigating the impact of participatory watershed management and soil and water conservation. The views expressed in this book should not be construed as representing any of these institutions.

1

SUSTAINABLE AGRICULTURE

One doesn't discover new lands without consenting to lose sight of the shore for a very long time.

André Gide, 1925, Les Faux-Monnayeurs

A VISION FOR AGRICULTURE

This book is about a vision of what can be achieved to make agriculture productive, environmentally sensitive and capable of preserving of the social fabric of rural communities. It is a book about the skills and ingenuity of local people and communities; about innovative agriculturalists who have sought to create an alternative agriculture; about the struggle for success in the face of huge odds.

There is now strong evidence that regenerative and resource-conserving technologies and practices can bring both environmental and economic benefits for farmers, communities and nations. The best evidence comes from the countries of Africa, Asia and Latin America, where the emerging concern is to increase food production in the areas where farming has been largely untouched by the modern packages of externally supplied technologies, such as pesticides, fertilizers, machinery, and modern crops and livestock. In these complex and remote lands, some farmers and communities adopting regenerative technologies have substantially improved agricultural yields, often only using few or no external inputs.

But these are not the only sites for successful sustainable agriculture. In the high input and generally irrigated lands, farmers adopting regenerative technologies have maintained yields while substantially reducing their use of inputs. And in the very high input lands of the industrialized countries, some farmers have maintained profitability, even though input use has been cut dramatically and yields have fallen. These improvements have occurred in initiatives focusing on a wide range of technologies, including pest and predator management, nutrient conservation, soil and water conservation, land rehabilitation, green manuring, water management and many others.

Such sustainable intensification emphasizing internal or available resources has been accompanied by indirect social and economic benefits. There is less need for expansion into non-agricultural areas, so ensuring that valuable wild plant and animal species are not lost. There is reduced contamination and pollution of the environment, so reducing the costs incurred by farming households, consumers of food and national economies as a whole. There is less likelihood of the breakdown of rural culture. There is local regeneration, often with the reversal of migration patterns as the demand for labour grows within communities. And, psychologically, there is a greater sense of hopefulness towards the future.

The Scale of the Challenge

As yet, the benefits of sustainable agriculture have been experienced by perhaps only a few thousand communities worldwide (see Chapter 7). They are potentially the pioneers for a new age in agriculture, pointing the way for the rest of the world. The scale of the challenge is huge and differs for different types of agriculture.

In the industrialized countries of the OECD and eastern Europe, some 1.2 billion people are supported by agriculture relying heavily on external inputs. In these regions, agriculture is highly productive, but also too often degrades and damages natural resources. An alternative and sustainable agriculture is unlikely to match current yield levels, but its reduced use of inputs lowers costs which means it can be financially viable. A subset of this type of industrialized agriculture, similar to the Green Revolution lands in terms of yield levels, occurs in the former countries of the eastern bloc. There is considerable reliance on modern technologies, but yields are relatively poor compared with countries of the OECD.

In the countries of the Third World, some 2.3–2.6 billion people are supported by agricultural systems characterized by modern technologies brought by the Green Revolution. The Green Revolution was characterized by the advent of new, high-yielding cereal varieties which, when cultivated with modern fertilizers and pesticides, transformed many agricultural systems. These systems have good soils and reliable water, and are close to roads, markets and input supplies. The area of these lands is some 215 million ha, and they currently produce 60 per cent of the grain in Third World countries. Alternative sustainable systems in these regions have been able to match their yields and profitability.

After these types of agriculture, this leaves some 1.9–2.2 billion people largely untouched by modern technology (based on estimates from FAO and World Bank data). They tend to be in the poorer countries with little foreign exchange to buy external inputs. Their agricultural systems are complex and diverse, and are in the humid and semi-humid lowlands, the hills and mountains, and the drylands of uncertain rainfall. They are remote from services and roads, and they commonly produce one-fifth to one-tenth as much food per hectare as farms in the industrialized and Green Revolution lands. It is in these regions that sustainable agriculture has had the greatest impact on local food production so far, with yields doubling to trebling with little or no use of external inputs.

The emerging evidence now shows that regenerative agriculture is possible and can have wider benefits. But this does not in itself indicate how it may be adopted by farmers worldwide. It suggests there can be many winners, but it is not clear who will be the losers in the short term. All successes have had three elements in common and there is much to be learnt from these. First, all have made use of locally adapted resource-conserving technologies. Second, in all there has been coordinated action by groups or communities at the local level. Third, there have been supportive external (or non-local) government and/or non-government institutions working in partnership with farmers.

Almost every one of the successes has been achieved despite existing policy environments. Most policy frameworks still strongly favour 'modern' approaches to agricultural development and at the same time discriminate against sustainability. They also tend to have an anti-poor and pro-urban bias. When policies are reshaped so as to support a more sustainable agriculture, and all three local conditions are present, then sustainable agriculture will be set to spread widely.

The Record of Modernized Agriculture

The impact of modern agriculture has been remarkable. About half of the rice, wheat and maize areas in Third World countries are planted to modern varieties, and fertilizer and pesticide consumption has grown rapidly. Nitrogen consumption, for example, increased from 2 to 75 million tonnes in the last 45 years, and pesticide consumption in many individual countries has increased by 10–30 per cent during the 1980s alone. Farmers have intensified their use of external resources and expanded into previously uncultivated lands. As a result, food production per capita has, since the mid-1960s, risen by 7 per cent for the world as a whole, with the greatest increases in Asia, where per capita food production has grown by about 40 per cent (FAO, passim).

Between 70–90 per cent of the recent increases in production have been due to increased yields rather than greater area under agriculture (World Bank, 1993). This has been described by Donald Plunkett (1993), scientific adviser to the CGIAR, as *'the greatest agricultural transformation in the history of humankind, and most of it has taken place during our lifetime. The change was brought about by the rise of science-based agriculture which permitted higher and more stable food production, ensuring food stability and security for a constantly growing world population'*.

A major problem is that these benefits have been poorly distributed. Many people have missed out and hunger still persists in many parts of the world. In Africa, for example, food production per capita fell by 20 per cent between 1964–1992. Estimates by the FAO and WHO (1992) and the Hunger Project (1991) suggest that around 1 billion people in the world have diets that are *'energy insufficient for work'*, of whom 480 million live in households *'too poor to obtain the energy required for healthy growth of children and minimal activity of adults'* (The Hunger Project, 1991). The causes are complex and it is not entirely the fault of Green Revolution technology. These have had an undoubted positive impact on the overall availability of food. None the less, the process of agricultural modernization has been an important contributing factor, in that the technologies have been more readily available to the better-off.

Modern agriculture begins on the research station, where researchers have access to all the necessary inputs of fertilizers, pesticides and labour at all the appropriate times. But when the package is extended to farmers, even the best performing farms cannot match the yields the researchers get. For high productivity per hectare, farmers need access to the whole package: modern seeds, water, labour, capital or credit, fertilizers and pesticides. Many poorer farming households simply cannot adopt the whole package. If one element is missing, the seed delivery system fails or the fertilizer arrives late, or there is insufficient irrigation water, then yields may not be much better than those for traditional varieties. Even if farmers want to use external resources, very often delivery systems are unable to supply them on time.

Where production has been improved through these modern technologies, all too often there have been adverse environmental and social impacts (see Chapter 3). Many environmental problems have increased dramatically in recent years. These include:

- contamination of water by pesticides, nitrates, soil and livestock wastes, causing harm to wildlife, disruption of ecosystems and possible health problems in drinking water;
- contamination of food and fodder by residues of pesticides, nitrates and antibiotics;
- damage to farm and natural resources by pesticides, causing harm to farmworkers and public, disruption of ecosystems and harm to wildlife;
- contamination of the atmosphere by ammonia, nitrous oxide, methane and the products of burning, which play a role in ozone depletion, global warming and atmospheric pollution;
- overuse of natural resources, causing depletion of groundwater, and loss of wild foods and habitats, and of their capacity to absorb wastes, causing waterlogging and increased salinity;
- the tendency in agriculture to standardize and specialize by focusing on modern varieties, causing the displacement of traditional varieties and breeds;
- new health hazards for workers in the agrochemical and food-processing industries.

Agricultural modernization has also helped to transform many rural communities, both in the Third World and industrialized countries. The process has had many impacts. These include the loss of jobs, the further disadvantaging of women economically if they do not have access to the use and benefits of the new technology, the increasing specialization of livelihoods, the growing gap between the well-off and the poor, and the cooption of village institutions by the state.

Modernist Perspectives

Despite these problems, many scientists and policy makers still argue vigorously that modern agriculture, characterized by externally developed packages of technologies that rely on externally produced inputs, is the best, and so only, path for agricultural development. Influential international institutions, such as the World Bank, the FAO and some institutions of the Consultative Group on International Agricultural Research, have long suggested that the most certain way to feed the world is by continuing the modernization of agriculture through the increased use of modern varieties of crops and breeds of livestock, fertilizers, pesticides and machinery. Remarkably, these international institutions often appear unaware at policy level of what can be achieved by a more sustainable agriculture. However, there are some signs of change, mostly limited to small groups of individuals, plus some small modifications in policy.

Box 1.1 contains examples of the types of perspectives brought to debates on the future of agricultural development that support modernist, high-external input themes. All make the point that the route to food security is through modern agriculture and external inputs. Some, such as the Nobel Laureate, Norman Borlaug, do so by vigorously putting down all alternatives, suggesting that unrealistic claims are made on behalf of alternative or sustainable agriculture. Others predict that if widespread starvation is to be averted, then *'soil fertility management based on anything other than increased chemical fertilizer would lead to massive increases in food imports'* (Vlek, 1990). This is echoed by Yudelman (1993), who more recently said *'there is no way short of unforeseen technological breakthroughs [that] raising yields in the future can be achieved without substantial increase in agrochemicals'*. The FAO has estimated that over 50 per cent of future gains in food crop yields will have to come from fertilizers. This calls for massive increases in fertilizer consumption by poor countries.

At the same time, traditional agriculture is presented as environmentally destructive, so needing to be modernized; or as efficiently managed systems which have hit a yield ceiling, so again needing modern technologies. Even where there have been recent shifts in emphasis, both in rhetoric and substantive policy, the Green Revolution model tends to be widely believed to be the *'only way to create productive employment and alleviate poverty'* (World Bank, 1993).

Clearly these quotes do not necessarily represent the exact position of these individuals or institutions. People and institutions change and adapt, and there are distinct signs of change illustrated in the box. These quotes do, however, show how value-laden is the debate. It is also true that the modernist perspectives are not the only ones that are value-laden. Those who promote an alternative or sustainable agriculture are equally value-laden. This book, for example, is explicitly about how a sustainable agriculture might be achieved, and how such a vision is blocked by current thinking and practices. The intention is to demonstrate its worth and value. This is not hidden.

Box 1.1 Modernist perspectives on future strategies for agricultural development

Technical Assistance Committee of the CGIAR (1988):
'Indigenous farm populations have learned to manage their systems quite efficiently, making it difficult to increase their production without resort to external inputs'.

FAO (1991):
'It seems likely that much of the growth in agricultural production will take place through increased use of external inputs'.

Norman Borlaug (1992):
'Some agricultural professionals contend that small-scale subsistence producers can be lifted out of poverty without the use of purchased inputs, such as modern crop varieties, fertilizer and agricultural chemicals. They recommend instead the adoption of so-called 'sustainable' technologies that do not require fertilizers and improved varieties... The advent of cheap and plentiful fertilizers has been one of the great agricultural breakthroughs of humankind... The adoption of science-based agricultural technologies is crucial to slowing – and even reversing – Africa's environmental meltdown'.

Norman Borlaug (1992):
'Development specialists... must stop 'romanticizing' the virtues of traditional agriculture in the Third World. Moreover, leaders in developing countries must not be duped into believing that future food requirements can be met through continuing reliance on... the new complicated and sophisticated 'low-input, low-output' technologies that are impractical for the farmers to adopt'.

International Fertilizer Development Center (1992):
'Higher yields per hectare produced by fertilizers will be the most persuasive argument to coax developing country farmers to abandon their environmentally destructive farming practices'.

The World Bank (1993):
'Experience such as obtained with the Green Revolution in parts of Asia has shown that broad-based agricultural growth, involving small and medium sized farms and driven by productivity-enhancing technological change, offers the only way to create productive employment and alleviate poverty on the scale required'.

FAO (1993a):
'When managed well, external inputs can lead to greater yields and improved nutrient content. This reduces the pressure to convert land to agriculture and improves food security. Few developing countries can therefore afford to forgo the benefits of external inputs. Used incorrectly, however, they can result in environmental pollution, threats to human and animal health, greater volatility in production levels, and reduced production and incomes'.

Stagnating Capacity in Modern Systems

High input agricultural systems will clearly continue to be immensely important for many farmers and economies. However, there is great uncertainty about the potential for significant further improvements in cereal yields and their sustainability. All countries where the Green Revolution has had a significant impact have seen average annual growth rates in the agricultural sector fall during the 1980s, compared with the post-revolution period of 1965–1980 (World Bank, 1994a).

Many recent agriculture projects have performed poorly too. The World Bank's own Operations Evaluation Department rated only half of their agricultural projects completed in 1989–1991 as satisfactory, compared with 67 per cent in the 1974–1988 period (World Bank, 1993). In addition, the *'percentage of satisfactory agricultural projects averages 10–20 per cent points below satisfactory ratings obtained in other sectors'*. This has encouraged donors to turn away from agriculture: the World Bank's spending on agriculture fell from 30 per cent of budget in the early 1980s (US $5.4 billion/year) to just 20 per cent in the early 1990s (US $3.9 billion/year). The US Agency for International Development reduced its support to agriculture in the Third World by 50 per cent in real terms between 1988–1994. None the less, less spending might be better, particularly if it is combined with a more careful look at the reasons for poor past performance.

There is also growing evidence that cereal yields in the high-external input areas cannot be sustained at their current levels. Consecutively monocropped modern cereals appear to be unable to maintain their initial yield levels, whether fertilized or not. At IRRI and other research stations in the Philippines, the yields of the highest yielding entry rice variety in long-term fertility trials fell steadily between 1966 and 1988 (Pingali, 1991; Flinn and De Datta; 1984; Flinn et al, 1981; Ponnamperuma, 1979). Similar yield declines have been detected in other experimental stations in India, Thailand and Indonesia (Pingali, 1991). In India, cereal yields have also declined over 16-year periods of annual cropping on some research stations. At Barrackpore, for example, wheat yields have declined from 4.4 to 3.3 t/ha and at Patnagar, rice has fallen from 6.4 to 5.2 t/ha (Gaur and Verma, 1991). These declines have only been reversed by increasing fertilizer applications by an extra 50 per cent.

Why this is happening is unclear, though pests, diseases, chemical toxicity, changing soil carbon–nitrogen ratios and chemical deficiencies are all possible explanations. The result, though, of this changed micro-environment for cereal production has been the need for increased input requirements for sustaining current yields, with declining profitability of monoculture systems. As Peter Kenmore (1991) has put it: *'the degradation of the paddy environment, whether by micro-nutrient depletion, atmospheric pollution, pest pressure or accumulative toxic change in soil chemistry, is greater than the capacity for genetic improvements in yield potential that breeders can select'*.

None the less, it is still possible that new technologies, such as biotechnology and genetic engineering, will open up new frontiers. Scientists hope that these will produce crops and animals that are more efficient converters of nutrients, with better drought tolerance and pest and disease resistance. One dream has been the incorporation of nitrogen-fixing nodules into the roots of cereals, so making these crops self-sufficient in nitrogen. If such breakthroughs do occur, it will be important that ways are found to ensure their availability to poorer farmers. If they are still part of a package, or rely on hybrid seeds that must be repurchased after every replanting, then they are likely simply to encourage even greater dependency on external resources and systems, and open up gaps between wealthy and poor farmers. Those low-income countries that are currently poorly endowed with natural resources and infrastructure are unlikely to benefit (Hobbelink et al, 1990).

WHAT IS SUSTAINABLE AGRICULTURE?

Contrasting and Confusing Terminology

Many different terms are used to describe agricultural systems. None are entirely satisfactory, because they mean different things to different people. Agricultural systems are sometimes defined with respect to agricultural production possibilities (high or low potential; favourable or marginal); to technological concentration (Green Revolution or complex and diverse); to the readiness to adopt new externally induced or derived technologies (modern or traditional); to the quality of available natural resources (resource-rich or resource-poor); and to use of external inputs (high or low external input).

There are also many terms that are used to describe alternatives to modern agriculture. These include sustainable, alternative, regenerative, low external input, low input sustainable agriculture, balanced inputs sustainable agriculture, resource-conserving, biological, natural, eco-agriculture, agro-ecological, organic, biodynamic and permaculture. These are often presented in opposition to modernized agriculture, which is described by terms such as conventional, resource-degrading, industrialized, intensive or high external input. For the sake of simplicity, the term sustainable is mainly used in this book, though resource-conserving, low input and regenerative are taken to be largely interchangeable. What characterizes them is the greater use of local resources and knowledge.

It is important to clarify the position of organic agriculture at this stage. Organic agriculture eschews the use of all synthetic fertilizers and pesticides. The term 'synthetic' is commonly used to differentiate between naturally occurring substances, such as manure, phosphate rock or sulphur, and compounds that have been manufactured or 'synthesized', such as inorganic nitrogen fertilizer or herbicides. Some argue that the only form of sustainable agriculture is organic and would suggest that any agriculture using external inputs is fundamentally unsustainable.

But there are others who would point to the way that organic practices can also damage the environment. Nitrates may leach from fields under legumes, ammonia can volatilize to the atmosphere from livestock manures, and heavy metals can accumulate in soils from the use of copper-rich Bordeaux mixture. It is also possible to apply relatively small amounts of fertilizers in a way that minimizes leaching or nitrification losses, and to use synthetic pesticides that are safe to humans and do not kill predators. None the less, organic agriculture is generally a form of sustainable agriculture, though not all sustainable agriculture is organic.

Goals for Sustainable Agriculture

During the past 50 years, agricultural development policies have been remarkably successful at emphasizing external inputs as the means to increase food production. This has produced remarkable growth in global consumption of pesticides, inorganic fertilizer, animal feedstuffs, and tractors and other machinery.

These external inputs have, however, substituted for natural control processes and resources, rendering them more vulnerable. Pesticides have replaced biological, cultural and mechanical methods for controlling pests, weeds and diseases; inorganic fertilizers have substituted for livestock manures, composts and nitrogen-fixing crops; information for management decisions comes from input suppliers, researchers and extensionists rather than from local sources; and fossil fuels have substituted for locally generated energy sources (Table 1.1). The specialization of agricultural production and associated decline of the mixed farm has also contributed to this situation. What were once valued internal resources have often now become waste products.

The basic challenge for sustainable agriculture is to make better use of these internal resources. This can be done by minimizing the external inputs used, by regenerating internal resources more effectively or by combinations of both. A sustainable agriculture, therefore, is any system of food or fibre production that systematically pursues the following goals:

- a more thorough incorporation of natural processes such as nutrient cycling, nitrogen fixation and pest–predator relationships into agricultural production processes;
- a reduction in the use of those off-farm, external and non-renewable inputs with the greatest potential to damage the environment or harm the health of farmers and consumers, and a more targeted use of the remaining inputs used with a view to minimizing variable costs;
- a more equitable access to productive resources and opportunities, and progress towards more socially-just forms of agriculture;
- a greater productive use of the biological and genetic potential of plant and animal species;
- a greater productive use of local knowledge and practices, including innovative approaches not yet fully understood by scientists or widely adopted by farmers;
- an increase in self-reliance among farmers and rural people;
- an improvement in the match between cropping patterns and the productive potential and environmental constraints of climate and landscape to ensure long-term sustainability of current production levels; and
- profitable and efficient production with an emphasis on integrated farm management, and the conservation of soil, water, energy and biological resources.

Table 1.1 Internal and external resources for agro-ecosystems

	Internal resources and processes	*External resources and processes*
Sun	Main source of energy	Supplemented by fossil fuels
Water	Mainly rain and small irrigation schemes	Large dams, centralized distribution and deep wells
Nitrogen	Fixed from the air and recycled in soil organic matter	Primarily from inorganic fertilizer
Minerals	Released from soil reserves and recycled	Mined, processed and imported
Weed and pest control	Biological, cultural, mechanical and locally available chemicals	With pesticides and herbicides
Energy	Some generated and collected on farm	Dependence on fossil fuel
Seed	Some produced on farm	All purchased
Management-decisions and information	By farmer and community gathered locally and regularly	Some provided by input suppliers, researchers, extensionists – assumed to be similar across farms
Animals	Integrated on farm	Production at separate locations
Cropping system	Rotations and diversity	Monocropping
Varieties of plants	Thrive with lower fertility and moisture	Need high input levels to thrive
Labour	Labour requirement greater – work done by family living on farm and hired labour	Labour requirement lower – most work done by hired labour and mechanical replacement of manual labour
Capital	Initial source is family and community; any accumulation invested locally	Initial source is external indebtedness or equity; any accumulation leaves community

Sources: adapted from Rodale, 1990, 1983

When these components come together, farming becomes integrated, with resources used more efficiently and effectively. Sustainable agriculture, therefore, strives for the integrated use of a wide range of pest, nutrient, soil and water management technologies. These are integrated at farm level to give a strategy specific to the biophysical and socioeconomic conditions of individual farms. Sustainable agriculture aims for an increased diversity of enterprises within farms, combined with increased linkages and flows between them. By-products or wastes from one component or enterprise become inputs to another. As natural processes increasingly substitute for external inputs, so the impact on the environment is reduced.

Can Sustainable Agriculture be Defined?

Although it is relatively easy to describe goals for a more sustainable agriculture, things become much more problematic when it comes to attempts to define sustainability: *'everyone assumes that agriculture must be sustainable. But we differ in the interpretations of conditions and assumptions under which this can be made to occur'* (Francis and Hildebrand, 1989).

A great deal of effort has gone into trying to define sustainability in absolute terms. Since the Brundtland Commission's definition of sustainable development in 1987, there have been at least 70 more definitions constructed, each different in subtle ways, each emphasizing different values, priorities and goals. The implicit assumption is that it is possible to come up with a single correct definition, and each author presumably regards their effort as the best.

But precise and absolute definitions of sustainability, and therefore of sustainable agriculture, are impossible. Sustainability itself is a complex and contested concept. To some it implies persistence and the capacity of something to continue for a long time. To others, it implies resilience and the ability to bounce back after unexpected difficulties. With regard to the environment, it involves not damaging or degrading natural resources. Others see it as a concept that means developmental activities simply take account of the environment. Economies are sometimes said to be sustainable if they carry on growing at the same rate or only if growth does not reduce the natural resource base.

In any discussions of sustainability, it is important to clarify what is being sustained, for how long, for whose benefit and at whose cost, over what area and measured by what criteria. Answering these questions is difficult, as it means assessing and trading off values and beliefs. Andrew Campbell (1994a) says that *'attempts to define sustainability miss the point that, like beauty, sustainability is in the eye of the beholder ... It is inevitable that assessments of relative sustainability are socially constructed, which is why there are so many definitions'.*

None the less, when specific parameters or criteria are selected, it is possible to say whether certain trends are steady, going up or going down. For example, practices causing soil to erode can be considered to be unsustainable relative to those that conserve soil. Practices that remove the habitats of insect predators or kill them directly are unsustainable compared with those that do not. Planting trees is clearly more sustainable for a community than just cutting them down. Forming a local group as a forum for more effective collective action is likely to be more sustainable than individuals trying to act alone.

At the farm or community level, it is possible for actors to weigh up, trade off and agree on these criteria for measuring trends in sustainability. But as we move to higher levels of the hierarchy, to districts, regions and countries, it becomes increasingly difficult to do this in any meaningful way. It is, therefore, critical that sustainable agriculture does not prescribe a concretely defined set of technologies, practices or policies at these levels. This would only serve to restrict the future options of farmers. As conditions change and as knowledge changes, so must farmers and communities be encouraged, and allowed to change and adapt too. Again, this implies that definitions of sustainability are time-specific and place-specific. As situations and conditions change, so must our constructions of sustainability also change. Sustainable agriculture is, therefore, not a simple model or package to be imposed. It is more a process for learning.

What is important is to ensure that the opportunities exist for wide-ranging debate on the appropriate levels of external and internal resources and processes necessary for a productive, environmentally sensitive and socially acceptable agriculture.

Some Misconceptions About Sustainable Agriculture

In addition to the problems over definitions, there are other misconceptions about sustainable and regenerative agriculture (NRC, 1989; Parr et al, 1990; Pretty et al, 1992). Perhaps the most common characterization is that sustainable agriculture represents a return to some form of low technology, 'backward' or 'traditional' agricultural practices. This is manifestly untrue. Sustainable agriculture does not imply a rejection of conventional practices, but an incorporation of recent innovations that may originate with scientists, farmers or both. It is common for sustainable agriculture farmers to use recently developed equipment and technology, complex rotation patterns, the latest innovations in reduced input strategies, new technologies for animal feeding and housing, and detailed ecological knowledge for pest and predator management.

Another misconception is that sustainable agriculture is incompatible with existing farming methods. For the development of a sustainable agriculture there is a need to move beyond the simplified thinking that pits industrialized agriculture against the organic movement, or the organic movement against all farmers who use external inputs. Sustainable agriculture represents economically and environmentally viable options for all types of farmers, regardless of their farm location, and their skills, knowledge and personal motivation.

It is also commonly believed that low or no external input farming produces low levels of output and so can only be supported by higher levels of subsidies. Such subsidies could be justified in terms of the positive benefits to the environment brought by sustainable farming, which could therefore be valued and paid for. But this may not be necessary. Worldwide, many sustainable agriculture farmers show that their crop yields can be better than or equal to those of their more conventional neighbours. Even if their yields are lower, these may still translate into better net returns as their costs are also lower. Sometimes yields are substantially higher and now offer the opportunities for growth for communities that do not have access to, or cannot afford, external resources (see Box 1.2). Either way, this means that sustainable farming can be compatible with small or large farms and with many different types of technology.

Box 1.2 **A highly productive, traditional system of agriculture: rice terraces in the Philippines**

Some traditional, integrated systems are highly productive. The ancient rice terraces of Bontoc in the Philippines yield some 6 tonnes of rice per hectare without the use of modern varieties, fertilizers and pesticides. This is more than two and a half times the current national average. Farmers maintain these levels of output with high inputs of labour and management. Terraces have to be regularly maintained; the soil puddled; weeds and wild plants collected to keep the terraces clear for fear of rat infestation; soil fertility is maintained with the use of *Azolla*, pig manures composts, and the import of mountain soils; and irrigation water is managed communally. Success is a function of both the resource-conserving technologies used and the organized community action to make best use of them.

Source: Padilla, 1992

SCIENCE AND SUSTAINABILITY

The Tradition of Positivist Science

Although there exist successful applications of sustainable agriculture throughout the world, still relatively few farmers have adopted new technologies and practices. One reason is that sustainable agriculture presents a deeper and more fundamental challenge than many researchers, extensionists and policy makers have yet supposed. Sustainable agriculture needs more than new technologies and practices. It needs agricultural professionals willing and able to learn from farmers; it needs supportive external institutions; it needs local groups and institutions capable of managing resources effectively; and above all it needs agricultural policies that support these features. It also requires that we look closely at the very nature of the way we conceptualize sustainability and how it might be achieved.

Since the early seventeenth century, scientific investigation has come to be dominated by the Cartesian paradigm, commonly called positivism or rationalism. This posits that there exists an objective external reality driven by immutable laws. Science seeks to discover the true nature of this reality, the ultimate aim being to discover, predict and control natural phenomena. Investigators proceed in the belief that they are detached from the world. The process of reductionism involves breaking down components of a complex world into discrete parts, analysing them and then making predictions about the world based on interpretations of these parts. Knowledge about the world is then summarized in the form of universal, or time-free and context-free generalizations or laws. The consequence is that investigation with a high degree of control over the system being studied has become equated with good science. And such science is equated with 'true' knowledge.

It is this positivist approach that has led to the generation of technologies for farmers that have been applied widely and irrespective of context. Where it has been possible to influence and control farmers, either directly or through economic incentives or markets, agricultural systems have been transformed. But where neither have the technologies fitted local systems nor have farmers been controlled, then agricultural modernization has passed rural people by. As indicated above, some 1.9–2.2 billion people now rely on agricultural systems in which cereal yields have remained of the order of 0.5–1.2 tonnes per ha over at least the past 50 years.

What is the Nature of Truth?

The problem is that no scientific method will ever be able to ask all the right questions about how we should manage resources for sustainable agriculture, let alone find the answers. The results are always going to be open to different interpretations. All actors, and particularly those with a direct social or economic involvement and interest, have a different perspective on what is a problem and what constitutes improvement in an agricultural system. As Brian Wynne (1992) has put it: *'the conventional view is that scientific knowledge and method enthusiastically embrace uncertainties and exhaustively pursue them. This is seriously misleading'*.

The critical issue is that data, which appear at first as objective and value-free, are of course constructed within a particular social and professional context. This context affects the outcomes, but is usually ignored when assuming that data can be objective and 'true'. This can have a profound impact on policy and practice in agricultural development.

Michael Stocking (1993) describes just how the values of the investigators affect the end result when it comes to soil erosion data. Since the 1930s, there have been at least 22 erosion studies conducted in the Upper Mahaweli Catchment in Sri Lanka. These have used visual assessments of soil pedestals and root exposure, erosion pins, sediment traps, run-off plots, river and reservoir sediment sampling, and predictive models. Between the highest and lowest estimates of erosion under mid-country tea, there is an extraordinary variation of some 8000 fold, from 0.13 t/ha/yr to 1026 t/ha/yr (Krishnarajah, 1985; El-Swaify et al, 1983; NEDCO, 1984). The highest estimate was in the context of a development agency seeking to show just how serious erosion is in the Third World; the lowest was by a tea research institute seeking to show how safe was their land management. As Stocking put it: *'The researchers were not lying; they were merely selective'*.

Similar differences are described by Thompson et al (1988) over forest issues in Nepal, where scientific estimates of fuelwood consumption vary by a factor of 67 and those of sustainable forest production by a factor of 150. Again, these differences are caused by the values of the actors involved in the analysis. A similar case is described by Jerome Delli Priscoli (1989) with regard to water use and energy needs in the north-west of the USA. One projection showed a steady growth in energy needs to the year 2000; this was conducted by the utility company. Another showed a steadily downward trend; this was conducted by environmental groups. Other projections by consultancy groups were found in the centre.

What does this say about the data? *'Each projection was done in a statistically 'pedigreed' fashion. Each was logical and internally elegant, if not flawless. The point is, once you know the group, you will know the relative position of their projection. The group, organisation or institution embodies a set of values. The values are visions of the way the world ought to be'* (Delli Priscoli, 1989)

Stocking's point about the erosion estimates is that science is not the neat, objective collection of facts about the nature and processes of the surface of the earth: *'The principal danger is whether in the end we delude ourselves and our clients into believing that our results are anything other than the product of our ideology, the audience to whom we play, and the measurement techniques we choose to use'*.

Here is the clear recognition that data are constructed by people with values and human foibles. The challenge Delli Priscoli identifies is not so much whether these differences have to be recognized, but that the competing values need to be mediated so as to produce agreements between these actors with very different agendas. This calls for some form of active participation between the competing actors and institutions in order to explore these different perspectives and values. Indeed, it suggests that the very framework of positivist science, in which it is believed that single, correct truths exist, is itself only a partial picture of the world's complexity.

This has profound implications for normal science. It may not yet have reached the crisis that Thomas Kuhn (1962) suggests is a fundamental element of any shift from one paradigm to another. However, as Michael Thompson and Alex Trisoglio (1993) suggest, *'the science that supplies us with facts that enable us to define the problems is in considerable disarray. The argument is that the science on which we have depended has missed out all the squiggly bits and, unfortunately, it is the squiggly bits that matter'*.

Alternatives and Additions to the Positivist Paradigm

The problem with the positivist paradigm is that its absolutist position appears to exclude other possibilities. Yet the important point about positivism is that it is just one of many ways of describing the world, and what is needed is pluralistic ways of thinking about the world and acting to change it (Kuhn, 1962; Feyerabend, 1975; Vickers, 1981; Reason and Heron, 1986; Habermas, 1987; Giddens, 1987; Maturana and Varela, 1987; Rorty, 1989; Bawden, 1991; Uphoff, 1992a; Wynne, 1992; Röling, 1994). Recent years have seen the emergence of a remarkable number of advances in a wide range of disciplines and fields of investigation. The sources include the so-called 'hard' sciences, such as physics, biology and mathematics, as well as the 'soft' sciences of philosophy and sociology. A summary is listed in Table 1.2.

Despite this wide-ranging list, those arguing for the seriousness and importance of developing alternatives to positivism are still in the minority. Many scientists argue strongly that information is produced by science, and then interpreted and applied by the public and policy makers. It is this process of interpretation that introduces values and confuses certainties.

The advances in alternative paradigms have important implications for how we go about finding out about the world, generating information and so taking action. All hold that *'the truth is ultimately a mirage that cannot be attained because the worlds we know are made by us'* (Eisner, 1990). All suggest that we need to reform the way we think about methodologies for finding out about the world. This should not be surprising, as *'the language of reductionism and positivism does not entertain the very complex and dynamic phenomena associated with the quest for sustainable practices'* (Bawden, 1991). Five principles set out the main differences between these emerging paradigms and positivist science (Pretty, 1994).

Table 1.2 Alternatives and challenges to the positivist paradigm from a range of disciplines

Alternatives and challenges to the positivist paradigm	*Key sources*
Chaos theory and non-linear science,	Gleick, 1987; Prigogine and Stengers, 1984; Gould, 1989
Fractal geometry and mathematics	Family and Vicsek, 1991; Lorenz, 1993
Quantum physics	See many sources, but especially theories of Schrödinger and Heisenberg
Neural networks	Holland et al, 1986
Soft-systems science	Checkland, 1981; Checkland and Scholes, 1990; Röling, 1994
Philosophy of symbiosis	Kurokawa, 1991
Historical sociology	Abrams, 1982
Morphic resonance	Sheldrake, 1988
Complexity theory	Waldrop, 1992; Santa-Fe Institute, passim
Gaia hypothesis	Lovelock, 1979
New and alternative economics	Ekins, 1992; Douthwaite, 1992; Daly and Cobb, 1989; Arthur, 1989
Post-positivism	Phillips, 1990
Critical systems theory	Jackson, 1991; Popkewitz, 1990; Tsoukas, 1992
Constructivist inquiry	Lincoln and Guba, 1985; Denzin, 1989; Guba, 1990; Röling and Jiggins, 1993
Communicative action	Habermas, 1987
Post-modernism	Harvey, 1989
Adaptive management and operationality in turbulence	Holling, 1978; Norgaard, 1989; Mearns, 1991; Roche, 1992; Uphoff, 1992a
Learning organizations and clumsy institutions	Peters, 1987; Handy, 1989; Argyris et al, 1985; Shapiro, 1988; Thompson and Trisoglio, 1993
Social ecology	Bawden, 1991, 1992; Woodhill, 1993

Note: The terms used in this table are not necessarily those that the listed authors or others would themselves use. What constitutes a theory, an alternative or a coherent paradigm is, of course, open to different interpretations.

The first is that any belief that sustainability can be precisely defined is flawed. It is a contested concept and so represents neither a fixed set of practices or technologies, nor a model to describe or impose on the world. The question of defining what we are trying to achieve is part of the problem, as each individual has different values. Sustainable agriculture is, therefore, not so much a specific farming strategy as an approach to learning about the world.

The second is that problems are always open to interpretation. All actors have uniquely different perspectives on what is a problem and what constitutes improvement. As knowledge and understanding is socially constructed, what each of us knows and believes is a function of our own unique contexts and pasts. There is, therefore, no single 'correct' understanding. What we take to be true depends on the framework of knowledge and assumptions we bring with us. Thus it is essential to seek multiple perspectives on a problem situation by ensuring the wide involvement of different actors and groups.

The third is that the resolution of one problem inevitably leads to the production of another 'problem-situation', as problems are endemic. Mao Tse-Tung once observed that *'every solution creates its own problems'*. The reflex of positivist science is to seek to collect large amounts of data before declaring certainty about an issue or problem. As this position is believed to reflect the 'real world', then courses of action can become fixed and actors no longer seek information that might give another interpretation. Yet in a changing world, there will always be uncertainties.

The fourth is that the key feature now becomes the capacity of actors continually to learn about these changing conditions, so that they can act quickly to transform existing activities. They should make uncertainties explicit and encourage rather than obstruct wider public debates about pursuing new paths for agricultural development. The world is open to multiple interpretations, so it is impossible to say which one is true. Different constructed realities can only be related to each other.

The fifth is that systems of learning and interaction are needed to seek the multiple perspectives of the various interested parties and encourage their greater involvement. The view that there is only one epistemology (that is, the scientific one) has to be rejected. Participation and collaboration are essential components of any system of inquiry, as any change cannot be effected without the full involvement of all stakeholders, and the adequate representation of their views and perspectives. As Sriskandarajah et al (1991) put it: *'ways of researching need to be developed that combine 'finding out' about complex and dynamic situations with 'taking action' to improve them, in such a way that the actors and beneficiaries of the 'action research' are intimately involved as participants in the whole process.'*

The positivists' response to these principles is often to suggest that they are all a recipe for chaos. If information is changeable, locally valid, value-laden and entirely open to interpretation, how can it be trusted? Whose illusion are we going to believe today? Where is the order? Does this not suggest that science is unbelievable and that 'anything goes'? Is there no more justification for scientific claims?

Few non-positivists would say that science does not work. They point out that what positivist science wants are ways of predicting and controlling nature, and so a good scientific theory simply gives better control and prediction. A more realistic way of thinking about science is as a human tool and not because it is in touch with some absolute reality. This simply means that *'no longer can it be claimed there are any absolutely authoritative foundations upon which scientific knowledge is based ... The fact is that many of our beliefs are warranted by rather weighty bodies of evidence and argument, and so we are justified in holding them; but they are not absolutely unchallengeable.'* (Phillips, 1990)

What is important to recognize is that language is the window through which people construct and interpret the world. We can only get a human idea of what is in the world and so science itself can only be a human picture of the world. How we see the world depends on what matters to us. As different people have different values, this raises critical issues for the methodologies we use for finding out about the world (see Chapter 6). What should become central is the people themselves, rather than the 'tools' or 'instruments'. How do their values affect the way we go about learning about the world? Why do they need the information? Why do they think it is important? How will they judge whether it is useful or good?

It is clearly time to add to, or even let go of, the old paradigm of positivism for science, and explore the new alternatives. This is not to suggest that there is no place for reductionist and controlled science. This will continue to have an important role to play. But it will no longer be seen as the sole type of inquiry. The process will not be without conflict. Thomas Kuhn's (1962) hugely influential analysis of paradigm changes in science describes this process of revolution for case after case. It inevitably means some huge transformations. But the result can be fundamental shifts in understanding: *'During revolutions scientists see new and different things when looking with familiar instruments in places they have looked before'.*

None the less, for the pioneers, this process will be extraordinarily difficult. When Richard Bawden quotes Thomas Kuhn (1962) *'I am quite aware that I risk fierce controversies, international name-calling, and dissolutions of old friendships'*, he says this has all happened during the changes initiated in recent years at Hawkesbury College in Australia: *'my Hawkesbury experience confirmed that all of this occurred in reality [to me] in our attempts to do things profoundly differently'.* Charles Darwin, at the end of his Origin of Species (1859) perceptively wrote: *'Although I am fully convinced of the merit of the views given in this volume,... I by no means expect to convince experienced naturalists whose minds are stocked with a multitude of facts all viewed, during a long course of years, from a point of view directly opposite to mine.... But I look with confidence to the future – to young and rising naturalists, who will be able to view both sides of the question with impartiality'.*

It is only when some of these old professional norms and practices are challenged and new ones in place that widespread change in the livelihoods of farmers and their natural environments is likely to be achieved. This has important implications for the whole process of transition towards a more sustainable agriculture.

CONDITIONS FOR SUSTAINABLE AGRICULTURE

The Transition to Sustainable Agriculture (see Chapter 7)

The modernization of agriculture has resulted in the development of three distinctly different types of agriculture. The first two types have been able to respond to the technological packages, producing high output systems of agriculture in the industrialized countries and in the Green Revolution lands. The third type comprises all the remaining agricultural and livelihood systems which are the low input systems, complex and diverse, with considerably lower yields.

The basic challenge for sustainable agriculture in each of these areas is quite different. In the industrialized agriculture of Europe and North America, it is to reduce substantially input use and variable costs, so as to maintain profitability. Some fall in yields would be acceptable given current levels of overproduction. In the Green Revolution areas, the challenge is to maintain yields at current levels while reducing environmental damage. In the diverse and complex lands, it is to increase yields per hectare while not damaging natural resources.

The evidence from farms and communities from all over the world shows that sustainable agriculture can be achieved in all three of these regions (see Chapter 7) (Figure 1.1):

- in the diverse, complex and 'resource-poor' lands of the Third World, farmers adopting regenerative technologies have doubled or trebled crop yields, often with little or no use of external inputs;
- in the high input and generally irrigated lands, farmers adopting regenerative technologies have maintained yields while substantially reducing inputs;
- in the industrialized agricultural systems, a transition to sustainable agriculture could mean a fall in per hectare yields of 10–20 per cent in the short term, but with better levels of financial returns to farmers.

But farmers do not get more output from less input. They have to substitute knowledge, labour and management skills to make up for the forgone added value of external inputs.

All of these successes have three elements in common (Figure 1.2). They have made use of resource-conserving technologies, such as integrated pest management, soil and water conservation, nutrient recycling, multiple cropping, water harvesting, waste recycling and so on. In all there has been action by groups and communities at local level; and there have been supportive and enabling external government and/or non-government institutions. Most though, are still localized. They are simply islands of success. This is because a fourth element, a favourable policy environment, is missing. Most policy frameworks still actively encourage farming that is dependent on external inputs and technologies. It is these policy frameworks that are the principal barriers to a more sustainable agriculture.

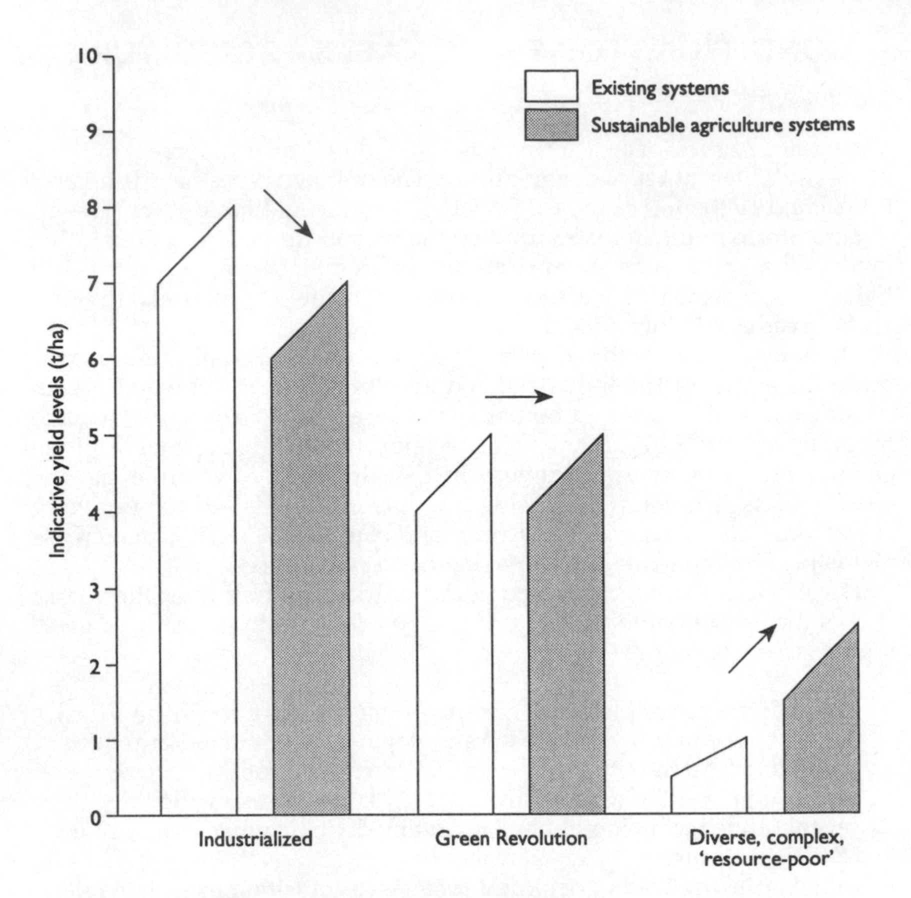

Note: Yields do not represent economic levels (see Tables 7.1–7.5)

Figure 1.1 The impact of sustainable agriculture on yields in three types of agricultural system

Resource-Conserving Technologies and Processes (see Chapter 4)

There is a wide range of proven and promising resource-conserving technologies and practices in the areas of pest, predator, soil, nutrient and water management. Many of these individual technologies are multi-functional, implying that their adoption will mean favourable changes in several components of a farming system at the same time.

A classic case of integrated farming is the rice–fish–*Azolla* system common in many parts of Asia. *Azolla* is a fern that grows in the water of rice fields, on the leaves of which live nitrogen-fixing algae. There are some 36,000 ha in the Fujian Province of China alone, in which all three components perform better than if they had been utilized alone (Box 1.3).

Integrated pest management involves the conjunctive use of a number of pest control strategies in a way that not only reduces pest populations to acceptable levels but can also be sustainable and non-polluting. Similarly, integrated nutrient conservation is the coordinated use of a

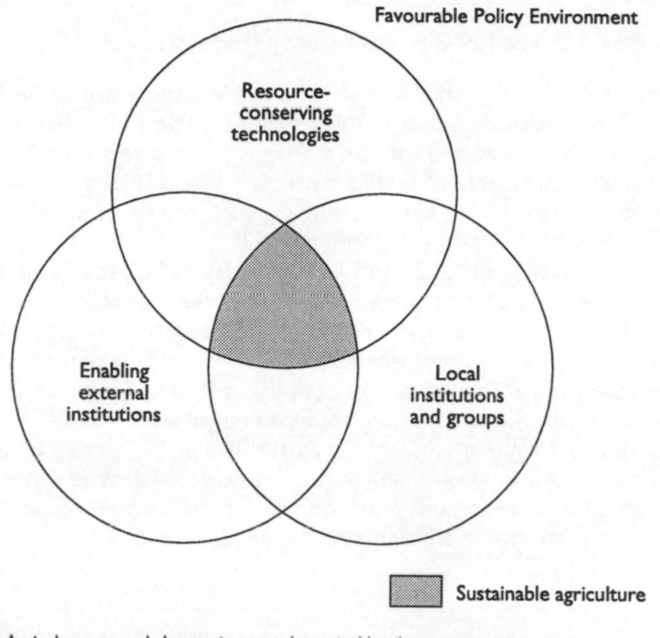

Note: Agriculture can only be persistent and sustainable when resource-conserving technologies are developed and used by local institutions and groups, who are supported by external research, extension and development institutions acting in an enabling way. For sustainable agriculture to spread, the wider policy environment must too be enabling.

Figure 1.2 The conditions for sustainable agriculture

range of practices to ensure sufficient local nutrient availability. Inevitably these are more complex processes than, say, relying on regular calendar spraying of pesticides or applications of inorganic fertilizer. They require a level of analytical skill and certain basic training, and the capacity to monitor on-farm ecological processes. However, evidence suggests that, with the appropriate incentives, farmers are willing and capable of adopting such management practices.

Although a range of resource-conserving technologies and practices have been widely proven on research stations to be both productive and sustainable, the total number of farmers using them is still small. This is because these technologies involve the substitution of management skills, knowledge and labour for previously used external inputs. Farming households face very real adjustment costs and so, in the short term, they may not see sufficient benefits from these practices. As current policies have tended to promote specialized systems, so farmers will have to spend time learning about a greater diversity of practices and measures. Lack of information and skills is thus a barrier to the adoption of a more sustainable agriculture.

The internal biological processes that make sustainable agriculture work also take time to become established. These include the rebuilding of stock of natural predators and wild host plants; increasing the levels of

Box 1.3 The rice–fish–*Azolla* integrated system of south-east China

In Fujian Province, in the south-east of China, there are some 36,000 ha of rice–*Azolla*–fish integrated farming systems. The *Azolla* fern grows on the surface of the water and fish live among the rice plants. The fish feed on the *Azolla* and fish manures fertilize the rice plants. More nitrogen becomes available to rice plants after passing through the fish compared with direct incorporation of *Azolla* as a green manure.

The presence of *Azolla* and fish suppresses the growth of weeds, while the movement of fish increases the oxygen content in the water, so promoting rice root growth. Fish also control some rice pests and diseases. The presence of fish and rice also prolongs the duration of *Azolla* growth. The result is a system that produces up to 750 kg fish/ha as well as slightly improved rice yields (105 per cent compared with non-fish fields). Recent data have shown that with some adjustments in the spacing of plants, choice of *Azolla* and fish species, and improved field construction, the applications of inorganic fertilizer and pesticide can be reduced by 66 per cent and 50 per cent respectively with no adverse effect on yields.

Source: Liu and Weng, 1991

nutrients and improving soil structure; and the establishment and growth of trees. In some cases, investment costs must be incurred for some time before returns increase, such as for labour to construct soil and water conservation measures; for pest and predator monitoring; for tree planting; and for purchase of livestock.

Local Groups and Institutions (see Chapter 5)

A sustainable agriculture cannot be realized without the full participation and collective action of farming households. This is for two reasons. First, the external costs of resource degradation are often transferred from the conventional farmer to the sustainable farmers. Second, one sustainable farm situated in a landscape of high input, resource-degrading farms may produce environmental goods which are undermined or diminished by the lack of support from neighbouring farmers.

A necessary condition for sustainable agriculture is, therefore, the motivation of large numbers of farming households for coordinated resource management. This could be for pest and predator management; nutrient management; controlling the contamination of aquifers and surface water courses; coordinated livestock management; conserving soil and water resources; and seed stock management. The problem is that, in most places, platforms for collective decision making have not been established to manage such resources (Röling, 1994).

The success of sustainable agriculture depends, therefore, not just on the motivations, skills and knowledge of individual farmers, but on action taken by groups or communities as a whole. This makes the task more challenging. Simple extension of the message that sustainable

producing extra benefits for society as a whole, will not suffice. What will also be required is increased attention to community-based action through local institutions and users' groups.

Six types of local group or institution are directly relevant to the needs for a sustainable agriculture:

- community organisations;
- natural resource management groups;
- farmer research groups;
- farmer to farmer extension groups;
- credit management groups;
- consumer groups.

Enabling External Institutions (see Chapter 6)

It is being increasingly well established that the benefits to local and national systems of engaging local people in analysis, decision making and implementation can be substantial. When people are already well organised or are encouraged to form groups, and when their knowledge is sought and given value during planning and implementation, then they are more likely to contribute financially and to continue activities after project completion.

The terms 'people's participation' and 'popular participation' are now part of the normal language of many development agencies. But there are many different types of participation. In passive participation, people are told what is going to happen. They may participate by giving information. They may participate in a consultative process. They may participate by providing resources, such as labour, in return for food, cash or other material incentives. It is common for all of these to be called 'participation'. But in none of these cases are local people likely to sustain activities after the end of the project or programme. As little effort is made to build local skills, interests and capacity, local people have no stake in maintaining structures or practices once the incentives stop.

If the objective is to achieve sustainable agriculture, then nothing less than interactive participation will suffice. In such participation, people engage in joint planning and implementation, which leads to the formation of new institutions or the strengthening of existing ones. It tends to involve interdisciplinary methodologies that seek multiple perspectives and make use of structured learning processes.

Central to sustainable agriculture is that it should enshrine new ways of learning about the world. Learning should not be confused with teaching. A move from a teaching to a learning style has profound implications for agricultural development institutions. The focus is less on *what* we learn, and more on *how* we learn and *with whom* (Bawden, 1991). Inquiry for sustainable agriculture, therefore, implies new roles for development professionals, and these all require a new professionalism with new concepts, values, methods and behaviour (Pretty and Chambers, 1993a).

Supportive Policies (see Chapters 8 and 9)

As identified in Figure 1.2, it is critical that the state should play a supportive role in the development of a more sustainable agriculture. Although many countries have taken some action in individual sectors, there is still a need for more integration if a sustainable agriculture is to be widely promoted and adopted. What has been achieved so far at community level represents what is possible in spite of the existing constraints (see Chapter 7). To date, policies have not been used with a view to directing agricultural practices towards greater sustainability. Indeed, sometimes they have had the opposite effect. The balance between high and low input systems is currently severely distorted in favour of high inputs through the inappropriate application of economic instruments.

So, for the transition to a more sustainable agriculture to occur, governments must facilitate the process with an appropriate range and mix of policy instruments and measures. They could decentralize administrations to reach down to local people. They could reform land tenure to individuals and give communities the right to manage their local resources. They could develop economic policy frameworks that would encourage the more efficient use of resources. They could encourage new institutional frameworks that would be more sensitive to the needs of local people.

Governments wishing to support the spread of sustainable agriculture can either offer incentives to encourage resource conservation, and/or penalize those polluting the environment. Although there are increasing numbers of policy components and areas of action relating to agricultural policy and environmental management, a more sustainable agriculture can only be achieved by integrated action at farm, community and national levels. For it to succeed, this will require the integration of policies too. As illustrated in Chapter 9, though, there are policies that are already known to work. Some of these could be implemented immediately.

Obstacles, Threats and Uncertainties

Despite this vision of a 'win-win' sustainable agriculture, in which farmers, rural communities, environments and national economies could all benefit, there are still many obstacles, threats and uncertainties. Many existing power structures will be threatened by change and it may not be possible for all to benefit in the short term. In addition, there are always winners and losers. The greatest challenge will be to ensure that the biases of the current development paradigm are not repeated, with the poorest and disadvantaged once again marginalized by improvements. The threats occur from international to local levels.

At international level, markets and trade policies have been tending to depress commodity prices, so reducing returns to farmers and economies. In the past ten years alone, commodity prices have fallen on average by 50 per cent. Agrochemical companies, too, will be seeking to protect their markets against alternatives that imply reduced use of their products.

At national level, macro- and micro-economic policies that still hinder the development of a more sustainable agriculture have to be targeted and changed. In some cases, this will be politically very difficult, particularly when it comes to implementing promised land reforms, which would give farmers the security to invest in sustainable practices.

The bureaucratic nature of large institutions is a further threat. They face difficulties in trying to work in a way that empowers local communities, as this implies giving up some power. Similarly, the conservative nature of universities and teaching institutions is an obstacle to the needs of a new professionalism. Most are unwilling or simply unable to train agricultural professionals capable of working with and for farmers.

Finally, farmers themselves face transition costs in the process of adopting sustainable agriculture practices and technologies, and acquiring new management and learning skills.

None the less, much can be achieved despite these structural problems, as illustrated by the 20 cases in Chapter 7.

OUTLINE OF THE BOOK

The next two chapters take a look at the modernist processes of agricultural development, the successes, and the social and environmental costs. Chapters 4 to 6 consider in detail the three essential conditions for a more sustainable agriculture, namely resource-conserving technologies and practices, local groups and institutions, and enabling external institutions working in partnerships with rural people. Chapter 7 presents empirical evidence of impacts following the transition to a more sustainable agriculture, and contains details of 20 cases from 12 countries of Africa, Asia and Latin America. Most of these successes have not arisen because of deliberate support from national policies, so Chapter 8 deals with the policy environment and what can be done systemically to support a more sustainable agriculture. The book finishes with 25 policies that are known to work and lays down a challenge to find ways to implement these in a new, more integrated and participative fashion.

2

THE MODERNIZATION OF AGRICULTURE

'Modernity is the transient, the fleeting, the contingent; it is the one half of art, the other half being the eternal and the immutable.'
C Baudelaire, 1863, from The Painter of Modern Life
(quoted in Harvey, 1989)

AGRICULTURAL AND RURAL MODERNIZATION IN THE TWENTIETH CENTURY

The Process of Modernization

Agriculture has had many 'revolutions' throughout history, from its advent some 8–10,000 years ago to the renowned seventeenth-to-nineteenth century agricultural revolution in Europe. In the past century, rural environments in most parts of the world also have undergone massive transformations. In some senses, these have been the most extraordinary in their speed of spread of new technologies, and the far-reaching nature of their impacts upon social, economic and ecological systems.

Two guiding themes have dominated this period of agricultural and rural development. One has been the need for increased food production to meet the needs of growing populations. Governments have intervened to transform traditional agricultural systems by encouraging the adoption of modern varieties of crops and modern breeds of livestock, together with associated packages of external inputs (such as fertilizers, pesticides, antibiotics, credit, machinery) necessary to make these productive. In addition, they have supported new infrastructure, such as irrigation schemes, roads and markets, guaranteed prices and markets for agricultural produce, as well as a range of other policies (see Chapter 8).

The other theme has been the desire to prevent the degradation of natural resources, perceived to be largely caused by growing populations and their bad practices. To conserve natural resources, governments have encouraged the adoption of soil and water conservation measures to control soil erosion. They have established grazing management schemes to control rangeland degradation. They have excluded people from forests and other sites of high biodiversity to protect wildlife and plants.

According to just these two themes, it would appear that agricultural and rural development has been remarkably successful. Both food production and the amount of land conserved have increased dramatically. Although often seen as mutually exclusive, both have been achieved with largely the same process of modernization. The approach is firmly rooted in and driven by the enlightenment tradition of positivist science (Kurokawa, 1991; Harvey, 1989; Hassan, 1985; Huyssens, 1984). Scientists and planners first name the problem that needs solving, in these cases too little food or too much degradation. Their concern is to intervene so as to encourage rural people to change their practices. Rational solutions are proposed and technologies developed. These technologies, known to work in a research station or other controlled environments, are assumed to work elsewhere. So they are passed on to the mass of rural people and farmers, and the benefits awaited.

Towards Coercion with Technologies

Central to this process of modernization is the assumption that technologies are universal. During the Green Revolution of the 1960s onwards, it was widely believed by scientists that they would be able to transform agricultural systems without affecting social systems (Palmer, 1977; Dahlberg, 1990). It was assumed that technologies existed independently of social context.

The terrible contradiction, however, is that modern agriculture has often had to resort to coercion and enforcement to ensure adoption of the new technologies and practices. In the early stages, it was common to find researchers and extensionists working closely and sensitively with farmers for a period (see Indonesia, Mexico, and soil and water conservation cases below). Where farmers' conditions are similar to those where the technology was developed and tested, then the technology spreads.

But most farmers have differing conditions, needs, values and constraints to those of researchers. When they reject a technology, say because it does not fit their needs or is too risky, modern agriculture can have no other response but to assume it is the farmers' fault. Rarely do scientists, policy makers and extensionists question the technologies and the contexts that have generated them. Rather they have blamed the farmers, wondering why they should resist technologies with such 'obvious' benefits. It is they who are labelled as 'backward' or 'laggards'. The problem is that *'technology does not take root when it is cut off from culture and tradition. The transfer of technology requires sophistication: adaptation to region, to unique situations and to custom'* (Kurokawa, 1991).

But technology transfer does not have sophistication. Central to this is the notion that new technologies are better than those from the past and so represent 'progress'. Such a process is usually depicted as linear, with the new replacing the old with no coexistence. This iconography is powerful in many disciplines, especially in evolution and natural selection

(Gould, 1989), and usually implies that what has gone before is not as good as what we have now.

Inevitably, there follows a qualitative shift in practice after the early sensibility, with intervention focusing on trying to change the local social and economic environment to suit the technology. Of course, it is possible to take the opposite stance and try to change the technology to suit the environment. But it has been a core characteristic of modernity not to do this. Indeed, adaptations are neither sought nor permitted. The technologies and practices are known to be effective; a great deal of money and effort has been expended on developing them; so why should they need adapting? This arrogance is reflected by the modernist architects. David Harvey (1989) describes the implications for some people who have had to live in modern buildings: '*this meant the eschewing of ornament and personalized design – to the point where public housing tenants were not allowed to modify their environments to meet personal needs, and the students living in Le Corbusier's Pavillion Suisse had to fry every summer because the architect refused, for aesthetic reasons, to let blinds be installed*'.

It has not been so different in agricultural development. Irrigation systems, for example, designed and 'perfected' by engineers, have long been put in place with no adaptations to local context. It is only recently that those concerned with irrigation have been able to voice the fact that much of the poor performance in irrigation systems stems from fundamental weaknesses in human processes of planning and management, which no amount of investment in technological hardware can overcome on its own (see Uphoff, 1992a; Bagadion and Korten, 1991).

A major part of the process of changing the local social and economic environment is the offering of financial incentives, such as directly through cash or food for work, or indirectly through subsidising the prices of fertilizers, pesticides or credit. In addition, regulations have been established to prevent certain practices, such as the collection of wild foods, hunting of wild animals, growing of traditional crop varieties, or to seek farmers' compliance by linking one practice or technology to the receipt of other benefits, such as the cultivation of new varieties as a precondition for access to credit.

The Erosion of Diversity

The assumption of the universality of technologies has inevitably led to greater standardization. As farmers are made to comply 'in their own best interests', they can do so only by completely changing their own livelihoods, simplifying their practices to incorporate the new technology. In addition, local institutions are coopted, or lose power and wither away. The state is acting as if it alone knows best.

Such universality of approach or technology leads to homogenization of environments. Where farmers used to grow tens or even hundreds of crop varieties, now they might only grow one or two. Where they used to use a range of biological and physical measures to control soil erosion, now they might only have terraces. Where they used to rely on wild plants and animals for food, medicine and fuel, now they might only rely on markets for these products. Modernization has brought with it the steady erosion of cultural and biological diversity (see Chapter 3).

This notion is not new. Modernity has always sought to sweep away the confusion of diverse local practices and pluralistic functions accumulated over the ages, so as to establish a new order. This order brings freedom from the constraints of history, and liberty in the new technologies and practices. One of the slogans of the modernist architect, Le Corbusier, was *'by order, bring about freedom'*. As Kisho Kurokawa (1991) put it *'the nail that sticks out is hammered down'*. Le Corbusier's other famous dictum that the home is a *'machine for modern living'* further celebrates the benefits of new technologies and universal benefits.

But this loss of diversity and increase in imposed and often bland order has done something to us too. We have lost something important, and probably do not know it: *'the very process of development even as it transforms the wasteland into a thriving physical and social space, recreates the wasteland inside of the developer himself'* (Berman, 1982). Modernity does not result in the triumph of rationality: rather it creates an *'iron cage of bureaucratic rationality from which there is no escape'* (Bernstein, 1985).

The world so created by this universal modernism is inevitably monotonous. It is associated with *'the belief in linear progress, absolute truths, the rational planning of ideal social orders, and the standardisation of knowledge and production'* (PRECIS 6, in Harvey, 1989). The contrast with what is required for a more sustainable agriculture is fundamental. Called postmodernism by some (coming after or contrasting with modernism), it favours heterogeneity and difference as the forces of liberation. What postmodern traditions have in common is the rejection of 'meta-narratives', or large-scale theoretical interpretations, plans or technologies that purport to have universal application. Those concerned with the development of a more sustainable agriculture, if it is to succeed, must not fall into the same traps. They must not make new grandiose claims to have the sole answer. They must be more modest, learning the lessons of modernization, and so establish a new tradition of science, policy and practice. The idea that all groups have a right to speak for themselves, in their own voices, and have their voice accepted as authentic and legitimate, is central to this new tradition (Harvey, 1989).

This chapter is concerned with themes of modern agriculture, and how these have influenced farmers and professional institutions. The hidden environmental and social costs of modern, technology-led agricultural transformations are explored in Chapter 3.

THE PURSUIT OF INCREASED PRODUCTIVITY

Industrialized Agriculture and the Green Revolution

The process of agricultural modernization has produced three distinct types of agriculture: these are industrialized, Green Revolution and all that remains – the low-external input, traditional and 'unimproved' (Chambers et al, 1989). The first two types have been able to respond to the technological packages, producing high input, high output systems of agriculture. Their conditions were either like those where the technologies were generated or else their environments could easily be changed and homogenized to suit the technologies. These tend now to be endowed with most of the following attributes:

- access to roads, urban markets, ports, and with these ready access to inputs, machinery, marketing infrastructure, transport, agroprocessing facilities and credit;
- good soils;
- adequate supply of water, either through stable rainfall or irrigation systems;
- access to modern crop varieties and livestock breeds; and
- access to petroleum-based products and machinery.

Most agricultural systems in industrialized countries count as high external input systems, save for the relatively small number of organic farmers. In the countries of the Third World, high-external input systems are found in the large irrigated plains and deltas of South, South-East and East Asia, and parts of Latin America and North Africa, but also in patches in other regions. They tend to be monocrop and/or monoanimal enterprises, geared for sale, and so include lowland irrigated rice, wheat and cotton; plantations of bananas, pineapples, oil palm, sugar cane; market gardening near to urban centres; and intensive livestock rearing and ranching.

These are the lands of the Green Revolution, the success of which lay in its simplicity. Agricultural scientists bred new varieties of staple cereals that matured quickly, so permitting two or three crops to be grown each year; that were day-length insensitive, so could be extended to farmers at a wide range of latitudes; and that were producers of more grain at the expense of straw. They were also much more nitrogen-responsive than traditional varieties. These modern varieties (MVs) were distributed to farmers together with high cost inputs, including inorganic fertilizers, pesticides, machinery, credit and water regulation. These technical innovations were then implemented in the best favoured agroclimatic regions and for those classes of farmers with the best expectations of and means for realizing the potential yield increases (Conway and Barbier, 1990). Now, some 50 per cent of the area of wheat, rice and maize is under MVs, though the uptake varies considerably across continents (Table 2.1). As a result, average cereal yields have roughly doubled in 30 years. This has led to an improvement of about 7 per cent in the total food produced per capita over the same period. This average does, however, hide significant regional differences: in South-East Asia, food production per capita has increased by about 30 per cent, but in Africa it has fallen by 20 per cent (FAO, passim).

A similar revolution occurred in industrialized countries. Farmers have modernized by introducing machinery, replacing labour, specializing operations, and changing practices to ensure greater aggregate production. The pressure to increase economies of scale, by increasing field and farm size, has meant that the traditional mixed farm, a highly

Table 2.1 Differences in area under modern varieties of wheat, maize and rice, late 1980s

Regions	Area under MVs of wheat (%)	Area under MVs of rice(%)	Area under MVs of maize(%)
Asia (not China)	76%	45%	36%
China	31%	81%	71%
Africa	51%	5%	51%
Latin America	78%	18%	54%
All Third World countries	52%	54%	45%

Source: Lipton and Longhurst, 1989

integrated system in which few external impacts are generated, has largely disappeared. In Britain, for example, the impact since the 1940s has been increased wheat yields from an average of 2.1 to 7 t/ha; barley from 2.1 to 5.9 t/ha; potatoes from 17 to 38 t/ha; and individual milk yields from about 6 to 14 litres per day (MAFF, MMB, passim). On the most productive farms, these averages may be almost doubled. Similar transformations in yields have occurred elsewhere.

The 'Forgotten' Agriculture

The third type of agriculture comprises all the remaining agricultural and livelihood systems. These represent a largely forgotten agriculture. These are the low-external input systems, and are located in drylands, wetlands, uplands, savannahs, swamps, near-deserts, mountains and hills, and forests. Farming systems in these areas are complex and diverse, agricultural yields are low, and rural livelihoods are often dependent on wild resources as well as agricultural produce. They are remote from markets and infrastructure; they are located on fragile or problem soils; and less likely to be visited by agricultural scientists and extension workers or studied in research institutions.

They also have low productivity, with cereal yields typically only 0.5 to 1 t/ha. The poorest countries tend to have higher proportions of these agricultural systems. Most of the food production in Africa comes from these low external input systems of agriculture (Table 2.2).

The extraordinary thing is that in the mid-1990s some 30–35 per cent of the world's population, about 1.9–2.1 billion people, are directly supported by this third and 'forgotten' agriculture (based on estimates from FAO and World Bank production and distribution data). Yet these people are currently excluded by development assistance and agricultural policies that focus on the high potential lands. A 1988 US Office of Technology Assessment study has said *'most agricultural development assistance ... has emphasised external resources'* (OTA, 1988). But people can rarely afford to sustain the use of external resources. Their only immediate

Table 2.2 Contribution of low-external input systems to food production in Africa according to various crops and animals

Crops or livestock	*Low-external input systems contribution to total production (%)*	*Proportion of crop receiving fertilizers (%)*	*Proportion of crop under modern varieties (%)*
Millet	72%	approx 0	very little
Sorghum	61%	approx 0	little
Maize	37%	60–75%[1]	25%
Rice	76%	75%[1]	5%
Cowpeas, pigeon peas, beans	49%	0	0
Groundnuts	55%	50%	little
Roots and tubers	90–100%	0–5%	0–5%
Cattle	85–90%	–	–
Sheep, goats, camels	95–100%	–	–
Fish	85–95%	–	–

1 Very little at recommended doses
Source: Horwith et al, 1989

alternative lies in low-external input systems using resource-conserving technologies.

Although there have been some recent changes in the recognition of what can be achieved by a more sustainable agriculture, many institutions still believe in the modernization paradigm. As the irrigated and well-rainfed lands provide the bulk of the world's food, so they are assumed to be the only places where further improvements can be made. To some, this implies that activities should clearly concentrate on high potential areas.

The remaining lands contribute small amounts in comparison as yields are so low. The common assumption is that improvements elsewhere will encourage people to migrate from the 'marginal' areas to the 'higher potential' ones, where they will have access to jobs and food. This will reduce the population pressure on marginal areas. The wider strategy would therefore to be to create employment and income opportunities outside agriculture in the marginal lands, and to *'boost productivity in the better endowed areas so that populations can move eventually from the marginal lands'* (Morse, 1988). Such an approach ensures that existing low input and traditional agricultural systems continue to be ignored.

None the less, it is increasingly apparent that both national and international agricultural research systems are making this forgotten agriculture a focus of attention (see Chapters 6 and 8).

Strategic Implications

Despite past improvements in food production, the most difficult challenges are just beginning. The world population will probably reach 8 billion by the year 2025 and may not stabilize until there are some 13 billion people. Even at the lowest estimates, and given current inequitable access and rights to resources, there will be a need for agricultural production to increase substantially if current levels of nutrition are to be maintained. Without very considerable growth the prospects for many people in poor countries and regions of the world are bleak. Given the extraordinary success of the Green Revolution, many argue that this model of development continues to provide the most effective and efficient prospect for all people of the South.

But though these agricultural revolutions have produced enormous benefits for farmers, consumers and economies, many of the poorest rural people are yet to benefit. They live in regions with poor quality soils and unpredictable rainfall. They are either remote from agricultural services that promote the package of inputs necessary to add value or they simply cannot afford to take a risk by adopting the whole package. In addition, inappropriate use of inputs imposes costs both in terms of economic efficiency, and the external costs imposed on others from agricultural pollution and environmental degradation.

As suggested in Chapters 1 and 7, the impacts of a more sustainable agriculture in each of these three agricultural types will differ. In the industrialized regions, it is possible that yields will fall by some 5–10 per cent; in the Green Revolution lands, it seems that yields can be maintained at the same level as conventional, high input systems or even slightly improved. It is in the remaining areas that significant growth is possible, with yields possibly doubling or trebling (see Figure 1.1).

THE PURSUIT OF ENVIRONMENTAL CONSERVATION

Conservation Themes

The pursuit of environmental conservation has also been a significant theme in rural development in the twentieth century. Conservationist beliefs have generally held that there is an inverse relationship between human actions and the well-being of the environment. The problems have been widely agreed upon by professionals: soil erosion, degradation of rangelands, spreading 'desertification', loss of forests and the destruction of wildlife. All of these problems have appeared to require action to prevent further deterioration and policies have consistently defined poor local management of resources as the principal cause of destruction.

As with any example of a rationalist tradition, technological solutions are developed once the problems are clearly defined, the solutions are technological and the barriers to adopting the solutions are placed squarely with the local communities. Recent examinations of all these areas have shown that the technologies and models of intervention arose in particular historical settings in industrialized countries. All the models were transferred to completely different contexts with little or no regard for the receiving environments. These historical studies include analyses of soil and water conservation with its beginnings in the USA; of the establishment of national parks for the preservation of biodiversity; and of rangeland management, and its origins in the USA and Australia.

Parks and Protected Areas

Parks and nature reserves have long been thought of as the best way of preserving wildlife. At the turn of the nineteenth century, governments in industrializing countries began to set aside areas of particular scenic beauty or uniqueness for conservation. The creation of most protected areas in North America during the last century involved the exclusion of local people. The famous tract of hot springs and geysers in north-western Wyoming, for example, was established as the Yellowstone National Park in 1872. The inhabitants of Yellowstone, mainly Crow and Shoshone native American, either left for reservations or were driven out by the army (Morrison, 1993). In the African colonies, authorities gazetted the first protected areas between the two world wars. The conservation philosophy emphasized that *'the public good was best served through the protection of forests and water resources, even if this meant the displacement of local communities'* (McCracken, 1987).

There are now close to 8500 major protected areas throughout the world today, increasing from some 2000 since the early 1970s (Pimbert and Pretty, 1995). Protected areas now exist in 169 countries, covering about 7.7 million km^2 or some 5.2 per cent of the world's land area. Following the Earth Summit in Rio de Janeiro in 1992, and the ratification of the Biodiversity Convention, many countries are seeking to transform 'as much land as possible' to strictly protected regimes. Many countries have already allocated more than 10 per cent of their territories as protected areas (Table 2.3).

The central concept has been of conserving the natural state of pristine environments and wildernesses that are rich in biodiversity, and where rare or unique species exist. But this concept of the wilderness as the *'untouched or untamed land is mostly an urban perception, the view of people who are far removed from the natural environment they depend upon'* (Gómez-Pompa and Kaus, 1992). It does not recognize the importance of local management and land-use practices in sustaining and protecting biodiversity. Traditional conservationists see the aesthetic and biological value of, say, a rainforest, but do not see the people. The notion that these can only be maintained without people has long influenced conservation policy (Manning, 1989; Gómez-Pompa and Kaus, 1992; West and Brechin, 1992; Pimbert and Pretty, 1995).

This neglect of resident people in or near parks and reserves strongly persists today. Until quite recently, plans for protected area management rarely mentioned the people living inside forests, coastal strips, wetlands and other biodiversity-rich areas earmarked for conservation. Yet virtually every part of the world has been inhabited and modified by people in the past, and apparent wildernesses have often supported high densities of

Table 2.3 A selection of countries with more than 10% of land under protected area regimes

Costa Rica	29%
Honduras	22%
Bhutan	20%
Botswana	18%
Panama	18%
Guatemala	16%
Nicaragua	14%
Central African Republic	12%
Malaysia	11.6%
Benin	11.5%
Tanzania	11.5%
Zimbabwe	11.3%
Senegal	10.8%
Rwanda	10.4%

Sources: Utting, 1993; CNPPA, 1993; MacKinnon et al, 1986

people. Many agricultural and forest people value and utilize wild resources, and there is good evidence from many different environments for effective and sustainable local management (Scoones et al, 1992; Nabhan et al, 1982; Alcorn, 1984). In India, for example, a study of 171 national parks and sanctuaries found that there were 1.6 million people living in the 118 parks that were inhabited (Kothari et al, 1989). Yet, by 1993, protected areas in India had already displaced some 600,000 tribal people, some 20 per cent of the country's tribal people.

Soil Conservation

The knowledge that soil erosion was both costly and damaging was first appreciated on a wide scale by agricultural authorities in the USA, colonial Africa and India in the early part of the twentieth century. They took the view that farmers were mismanagers of soil and water, and so had to be encouraged to adopt conserving practices. Erosion was considered a technical problem requiring technical action, and so authorities have encouraged farmers to construct terraces, bunds (embankments of soil), ditches and drains, and to adopt alternative cropping patterns and contour planting. They have also resettled people to discourage the use of certain lands and destocked regions of livestock to reduce grazing pressure (Pretty and Shah, 1994).

The style of intervention was first established in the USA and particularly followed the period of severe wind erosion and dust storms that came to be known as the Dust Bowl of the early 1930s. Even though there were subsidies to encourage farmers to adopt new measures, authorities were granted wide-ranging powers to enforce land-use regulations. The powers to overcome non-compliance were extensive, and included fines, the authority to gain access to farms to conduct conservation work and the capacity for direct tax-billing to pay for such remedial work (Headley, 1985; Trimble, 1985).

This pattern of intervention was repeated by colonial authorities in Africa and Asia. Early regulations had been adapted to local conditions, and were grounded in farming and grazing practice (Stocking, 1985; Gichuki, 1991). But later, administrators travelling to the USA saw the devastation, and brought back recommendations for large-scale bunding and ridging, combined with contour ploughing and planting. Locally adapted practices were largely ignored, even though they were more effective in droughts (Beinart, 1984). These measures were imposed on farmers, who were then monitored closely to ensure their compliance. In some countries, this meant the compulsory resettlement of many people to new villages.

This has been the style for soil conservation ever since. Technologies known to work under certain conditions are widely used or recommended, and backed up by local and national policies that give powers to the state to execute specified improvements on farmers' fields, and to allocate the costs of these improvements between the farmers and the state. In many places, provisions have been made for compulsory treatment of the fields of farmers refusing land treatment. This has led to increased alienation with, for example, people uprooting plantations, and destroying fencing and conservation measures.

The quantitative achievements of conventional soil conservation programmes can appear impressive. Throughout the world, terraces have been built, trees planted and farmers trained on a massive scale. Yet these results have mostly been short lived, tending to occur only within project boundaries and before project completion. If performance is measured over long periods, the results have been extraordinarily poor for the amount of effort and money expended: technologies have neither persisted nor spread independently into non-project areas (Shaxson et al, 1989; Hudson, 1991; Reij, 1991; Pretty and Shah, 1994).

Rangeland Management in the Drylands

Rangeland science and management has had a remarkably similar history. The concept of a 'sustainable yield' and the goal of improved productivity had its origins both in North America and in Australia. The approach was well adapted to the social and ecological context of these rangelands. A central feature of range management is that it has evolved to meet the needs of a system based on privately owned land. As Russell and Ison (1991) put it: *'so pervasive is this history, which constitutes this particular 'tradition of understanding', that it is difficult for those involved in it to see range management in any way other than their own way'*.

For at least 50 years, policy makers have consistently defined the major concern of pastoral regions as overstocking leading to ecological disaster. The problem was clear (too many cattle), as was the technical solution (destocking of the cattle). The central assumption is that pastoral ecosystems are potentially stable and equilibrial systems, which can become destabilized by overstocking and overgrazing. This led to the establishment of group ranches, grazing blocks and grazing associations. But these have never worked and the ground assumptions behind this modernizing tradition of range management are being increasingly discredited (Benkhe and Scoones, 1992; Russell and Ison, 1991; Ellis and Swift, 1988; Sandford, 1983).

What recent studies are showing is that pastoral systems are non-equilibrial, with system dynamics affected more by social than biological factors. Pastoralists are opportunistic, continually adapting to varying conditions and their persistence is a function of their capacity to change. Ironically, it is the now conventional rangeland practices themselves that are the destabilizing influences on pastoral systems, as they have prevented traditional adaptive systems from succeeding. These are more productive than the imposed models because *'the producer's strategy within non-equilibrium systems is to move livestock sequentially across a series of environments each of which reaches peak carrying capacity in a different time period. Mobile herds can then move from zone to zone, region to region, avoiding resource-scarce periods and exploiting optimal periods in each area they use'* (Benkhe and Scoones, 1992).

In this way livestock producers can maintain a total livestock population with levels of productivity in excess of that which could be sustained by several separate herds confined to their individual areas or ranches.

THE MODERNIZATION OF INDONESIAN RICE FARMING

Early Themes and Processes

The modernization of Indonesian rice farming illustrates how the Green Revolution came to represent for a while all that was good about modern agriculture. Rice farming has been influenced since the early 1960s by a sequence of government programmes that have intervened heavily in farmers' practices. Farmers have been encouraged to adopt the new technology packages by the presence of extension services, subsidies on inputs, price floors on rice and price stabilization schemes, credit provisions and the establishment of village organizations.

Although Indonesia has always been a big rice producer, substantial imports of rice had always been necessary. The import bill in 1963 alone was US$133 million. In aiming to raise rice output and increase rural employment, efforts to intensify rice production were directed from the start to the best endowed and irrigated areas. The impact on production has been remarkable. Annual milled rice production grew from 9.8 to nearly 31 million tonnes from the early 1960s to 1990. Self-sufficiency was reached in 1985 (Kartasubrata, 1993; van der Fliert, 1993).

There have been four major national initiatives, beginning with the *Demas* programme of 1964–5, leading to the current *New Supra Insus* effort of the late 1980s and early 1990s (Palmer, 1976, 1977; Soetrisno, 1982; Sawit and Manwan, 1991; Fox, 1992; Winarto, 1993). Intervention started with genuine partnerships between extension workers and farmers. In 1963, a new approach to extension was tried in three villages of West Java.

Student extensionists from Bogor Agricultural College lived in the villages, and each worked with small groups of farmers on implementing new technologies for rice cultivation. They focused on 'five endeavours': the use of modern varieties (MVs), fertilizers, eradication of diseases, improvement of methods of cultivation, and irrigation.

The extensionists established close working relationships with farmers and their personal commitment did much to persuade farmers to adopt the new practices. As a result, yield increases of 40 to 150 per cent were achieved over non-participating villages. This was so successful that the initiative was given the name *Demas* (from *Demonstrasi Massal*: mass demonstration), and expanded to some 10,000 ha. The students were seen as willing to work with farmers and, unlike normal bureaucrats, did not work in expectation of returned favours or loyalty: *'their willingness to work every day in the fields and to submerge themselves in discussion between the farmers was bound to increase the interest of the farmers in new methods'* (Palmer, 1977).

Towards Coercion

This rapid success was its downfall. It was soon replaced by the *Bimas* (mass guidance) programme in 1965. This marked the beginning of the formal transfer of new rice technology to farmers. It also marked the first erosion from partnerships with farmers towards enforcement. In the search for economies of scale, the programme was expanded too quickly. Some 1200 student extensionists were put in the field in 1965–6, each now having to work with 512 farmers. As they no longer lived in the villages, their contact was more with village leaders, who were then expected to instruct farmers. This resulted in an inevitable dilution of their impact and a substantial change in work ethic.

As adoption was too slow, *Bimas* then obliged all farmers in a contiguous area to adopt new practices, willingly or not. Many farmers were reluctant to adopt MVs of rice, but they were *'instructed and enforced'* by extensionists (Hansen, 1978) or *'influential farmers enticed, bullied, cajoled and blackmailed weaker farmers into taking their place in the scheme'* (Palmer, 1976). To ensure they met their targets, some extension officers even rented farms and then employed the owners as wage labourers.

The reluctance to participate did not diminish, causing the government to establish the *Bimas Gotong Royong* (*Bimas GR*) programme in the late 1960s. But enforcement was to be yet stronger. A multinational corporation was contracted to supply a package of fertilizers, pesticides and cash credit directly to village level on some 85,000 ha of Java. The government paid $52.50/ha for this service. The package was distributed and *'in some villages, farmers were given only two weeks advance notice to undertake a significant change in cultivation practice with no alternative but to accept'* (Palmer, 1977). The payment for inputs was initially one-sixth of the yield, but as farmers were suspected of under-reporting their yields, a fixed payment of 1 tonne was set in 1969.

The pesticides were sprayed from the air, regardless of whether or not all farmers were part of the programme. The operators found it impossible to plan cut-outs when flying over farms which had successfully objected to being sprayed. So farmers were obliged to plant the same rice variety at the same time in whole blocks of farms (although many still found ways to cultivate their own local varieties). These obligations, along with the fixed composition of inputs imposed regardless of local conditions and practices, aroused yet more hostility.

Village officials and other authorities reinforced these actions at local level. Elske van der Fliert (1993) describes the process in Sugihsari, Central Java: '*the police officer commanded that everybody had to take an input package. Seed was supplied by the subdistrict, and fertilizers and pesticides by a local irrigation project. Initially, many farmers were reluctant, but they were forced to buy the new seeds, fertilizers and chemicals... The farmers were threatened that the irrigation water would be closed, and that their fields would be harrowed over if they tried to plant traditional varieties*'.

But in 1970, the 550,000 ha of *Bimas GR* was abandoned one month after an incognito visit by the president to villages to establish for himself the seriousness of the situation.

The improved *Bimas* programme began with an emphasis on local institutions. Through these *BUUDs* (*Badan Usaha Unit Desa*), supplies of seeds, fertilizer and pesticides were guaranteed at subsidized prices. A guaranteed floor price for paddy and local storage facilities were put in place to encourage further adoption. To have access to these incentives, farmers had to join the *BUUD*, and being a member meant compulsory acceptance of the new technology, inputs and guidance of extension workers. And although farmers were given vouchers for the purchase of subsidized fertilizer, these were not necessarily a guarantee of getting the required amounts at the right time. Irregular supplies meant farmers often had to buy in shops at a higher price and then later sell their *Bimas* fertilizer at a lower price.

During the wider and more rapid adoption of MVs, and the package of associated technologies, significant numbers of farmers continued to cultivate traditional varieties. Some did not adopt the new package and many others rejected it soon after adoption. This is despite the fact that yields from MVs were consistently higher than from other varieties, and that almost all types of farmers, irrespective of size and tenurial status, could average 20–80 per cent higher yields when planting them. It was also during the late 1960s and 1970s that it became increasingly apparent that widespread social change was occurring as a result of rice modernization. Rural jobs were being lost, women isolated from the production process, and complex ecological and cultural relationships undermined (see Chapter 3).

The next phases brought the *Insus* and later *Supra Insus* (SI) programmes. These promoted rice intensification through a ten-point technological package implemented through the 'training and visit' system of extension adopted in the early 1980s. This involved extension agents forming farmer groups, subsidized credit, establishment of cooperatives for the distribution of inputs, synchronization of planting and varietal rotations, and direct support for input supply firms. Extension workers continued to promote MVs and transferred information on pesticide products in farmers' meetings. Local and regional administrations, together with extension workers, continue to play an important role in guiding farmers' activities (Sawit and Manwan, 1991; Winarto, 1993). In particular, they influence decisions on components of the credit packages; they refuse requests for partial packages; and suggest particular rice varieties for particular seasons. Farmers do have the opportunity to continue to be creative, but most of the external knowledge and information has tended to reinforce the notion that there is only the one way of farming rice.

Fundamental Shifts in Policy

The greatest policy change in Indonesia has occurred, though, as a result of pest resurgences caused by the overuse of pesticides. As early as 1970, research had shown that population densities of brown planthopper (BPH) were highest where farmers had applied diazinon and endrin. But these products continued to be applied, leading to severe outbreaks of BPH and other pests (see Chapter 3). This led, in November 1986, to the banning by presidential decree of 57 broad spectrum brands of pesticide. At the same time, integrated pest management was declared as a national strategy and 1500 new pest observer positions created. Subsidies on pesticides were progressively cut from an 85 per cent level in 1984 to zero by January 1989, saving the country some US$130–160 million each year. Pesticide production in Indonesia fell by nearly 60 per cent between 1985–90, from some 57,000 to 24,000 tonnes (van der Fliert, 1993).

Extension workers are now seen more as facilitators, aiming to transmit knowledge and ways of learning rather than technologies. Farmers learn in Farmer Field Schools about predator–prey relationships and how to control pests while cutting pesticide use. Rice yields have been maintained or even improved, while pesticide applications, and so costs to farmers, have been cut substantially (see Case 11, Chapter 7). Although this does not necessarily signal the end of modernization, new partnerships between extensionists and farmers are being established. Varietal rotation is one strategy where collective action is needed. It is designed to avoid the build up of resistant insects and virulent strains of tungro virus, but success depends on the strategy adopted by the extension authorities and the responses of farmers (Winarto, 1994; Sawit and Manwan, 1991).

The rice variety IR64 was widely favoured by farmers, as it was palatable, productive and resistant to BPH. Its use spread rapidly, but then it was severely damaged by white rice stem borer in 1989–90. Farmers in some communities then responded by resorting to growing a greater diversity of traditional varieties (Winarto, 1993). In West Java in the late 1980s, it was recommended that the Cisadane variety of rice be planted on no more than 30 per cent of the area. It is vulnerable to brown plant hopper if planted continuously, but it has a good flavour and fetches a high price. But in 1987–8, 43–80 per cent of the area of different districts was planted to Cisadane. What was lacking at the time were appropriate local institutions in which group plans for rice cultivation could be discussed and agreed. The rice-IPM programme seems now to be offering such new opportunities (Sawit and Manwan, 1991; Winarto, 1994).

THE MODERNIZATION OF MEXICAN WHEAT AND MAIZE FARMING

Early Research Themes

Another country where notable advances in modern agriculture have led to increased food production is Mexico. Over the last 50 years, farmers have been encouraged to adopt modern methods of production. Research on increasing food production was first coordinated with the establishment of the office of experimental stations within the Ministry of Agriculture during the 1930s. It was staffed by young Mexican scientists, who were closely associated with the philosophy of development represented by the government of Lazaro Cardenas. They searched rural areas for local varieties of maize and wheat with potential for high yields. They were *'little interested in importing technology from abroad, preferring to work slowly at the local level'*, particularly through the regional Practical Schools of Agriculture (Hewitt de Alcantara, 1976).

The approach of these existing institutions, later merged as the Institute for Agricultural Investigation (IIA) in 1947, was to build on existing farming practice so as to develop technologies grounded in local practices and conditions. The position of agricultural scientists was described at the time by Edmundo Taboada in this way: *'it is possible that a discovery can be made in a laboratory, in a hothouse, at an experiment station, but useful science, manageable, operable science, must grow out of the local communities of... small farmers, ejidatarios, and indigenous communities'* (in Hewitt de Alcantara, 1976).

These institutions were, however, to be overshadowed from 1943 by the Office of Special Studies (*Oficina de Estudios Especiales*, or *OEE*). This semi-autonomous body within the Ministry of Agriculture was established following a Rockefeller Foundation funded mission in 1941 (Hewitt de Alcantara, 1976; Stakman et al, 1967). The new government of Camacho persuaded the foundation to begin a programme of foreign assistance to improve agricultural productivity, and so the *OEE* was set up to conduct agricultural research to increase wheat and maize production in the prosperous irrigated sector of Mexican agriculture. This would be achieved by importing proven technologies, and bringing scientific personnel from the USA to conduct the research. It was the *OEE* which generated the technology now associated with the Green Revolution. The demise of the *IIA* occurred later in 1961, when it was merged with the OEE to form the National Institute for Agricultural Investigation (*INIA*). This adopted wholesale the approach of the *OEE*, and so had no focus on small and subsistence farmers.

The Modern Approach

Norman Borlaug became director of the *OEE* in 1944, with the primary goal of a rapid increase in yields of cereals until national self-sufficiency was achieved (Borlaug, 1958). Plant breeders first collected wheat seed from Mexico and abroad, so as to breed new varieties with the preferred characteristics of high yield, resistance to stem and leaf rust, shortness of duration, drought resistance, adaptability to summer planting and resistance to lodging. By the late 1940s, five modern varieties were being multiplied for commercial use, and in 1949–50, some 50,000 ha were planted. There was a very rapid adoption of MVs and by the early 1960s all the wheat area was under MVs. But the commercial lifetime of each variety was only four years, as large areas under single varieties were regularly overcome by new rust races. This meant that strong lines of communication were needed between the scientists continually producing new technologies and the large-scale farmers of wheat.

At the same time, breeders were experimenting to produce higher yielding varieties of maize. The best yields come from hybrids, but farmers growing hybrids must have new seed every year as productivity declines markedly if they use second or third generation seed. Open pollinated varieties are less productive, but have the advantage of permanence as farmers can use the seed in subsequent years. But the *OEE* argued that maize yields were so low that every effort should be made to improve them through the most advanced scientific means possible, which meant the use of hybrids. Hybrids also need fertilizers and irrigation, both of which were not available to small farmers. The *IAA* opposed this breeding strategy of the *OEE*, but was not strong enough to prevail. In 1948, 80 per cent of all improved varieties planted were open pollinated but, by 1956, 96 per cent of research resources were being devoted to hybrid development and seed production.

As in Indonesia, the state began the process of modernization by simply making technologies available to farmers. But when they were not adopted at a rate felt necessary, so more incentives and encouragements were put in place. In 1953, after a disastrous harvest, the government decreed that the function of Directive Committees for each of the 30 irrigation districts was to select cropping patterns for farmers according to national goals; allocate water at precise times; and provide a direct link between national policy and farming practice. Irrigation water was subsidized as the costs of investment, maintenance and repair were not charged to farmers. Prices were guaranteed for the principal crops and credit made available. Crop insurance schemes were also established to protect farmers against natural disaster, with a federal subsidy covering 40–60 per cent of the costs to farmers.

There have also been programmes for small-scale producers on non-irrigated lands, the most notable being the Plan Puebla (Volke Haller and Sepulveda-González, 1987; Gómez-Pompa et al, 1993; Gladwin, 1979). This was started by *CIMMYT* in 1967, the aim being to breed improved maize hybrids to perform better than local varieties. When it found none, the project focused on recommendations for timely applications and the increased use of fertilizers, detailed recommendations on plant spacing for different types of maize, coupled with provision of credit and advice. Even though maize yields more than doubled on some farms between 1967 and the early 1980s. Less than 1 per cent of farmers adopted the complete system as recommended. Most just adopted one or two components of the package into their existing system. Some of the reasons for not following the recommendations were that farmers do not know enough about new technologies and perceived them as being too risky; that farmers had an aversion to credit; that there were delays in fertilizer deliveries; and that there was lack of understanding among professionals of the complexity of farms and farmers' reasons for decision making (Gómez-Pompa et al, 1993).

Impacts of Modernization

National yields of both wheat and maize have increased since the 1940s, wheat from 750 to 4400 kg/ha, and maize from 600 to 1750 kg/ha (FAO, *passim*). Maize is the crop of the small farmer and yields have stayed low because the majority of farmers did not have access to the full package of technology. Wheat farms tend to be larger and are more likely to be irrigated, and so commercial wheat farmers have received more support from government programmes.

But this success has not been persistent. The early growth rates have not been repeated, and crises in agriculture became a feature of the 1970s and 1980s. Even though *'proud government politicians were still presenting Mexican agriculture as an enviable model which had succeeded in combining accelerated development with political tranquillity'* (Bartra, 1990), protests and conflicts increased in the countryside. Mexico has increasingly had to rely on food imports rather than internal production. Since the mid-1960s, both maize and wheat imports have increased substantially (Calva, 1988).

In the early 1980s, the government initiated a programme for food self-sufficiency (the SAM: *Sistema Alimentario Mexicano*), which continued the emphasis on the provision of credit and fertilizers. Again, there were no provisions for small farmers. The creators of SAM *'recognised that some producers would fall outside their development plans and 'would not be able to survive as farmers'* (Bartra, 1990). SAM proposed that their plots of land be combined. SAM was happy to see these 'inefficient' producers leave the land. Despite some success with better-off farmers, SAM has had to be discontinued because of the heavy investment required from government (Gómez-Pompa et al, 1993).

What these past efforts for improved modern production have tended to ignore is knowledge and labour-intensive practices that exist on small farms. Indeed, traditional forms of agriculture, such as the maize-beans-squash complexes, the raised field chinampas and the *cajete* systems of canals and terraced fields, have been overtaken by modernized agriculture, becoming progressively marginalized in the process (Dewey, 1981; Gómez-Pompa et al, 1982; Wilken, 1987; Borowitz, 1989; García-Barrios and García-Barrios, 1990). External resources have substituted for internal resources and processes, and knowledge about the complex local linkages and relationships needed to keep these systems going has been lost. Yet these traditional systems make use of diversity to control pests, maintain nutrient flows and provide a range of products of varying nutritional value to the household (Box 2.1).

In some areas, rural culture has been completely undermined by modernization. Some argue that the increasing social conflict is a direct result of the disempowerment of the small farmers or 'peasants' (Bartra,

Box 2.1 The impact of the loss of traditional complex and diverse systems of agriculture in Mexico

A: Traditional farms in Tabasco

Farmers have traditionally cultivated maize, beans, squash, manioc, sweet potatoes, plantains, rice, 31 species of fruit and 19 species of vegetable and spices. Modernized agriculture has brought land drainage and resettlement, and a focus on the farming of sugar cane, rice and cattle. Rising incomes associated with this agriculture have not all been beneficial:

- the falling diversity of crops has decreased dietary diversity and nutrition;
- increased cash income has meant increased cash expenditure on soft drinks and snacks. *Pozol*, a traditional drink made from ground maize and cocoa beans and drunk by children, one litre of which gives 1000 kcal, 25g of protein and various vitamins and minerals, has been replaced by soft drinks, one litre of which gives 500 kcal, and no other value;
- there has been a decline in mutual labour exchange arrangements;
- there has been a changed division of labour, with men now dominating agricultural activities.

B: Upland farming in the volcanic highlands of Tlaxcala

In the highlands, farmers have long maintained systems of canals and terraced fields, with *cajetes* at the base of the terraces to capture soil and organic matter, and act as compost pits. A wide range of crops are grown in complex intercropping and rotational patterns. Soil fertility management involves use of manures and the *cajetes*, and trees on field borders provide various functions. Despite their proven success, many farmers are abandoning these practices by substituting external inputs. To qualify for credit, farmers must monocrop with MVs. Monocrops are vulnerable to pests, so they must use more pesticides. The government subsidises the construction of new terraces with tractors, but these new terraces lack *cajetes*, so farmers increasingly use nitrogen fertilizer.

Sources: Dewey, 1981; Wilken, 1987; Borowitz, 1989

1990). In the Yaqui valley, Yaqui Indians had long been farming maize, beans, squash, wheat and many vegetables, utilizing wild fruits and seafoods, and doing this with strong social organizations. But now traditional agriculture has all but disappeared and has brought serious disintegration of the social fabric (see Chapter 3).

The modernization of Mexican farming has brought new agricultural practices and higher yields for some farmers in some areas. But there has been considerable cost to this improvement, with the loss of local institutions and traditional technologies (see Chapter 3). Despite huge investment in technology generation and extension, Mexico is still not self-sufficient in food.

THE MODERNIZATION OF SOIL AND WATER CONSERVATION IN THE USA AND AFRICA

Beginnings in the USA

The third case of agricultural modernization comes from the field of soil and water conservation. This too has a history of coercion and control (Pretty and Shah, 1994). Rural development policies and practice have generally taken the view that erosion occurs because farmers are poor managers of soil and water. Farmers have thus been advised, paid and coerced into adopting soil and water conservation measures and practices new to them. Many have done so, and environments and economies have benefitted for short periods. But the various financial and legal incentives have brought only short-lived conservation, and farmers soon revert to their own practices. These efforts have thus been remarkably unsuccessful, often resulting in more erosion (see Chapter 3).

The style of intervention was first established in the USA, where there is still a marked contrast between the widespread and enduring success of indigenous soil and water conservation, and the approach adopted by soil conservation authorities. Indian farming cultures had been farming with soil and water conservation measures for at least 1500 years in the Greater South-West of North America. Farmers of Anasazi, Hohokam, Pueblo, Zuni, Hopi and Papago cultures located fields where water ran off hills, built earthen diversion dams and channels to conduct water, used contour bunds, stone terracing and contour hedges of agave cacti, sited silt traps to produce gully fields, grew crops in mounds and on ridged fields, and stored runoff in reservoirs (Rohn, 1963; UNEP, 1983; Fish and Paul, 1992).

These combined to produce complex, diverse and productive agricultural systems. At Point of Pines, for example, 2500 ha of cultivated land with contour terraces, check dams and bordered gardens supported at least 3000 people for 500 years. And in New Mexico, bordered gardens connected by ditches to vast rain catchment areas supported a population density of 700 per sq km. None the less, these were ignored by the modern conservationists.

By far the greatest boost to modern conservation ideology was given by the Dust Bowl that struck the southern and south-western states of Oklahoma, Arkansas, Kansas, Colorado, Texas and New Mexico during the 1930s. In the early years of this century farmers were being encouraged to expand westwards by favourable homestead policies and the high price of wheat (Worster, 1979). In the 1910s, 30,000 farmers each year registered new land holdings in these states and, in 1919 alone, some 4.5 million hectares of grassland were ploughed for the first time to grow wheat. By the time the dust storms began, much of the land had only been farmed for a generation. Eventually some 50 million hectares of farmland were said to be severely affected by erosion. Dust and earth blanketed houses and crops, and there were potent images of destruction, the landscape having become *'a vast desert, with... shifting dunes of sand'* where there had once been crops (Worster, 1979).

These images of erosion linked farmers' cropping and grazing practices to increased frequency of droughts. The message was clear. Farmers caused land degradation which could lead to national ruin. At the time, several influential writers suggested that whole civilizations had collapsed through neglect of the soil (Bennett, 1939; Jacks and Whyte, 1939). The head of the US Soil Conservation Service, Hugh Bennett, painted environmental catastrophe large by indicating that *'the ultimate consequence of unchecked soil erosion, when it sweeps over whole countries as it is doing today, must be national extinction'* (in Beinart, 1984). Over a relatively short period, policy makers came to treat the problem as so serious that widespread social and institutional action had to be taken.

As a result a federal Soil Conservation Service (SCS – formerly the Soil Erosion Service) was established in 1935 as a separate body to the existing extension service. Its agents were to conduct a national inventory of erosion, so they could *'help the farmer do things correctly'* (in Trimble, 1985). From the start, erosion was seen as a problem arising out of bad farming practices that had to be corrected. The perception of great urgency in the face of 'looming catastrophe' encouraged the adoption of large-scale engineering to protect as much land as possible. But to demonstrate the efficacy of the approach, Bennett needed large amounts of land to practise the new conservation measures. As insufficient agreements came from private farmers, they selected Navajo native American reservations on which to experiment (Kelly, 1985).

They constructed physical measures, and enforced compulsory destocking of sheep and goats. The projects used college graduates for the technical work and employed Navajos as labourers. But the project provoked an intense negative reaction, not only to soil conservation but also to all government programmes. Anthropologists employed to conduct surveys discovered that the native Americans were not against soil conservation but were opposed to the way it was being implemented (Kelly, 1985). They took exception to the locations of the devices, as these interfered with and hampered other activities. It was not a lack of interest that prevented them from maintaining or repairing the structures and earthen dams, but rather the dams had been constructed with heavy equipment over which they had no access or control.

Conflict over budgeting and approach was to continue to hamper the SCS approach. The SCS approach was vigorously opposed by the extension service, whose agents at county level and in land grant colleges had a good knowledge of the local diversity of conditions. The SCS applied terracing technology widely, while local agents argued for locally adapted and appropriate technologies. But the dissenting voices were ignored. Sauer was one of the few who indicated that construction without maintenance did more harm than good: *'the present erosion crisis is the result primarily of the introduction of terracing, originally thought of as protection against erosion'* (Sauer, 1934 in Trimble, 1985).

Transfer to Africa

The pattern of intervention was repeated by colonial authorities in Africa. Erosion was first recognized as a problem as early as the 1870s, though it was not until the early part of the twentieth century that concern grew over farming as practised by indigenous people. At first, farmers were persuaded to adopt soil conservation practices through publicity bulletins extolling the virtues of contour ploughing and grass strips, by establishing demonstration plots and enacting some legislation, such as prevention of tree removal near streams (Stocking, 1985; Gichuki, 1991). Few adopted the technologies, even though groups of farmers were taken to demonstration farms to see the benefits of the new farming practices.

New grazing management systems of enforced enclosure of grazing lands developed in Texas were also implemented (Huxley, 1960). Potent images of erosion spurred these efforts. In Kenya, Huxley described *'gullies 15–20 feet deep... in places, the landscape seems as dead as the moon's'* in the west, and elsewhere the *'land is gashed... scraped bare, pounded into dust by the hoofs of little cattle and greedy goats'* (Huxley, 1960). It was clear to officials that local people were to blame. They sought technical guidance from the USA and brought back recommendations for large-scale conservation intervention. There were occasional dissenting voices. Writing in 1930, Sampson drew attention to indigenous methods of cultivation designed to check erosion, particularly mounding and ridge-and-furrowing systems on the contour. He indicated that local farmers *'fully realise the losses caused by erosion and consequent soil exhaustion, and their methods are well worth studying not only for themselves, but as a guide to those who seek to improve on them'* (Sampson, 1930). But these sentiments were rare.

When some of the new soil conservation efforts proved to be too costly to sustain, particularly where mechanization was required, administrators increased the use of local labour rather than adapt the technologies (Anderson, 1984). They also put together the components of good conservation practice into farm plans. These were literally a layout on a blueprint chart showing what every field was to grow for ten years, with all contours marked, the locations for woodlots, paddocks and homestead, and where to plant cash and food crops.

All of this required the monitoring of farming practices to ensure compliance. The final stage of control was achieved by the compulsory resettlement of farmers to centralized linear settlements where they could be observed more easily. In Kenya, more than one million people were moved in the mid-1950s to some 850 new linear villages (Huxley, 1960). Officials, proud of the new neatness and order, commented that farms of one village in Zimbabwe (then Rhodesia), were *'all in lines and look very nice'* (Alvord in Beinart, 1984). This was a complete contrast to the traditional way villages in East and Southern Africa are arranged. Now many of the straight paths and tracks readily became gullies, as they concentrated water flow down slopes. The contrast again with traditional practices is significant, where paths were laid out in zig-zag patterns and *'if one developed a gully, it would be moved'* (Mr Chibidi in Wilson, 1989). Soil and water conservation had extended to the remoulding of all aspects of rural life.

Modernized soil and water conservation has continued in this style in post-colonial times (see Chapter 3). Many interventions have led to more, not less, erosion from farmers' fields. Even though the short-term achievements can appear impressive in virtually all these soil conservation projects, structures and practices tend not to persist. As farmers are treated at best as labourers for construction, they have few incentives to maintain structures or continue with practices that they neither own nor have had a say in designing. All too often, impressive new structures and practices quickly degrade and eventually disappear, leaving little evidence of interventions and institutions (Pretty et al, 1994; Kerr and Sanghi, 1992; Haagsma, 1990; Reij, 1988; Marchal, 1986).

CONSEQUENCES OF PACKAGED TECHNOLOGIES

The Whole Package or Nothing?

As has been indicated earlier, few farmers are able to adopt the whole modern packages of production or conservation technologies without considerable adjustments. Part of the problem is that most modern agricultural research still occurs mostly on the research station, where scientists experience quite different conditions to those experienced by farmers. Scientists have access to all the necessary inputs of fertilizers, pesticides, machinery and labour at all the appropriate times. So when farmers try new technologies, they rarely do as well as the researchers.

Many farming households also cannot adopt the package because they simply cannot afford to take risks. They know they must have the whole package if they are to get any benefits. Paying for just half the package will be a waste as it gives no returns. For farmers, it is all or nothing. If one element of the package is missing, the seed delivery system fails or the fertilizer arrives late, or there is insufficient irrigation water, then yields may not be much better than those for traditional varieties. Although it is gradually being recognized, particularly in Asia, that packages are not appropriate to the complexities of rural life and that ways have to be found that allow farmers to select the components that make sense for them, it is still common to find this approach.

In Malawi, agricultural research since the 1960s has been developing new varieties of maize. These have been dent varieties that have soft starch which is easier for modern rollers to handle for flour production (Barbier and Burgess, 1990; Conroy, 1990; Kydd, 1989). These yield about twice as much grain per kg of nitrogen added than the traditional flint varieties. But many farmers find it difficult to get fertilizers when they need them, as they are often delivered late or not at all. Moreover, in a drought year, the response of the MVs to nitrogen is no better and so farmers applying inorganic fertilizer are worse off than those who do not (Conroy, 1990). Rural people also prefer flint varieties for their taste and high starch content, and because they are less subject to insect damage during storage. Scientists knew of these drawbacks, but added further new technologies to solve the problems. These included the promotion of insecticides to control storage pests and mechanical mills to overcome the difficulties of hand milling dents. But still the package was too costly and risky to farmers, and by the beginning of the 1990s, only 5 per cent of the maize area was being planted to modern dent varieties.

Added to this is the problem of very different adoption across wealth classes, with the poorest missing out (Table 2.4). Nearly a quarter of all households have less than 0.5 ha of land, of which close to half are female-headed. Only 2 per cent of their land area is under MVs, only 15 per cent use inorganic fertilizers, only 8 per cent are members of credit groups and very few are visited by extension workers or attend meetings. The contrast with farmers with more than 3 ha is marked: it is they who are capturing most of the benefits.

Table 2.4 Comparison of the characteristics of three classes of farmers in Malawi and the relative rates of adoption of components of the Green Revolution package, 1984–8

	Ultra poor landholding class (< 0.5 ha)	*Poor landholding class (0.5–1 ha)*	*Relatively wealthy land-holding class (> 3 ha)*
Proportion of households falling in class	23%	33%	4%
Proportion of households in each class that are female headed	42%	34%	8%
Proportion of farm area in each class under MVs	2%	2%	25%
Proportion of households in each class using fertilizer	15%	22%	36%
Proportion of households in each class in credit groups	8%	17%	25%
Proportion of households in each class visited by extension worker in last year	5%	8%	23%
Proportion of households in each class attending extension meetings	13%	25%	37%

Sources: adapted from Kydd, 1989; Conroy, 1990; Barbier and Burgess, 1990

Despite the failures of Green Revolution packages in Africa, efforts still continue along these lines (SAA/Global 2000/ CASIN, 1991, 1993). The Sasakawa Global 2000 project, for example, with the high profile support of ex-president Jimmy Carter and Norman Borlaug, has been working since 1986 in a wide range of countries. The central principle is that agricultural development *'cannot be achieved unless farmers have greater access to the products of science-based agriculture, namely improved varieties, chemical fertilizers and crop protection products, and improved crop management practices'* (Dowswell and Russell, 1991). The president of the Sasakawa Foundation, Yokei Sasakawa (1993), indicated that *'we believe that many of the agricultural lessons of Asia's Green Revolution can indeed be applied to Africa as well'.*

The project's strategy for technology transfer is the management training plot (MTP), which is managed by the farmer and supervized by an extension worker. Farmers receive loans to pay for inputs, which are subsidized. In Tanzania, for example, fertilizer prices to farmers are just one-third of the international price. The project would appear to be a great success, as yields on these plots are at least two to three times greater than traditional yields. Yet, like other Green Revolution efforts, there must be concerns about the sustainability of the efforts. Farmers can participate for a maximum of three years, after which *'they are graduated from the program and must obtain inputs using their own resources'* (Quinones et al, 1991).

Despite the fact that the project has encouraged the involvement of some 200,000 farmers in training plots, there remain significant questions about the capacity of countries to support this high external input approach when the projects cease. Joseph Makwete (1993), the Tanzanian minister of agriculture, said *'Tanzania cannot sustain the purchases of fertilizers required'*, and Christopher Dowswell (1993), the director for Program Coordination, indicated that *'it is unlikely that national governments will – or can – adopt the full program, given their budget limitations'*. These sentiments must call into question the appropriateness of the modern packages being promoted. MTP farmers themselves have expressed serious concerns about the supplies of inputs once the project withdraws (SAA/ Global 2000/ CASIN, 1993).

One critical view of the Global 2000 approach has come from the Ghanaian scientist Elsie Ayeh (1990). She describes how farmers stopped using the local neem insecticide, instead adopting commercial pesticides costing close to 400 times as much: *'many times, the impression is given when external inputs are introduced that they are the only cost effective ones'*. She describes farmers registering mainly to get fertilizer on credit, but then reverting to original practices after their eligibility ceases. As she says: *'it cannot be denied that the programme has helped farmers increase yields in the short-term, but should not its longer sustainability be determined?'*

Packages and Sustainable Agriculture

Crop research, however, has no monopoly on developing and extending technology packages that are not adopted by farmers. The same is true of many sustainability-enhancing innovations. Even if resource-conserving technologies are productive and sustainable (see Chapter 4), if they are imposed on farmers, then they will not be adopted widely.

A recent study of upland agriculture projects in South East Asia has found that farmers have not adopted resource-conserving technologies on a significant scale (Fujisaka, 1991a). Contour hedgerows, bench terraces, earth bunds, multiple cropping, legumes, perennial crops, contour tillage and alley cropping have all been introduced to farmers, as they offer the opportunity for increased yields on a sustainable basis (Box 2.2).

Box 2.2 Ten reasons for non-adoption of resource-conserving technologies in upland projects in South-East Asia

Absence of a problem	No adoption of erosion-controlling technologies in the Philippines and Indonesia where soils were fertile and showed little signs of erosion.
Inappropriate innovation	Permanent cropping and terracing recommended in Laos did not address the main problems for farmers, namely weed infestation.
General unawareness	Actually rare for farmers to be unaware of soil nutrient and erosion issues.
Incorrect identification of adoption domains	Researchers misunderstood reasons for local practices and targeted innovations inappropriately.
Local practices better	Local practices were often as effective as contour hedgerows for erosion control and biomass recycling.
Generation of new problems	Project-recommended permanent terraces in Laos became so infested with weeds that farmers forced to shift to new fields; napier grasses planted in hedgerows in Indonesia and Philippines caused resurgences of white grubs, a pest of rice and maize.
High cost	Labour and materials costs are immediate, but benefits may accrue only in the future.
Poor extension	Incorrect demonstration of a technology loses interest of farmers and undermines credibility of work; tendency to work with 'progressive' farmers only, and so misunderstood constraints of others.
Insecure land tenure	Tenant farmers generally less likely to improve their land where investments are in trees or terracing.
Inappropriate incentives	Adopters in Laos constructed ditches along contours in return for food from the World Food Programme, but then refused to do further work when food payments stopped.

Source: Fujisaka, 1991a

Again and again, the imposed models look good at first and then fade away. Alley cropping, an agroforestry system comprising rows of nitrogen-fixing trees or bushes separated by rows of cereals, has long been the focus of research (Kang et al, 1984; Attah-Krah and Francis, 1987; Young, 1989; Lal, 1989). The result has been the development of a range of systems that are very productive and sustainable, needing few or no external inputs. They control erosion, produce food and wood, and can be cropped over long periods. But the problem is that very few, if any, farmers have adopted these alley cropping systems as designed. Despite millions of dollars of research expenditure over many years, systems have been produced suitable only for research stations.

Where there has been some success, however, is where farmers have been able to take one or two components of alley cropping and then adapt them to their own farms. In Kenya, for example, farmers planted rows of leguminous trees next to field boundaries or single rows through their fields; and in Rwanda, alleys planted by extension workers soon became dispersed through fields (Kerkof, 1990). But the prevailing view tends to be that farmers should adapt to the technology. Of the Agroforestry Outreach Project in Haiti, it was said that: *'Farmer management of hedgerows does not conform to the extension program... Some farmers prune the hedgerows too early, others too late. Some hedges are not yet pruned by two years of age, when they have already reached heights of 4–5 metres. Other hedges are pruned too early, mainly because animals are let in or the tops are cut and carried to animals... Finally, it is very common for farmers to allow some of the trees in the hedgerow to grow to pole size. These trees are not pruned but are harvested when needed for house construction or other activities requiring poles.'* (Bannister and Nair, 1990)

Farmers were clearly making their own adaptations according to their own needs.

There has been a similar long experience with integrated pest management which, until recently, has tended to follow a typical cycle after overdependence on a single pest control tactic, usually insecticides. Patricia Matteson and colleagues (1992) describe the depressing scenario: *'There ensues a crisis or disaster phase of crop unprofitability or failure. The spectre of frightening economic and political losses spurs government to espouse and invest meaningfully in IPM, and gets farmers' attention. During the IPM phases an array of tactics based on ecological crop protection principles... are promulgated and applied. Then, after enjoyment of a period without pest control emergencies, a deterioration phase sets in. Policy makers and farmers are no longer motivated to stick to ecological principles, backsliding to routine pesticide applications that require less attention, effort and management skill. IPM research and extension programmes falter'.*

Another chastening experience comes from Mexico, where Arturo Gómez-Pompa and colleagues (1993, 1989, 1982) worked in the 1970s and early 1980s on transferring the Chinampa system of cultivation to lowland areas of Tabasco and Veracruz (see Chapter 4 for description of system). It was known to be an intensive, high output system of great antiquity. Experimental chinampas were constructed in the swamps, with more than

20 species of staples and vegetables cultivated on the fields, and fish and turtles harvested from the canals.

But when the local farmers and fishermen took charge, they began to abandon techniques and species. Some became just fishermen. Others came into conflict with local authorities that controlled the land. Others had difficulties with marketing the produce. As Gómez-Pompa and Jimenez put it in 1989: '*After all the initial effort to disseminate the knowledge of the Chinampas farming system in Tabasco in the years 1975–80, not one functioning hand-made chinampa in Tabasco remains at this moment*'. The principal reason for the failure was because of problems over marketing of the new produce. None the less, these efforts at integrated farming have triggered community diversification into pig, fish and chicken farming.

Just occasionally, however, an environmentally beneficial technology is developed that appears to require no knowledge of farmers' conditions. The integrated pest management (IPM) programme to control cassava mealybug (CMB) (*Phenacoccus manihoti*) in west and central Africa is one example (see Chapter 4). CMB was first recorded in Africa in 1973, and an effective natural enemy, the wasp *Epidinocarsis lopezi*, was found in 1981. Since releases began, it has became established in 25 countries, providing good control of CMB. It is to some extent a 'perfect technology' for scientists, as it is released from the air without the knowledge of farmers. It is, however, not necessarily a perfect technology for farmers. The contrast with another IPM programme in West Africa is significant when it comes to issues of sustainability (Box 2.3).

THE FINANCIAL COSTS OF MODERNIZATION

International Stresses on Poor Countries

Major changes in the economic conditions for many poorer countries put these externally generated and costly technological packages out of reach of many of their farmers. These stresses come from the falling real prices of agricultural products on world markets, the increasing debt burdens and the harsh economic conditions brought by structural adjustment programmes. Poor countries receive less for their agricultural exports and they have less money to purchase external inputs, which have in turn become more costly because of currency devaluations.

Low and very low income countries (with per capita GNPs of less than US$500) generally derive some 30–60 per cent of total national income from agriculture alone (World Bank, 1994a). This compares with less than 4 per cent for countries with per capita GNPs of US$6–30,000. For many poor countries, export earnings are highly dependent upon agriculture, with some countries almost entirely relying upon a single crop or commodity (Barbier, 1991; Conway and Barbier, 1990). Coffee comprises 96 per cent of the total export earnings for Uganda, 76 per cent for Rwanda, 61 per cent for El Salvador, 54 per cent for Ethiopia, and 25–30 per cent for Kenya, Colombia and Costa Rica. Some 70 per cent of Comoros' earnings are from cloves and vanilla, 60 per cent of those of Honduras are from bananas and coffee, and 53 per cent of those of Sri

Box 2.3 Comparison of farmers' involvement in two IPM programmes

A: Cassava mealybug control with *E lopezi*

The programme has involved close collaboration between IITA and NARSs, involving training of local technicians to participate in releases. Now mass rearing of *E lopezi* is done in Benin, from where they are transported by air for air release. According to IITA, an important component of success has been that farmers and extension agents have not had to be involved. Farmers do not, therefore, know anything about the releases. One survey of farmers in Ghana and Cote D'Ivoire found that they recognized CMB and how it was a devastating pest. All those where *E. lopezi* had been introduced at least six months before had observed a significant decline in CMB. But as none of them knew about the programme, they attributed the decline to recent heavy rains and other climatic factors.

B: Mango mealybug control in Togo

The CMB programme contrasts with the successful introduction of the parasitoid *Gyranusoides tebyii* to Togo in 1987 to control the mango mealybug (*Rastrococcus invadens*). The parasitoid was found in India, and following testing, rearing and release, it rapidly spread over the whole of Togo. By 1989, no mango trees could be found on which mango mealybug was present without being parasitized. But success would be threatened without public interest, as any use of chemical control methods would kill the parasites. A great deal of publicity was given, using radio, TV and advisory leaflets. Considerable economic losses are now being prevented by the biological control system.

Sources: Kiss and Meerman, 1991

Lanka are from tea and rubber. Very few of these countries depend upon food staples for earnings, the most important commodities being beverages, sugar, cotton, fruit, rubber and spices.

But the biggest problem for all countries relying on exporting agricultural produce comes from the steady fall in international prices of agricultural commodities. Over the last ten years, food and agricultural commodity prices have fallen substantially in real terms, to about 50 per cent of their former levels (Table 2.5). It is only bananas that have held anything like their price levels of ten years ago. Timber, it should be noted, has maintained parity and countries with timber stocks are clearly going to be doing better extracting those than relying solely on agricultural commodities.

Debt is still a pressing problem for very many countries. External debt as a proportion of GNP and debt servicing as a proportion of GNP and exports have risen substantially in recent years. In Africa, 13 countries had more long-term public debt than their total GNP in 1991 and a further 17 had a debt burden of more than half of GNP. Just servicing these debts uses up a substantial proportion of income earned from exports. Some 49 countries throughout the world now pay more than 25 per cent of their

Table 2.5 Changes in commodity prices 1980–2 to 1990–2

Commodity	*Price 1980–2*	*Price 1990–2*	*Proportion of 1992 to 1982 (%)*
Cocoa ($/kg)	2.98	1.16	39%
Coffee ($/kg)	4.38	1.71	39%
Tea ($/kg)	2.87	1.91	68%
Rice ($/t)	571	288	50%
Maize ($/t)	170	104	61%
Wheat ($/t)	257	154	60%
Beef ($/kg)	3.54	2.49	70%
Soybeans ($/t)	385	234	61%
Bananas ($/t)	536	511	95%
Sugar ($/kg)	0.55	0.22	40%
Copra ($/t)	531	289	54%
Cotton ($/kg)	2.54	1.55	61%
Rubber ($/kg)	1.80	0.99	55%
Sisal ($/t)	929	615	66%
Wool ($/kg)	5.94	3.74	63%
Palm oil ($/t)	742	330	44%
Tobacco ($/t)	3278	1991	61%
Indices			
Agricultural commodities	171	94	55%
Food commodities	175	93	53%
Cereals	174	98	56%
Beverages	210	89	42%
Non-food agricultural products	157	80	58%
Timber	112	105	94%

Source: OECD, 1993a

income for debt servicing (World Bank, 1993, FAO, 1993a).

In most poor countries, yields of major staples are low: virtually no very low income country produces more than 2 t/ha for any of the four major cereal staples. The use of fertilizers is generally low, and few countries have significant amounts of agricultural land irrigated or with adequate and predictable rainfall. For poor countries, the need to increase food production while not degrading the resource base is increasingly pressing. Yet they must first meet debt repayments and agricultural production must grow simply to offset the falling earnings from international markets.

The High Cost of the Modern Packages

The modern packages of technologies for production and conservation improvements would appear to be now too costly for most poor countries and poor farmers. All of this suggests that importing fertilizers, pesticides and machinery, investing in irrigation or other infrastructure, and putting in place large-scale terracing, are not viable options for many countries.

The high cost of the soil and water conservation packages gives a clear example of the scale of the problem for poor countries. The financial costs for constructing earth bunds and water harvesting structures in semi-arid Niger and Kenya are US$600–800 per ha; for constructing terracing in China, India, Morocco and Indonesia some $350–600; and for rehabilitating existing terracing in Korea and Peru some $1900–2500 per hectare (Reij, 1988; Magrath and Doolette, 1990; Treacy, 1989). Their high cost makes replication into non-project regions virtually impossible. As a result, coverage is usually limited to the project areas. When these costs are matched with the scale of effort required to treat all degraded lands, the aggregate costs using the current project packages are staggering. It would cost, for example, $0.6 billion for Niger (which was some 25 per cent of GNP in 1991) and $1 billion for Peru (some 4.3 per cent of GNP). Such expenditure is clearly beyond the means of most countries. As Chris Reij (1988) put it: *'It will be evident that only low-cost soil and water conservation techniques could possibly be applied on a large scale'*.

Many countries also spend very large sums on subsidizing inputs (see also Chapter 8). In India, the cost of subsidizing inputs was 2.8 per cent of GDP in 1990 (Rajgopalan, 1993). Pesticide subsidies can be huge too, with many countries spending tens to hundreds of millions of dollars each year keeping prices low (Repetto, 1985). In Indonesia, subsidies on pesticides alone cost $1.21 billion between 1977–86, with an annual high of $160 million reached in 1981 (Kenmore, 1991).

The cost of new infrastructure is also beyond most countries. Irrigation has been an important component of the successes of the Green Revolution. Between 1960–90, the extent of irrigated land grew from 100 to 170 million ha. In recent years, expansion has slowed, partly because investment is increasingly having to be made in rehabilitating existing systems, but also because development costs are so high. If recent trends of growth are to continue, this would mean a growth rate in irrigated area of a further 130 million ha by 2025. But at current prices this would mean an investment of some $400–800 bn, which is 3–6 times more than all resource flows to Third World countries – including from official development assistance, direct foreign investment, from NGOs and export credits, and some 7–14 times greater than from official development assistance (OECD, World Bank, passim).

These financial costs suggest that it will be difficult in the immediate future for many countries to adopt or benefit from modernist strategies for agricultural development. But in addition to these costs, there are also many hidden environmental and social costs experienced by countries following modern strategies. These are the subject of Chapter 3.

SUMMARY

In the past century, rural environments in most parts of the world have undergone massive transformations. Two guiding themes have dominated: the pursuit of increased food production and the desire to prevent environmental degradation. As a result, both food production and the amount of land conserved have increased dramatically. The process of modernization has characterized both these achievements. Technologies and practices are assumed to be universal, and exist independently of social context. When they are not widely favoured or adopted, then the external response has been to blame the farmers rather than the technology. This in turn has led to modern agricultural development being characterized by coercion and enforcement at one time or another.

The process of agricultural modernization has produced three distinct types of agriculture: the industrialized, the Green Revolution, and the complex and diverse agriculture. Modernization has succeeded in the first two, leading to big increases in productivity. The third has largely been forgotten, even though it supports some 1.9–2.2 billion people. The specific cases of rice in Indonesia, and maize and wheat in Mexico, illustrate the process of modernization during the past 40–50 years. In both countries, there have been significant improvements in productivity, but in both farmers have either been obliged or coerced into adopting the package of modern technologies, or have been simply ignored. Many have become poorer as a result.

The pursuit of environmental conservation has been characterized by technologies developed in particular historical contexts and transferred irrespective of the receiving environments. Efforts to conserve soil and water, establish protected areas and manage rangelands have all been highly ineffective and costly, often eventually leading to increased alienation of local people. Despite widespread evidence of indigenous soil and water conservation, modern conservation ideology has imposed measures as a result of the Dust Bowl that struck the southern USA during the 1930s. In the name of conservation, farmers have been paid, coerced and enforced into adapting conservation technologies in Africa and Asia, yet few have maintained these measures.

The problem for many farming households is they cannot adopt modern packages without significantly changing their own practices. They either adopt the whole package or reject it entirely. Although it is gradually being recognized that the complete package is not appropriate for the complexities of most rural life, packages are still being used in programmes promoting both high external input and sustainable agriculture.

New processes are needed, as most modern, external packages are financially costly for countries and farmers, and it is difficult to see how many could benefit from the modernist strategies for agricultural development. Only low-cost technologies and practices will be able to be applied on a scale wide enough to improve the livelihoods of some 2 billion people. This will require the adoption of an entirely different approach to agricultural and rural development.

3

THE ENVIRONMENTAL AND SOCIAL COSTS OF IMPROVEMENT

'But when the motor of a tractor stops, it is as dead as the one it came from. The heat goes out of it like the living heat that leaves a corpse. Then the corrugated iron doors are closed and the tractor man drives home to town, perhaps twenty miles away, and he need not come back for weeks or months, for the tractor is dead. And this is easy and efficient. So easy that the wonder goes out of the work, so efficient that the wonder goes out of the land and the working of it, and with the wonder the deep understanding and the relation'.

John Steinbeck, The Grapes of Wrath, 1939

THE GENERAL COSTS OF IMPROVEMENT

The pursuit of increased productivity and conserved natural resources in the course of rural modernization has produced benefits in the form of improved food production and some improvements in resource conservation. The increases in food production have been significant. Despite the world population more than doubling in the past 50 years to some 5.6 billion, food production per capita has been able to keep pace. Over the same period, the amount of land conserved or protected has also increased. In the tropics alone, the land devoted to national parks and protected areas has grown from 58 to 174 million ha.

These improvements look so good that it is easy to be tempted to forget: 'What is the cost of this improvement?' 'Who benefits and who loses out?' Many would argue that the ends surely justify any reasonable means. Yet it is increasingly being recognized that the social and environmental costs of agricultural modernization cut deep into the fabric of society. Modernization in the urban environment has been characterized by alienation and conflict, increased individualism and a breakdown of communities. Much the same is true in rural environments. Jobs have been lost, environments polluted, communities broken up and people's health damaged.

All sectors of economies are affected. The drive for agricultural efficiency has drastically cut the numbers of people engaged in agriculture in industrialized countries. External inputs of machines, fossil fuels, pesticides and fertilizers have displaced workers in Green Revolution lands. Rural cultures have been put under pressure, as more and more people have been forced to migrate in search of work. Local institutions, once strong, have become coopted by the state or have simply withered away. Farms have become simplified and some resources, once valued on the farm, have become wastes to be disposed off the farm. Some external inputs are lost to the environment, so contaminating water, soil and the atmosphere. Agriculture has become more fossil-fuel intensive, so contributing to global warming. Overuse or continued use of some pesticides causes pest resistance and leads to pest resurgences, encouraging farmers to apply yet more pesticides.

Environmental Pollution and Contamination by Agriculture

The agricultural production increases brought about by high input packages have brought great benefits. Without them many people would be worse off than they are now; many others might have died of starvation. But in order to assess the true net benefits of high input packages, it is important also to understand some of the external costs.

The environmental problems caused by farming are a direct result of an increasingly intensive and specialized agriculture. The mixed farm can be an almost closed system, generating few external impacts. Crop residues are fed to livestock or incorporated in the soil; manure is returned to the land in amounts that can be absorbed and utilized; legumes fix nitrogen; trees and hedges bind the soil, and provide valuable fodder, fuelwood and habitats for predators of pests. In this way the components of the farm are complementary in their functions. There is little distinction between products and by-products. Both flow from one component to another, only passing off the farm when the household decides they should be marketed.

Over the last half century, many such highly integrated systems have disappeared. Farms have become more specialized with crop and livestock enterprises separated. Intensification of agriculture has meant greater use of inputs of pesticides, fertilizers and water, and a tendency to specialize operations. The inputs, though, are never used entirely efficiently by the receiving crops or livestock and, as a result, some are lost to the environment. Some 30–80 per cent of applied nitrogen and significant but smaller amounts of applied pesticides are lost to the environment to contaminate water, food and fodder and the atmosphere (Conway and Pretty, 1991). Water is often wasted or used inefficiently, leading to groundwater depletion, waterlogging and salinity problems. This is not only wasteful, but costly to those who want to use these resources and expect them to be uncontaminated.

Many environmental and health impacts have increased in recent years; others have continued to persist despite all efforts to reduce them (Conway and Pretty, 1991). Water systems have become increasingly contaminated. Nitrate in water can give rise to the condition methaemoglobinaemia in infants and is a possible cause of cancers. Pesticides contaminating water can harm wildlife and exceed drinking water standards. Nitrates and phosphates from fertilizers, and organic wastes from livestock manures and silage effluents all contribute to algal growth in surface waters, deoxygenation, fish and coral deaths, and general nuisance to leisure users. Eroded soil also disrupts water courses, and runoff from eroded land causes flooding and damage to housing, irrigation systems and natural resources.

Various pollutants also harm farm and local natural resources. Pesticides damage predator populations and other wildlife and induce resistance in target pests. Nitrates from fertilizers and ammonia from livestock waste disrupt nutrient-poor wild plant communities. Metals from livestock wastes raise metal content of soils, and pathogens in wastes can harm human and livestock health. The atmosphere is contaminated by ammonia, which plays a role in acid rain production; nitrous oxide derived from fertilizers, which plays a role in ozone layer depletion and global warming; and methane from livestock and paddy fields, which also affects global warming.

The consumer is most likely to be directly affected by eating food contaminated mainly by residues of pesticides, but also by nitrates and antibiotics. In the industrialized countries, the levels of pesticides in foods have been falling steadily since the 1950s (Conway and Pretty, 1991; WPPR, 1994; Gartrell et al, 1986a, b). None the less, there are occasional public scares over particular products and rare incidents of severe poisoning arising from the spraying of illegal products. But in Third World countries, daily intakes are often very high. These may be in cereals, such as in India (Kaphalia et al, 1985; Sowbaghya et al, 1983); in fish, such as from rice fields in Malaysia (Chen et al, 1987) or lakes and rivers of Kenya, Nigeria and Tanzania (Atuma, 1985; Atuma and Okor, 1985); and in milk from cows affected by spray drift from cotton plantations in Nicaragua and Guatemala (ICAITI, 1977).

But the major hazard lies in locally marketed food. Leafy vegetables are often sprayed twice a week and may come to market with a high degree of contamination, especially in the dry season. Over 50 per cent of green leafy vegetables collected around Calcutta during the dry winter months contained residues, though this fell to 8 per cent in the wet season (Mukherjee et al, 1980). In Indonesia, cabbages and mustard greens have been found to contain organophosphates many times in excess of human tolerance limits (Darma, 1984). Similar levels of contamination have been recorded from Africa (Atuma, 1985).

These costs of environmental damage are growing, and are dispersed throughout many environments and sectors of national economies. For a comprehensive review of the effects of agricultural pollution on natural resources, wildlife and human health, see Conway and Pretty (1991). What has characterized recent analyses has been the recognition that farmers themselves are suffering declining incomes or health effects from these modern approaches to agriculture. The following sections consider the issues of energy consumption by agriculture, pest resistance and resurgences, health impacts of pesticides and soil erosion.

Energy Consumption by Agriculture

A largely hidden cost of modern agriculture is the fossil fuel it must consume to keep outputs high. Modern agriculture has tended to substitute external energy sources for locally available ones (see Chapter 1). With the increasing use of nitrogen fertilizers, pumped irrigation and mechanical power, which are all particularly energy intensive, agriculture has become progressively less energy efficient. These three account for more than 90 per cent of the total direct and indirect energy inputs to farming in Third World countries (Leach, 1985, 1976). Mechanization reduces the labour required for agriculture and so can cut variable costs if energy is cheap, as it is in most industrialized countries. But for poorer countries, mechanization forces increased foreign exchange expenditure on fuel, oil, engines and spares.

There have been many approaches to energy accounting for agricultural systems (Leach, 1976, 1985; Stout, 1979; Stanhill, 1979; Pimentel, 1980; Smil et al, 1982; Dovring, 1985; Pimentel et al, 1989; OECD/IEA, 1992; OECD, 1993). These use such a variety of conventions that it is difficult to make direct comparisons. Some include only the direct fossil fuel energy consumed on farms; others seek comprehensive energy balances by including all the indirect energy consumed in manufacturing equipment and inputs, transporting produce to and from farms, and the energy required to feed human and animal labour on the farm. Direct energy represents what is immediately vulnerable to supply interruptions and so is of more immediate interest to farmers. In general, apart from nitrogen fertilizers, the manufacture of which is extremely energy intensive, direct energy costs far exceed indirect costs (Leach, 1985).

With the greater use of machinery, fuel and nitrogen fertilizers in modern high input agriculture, energy consumption is substantially greater than equivalent low input or organic systems (Table 3.1). In the Philippines, for example, a doubling of yields comes at the cost of an 8 to 30 fold increase in energy consumption. In India, a 10–20 per cent increase in yields following mechanization costs an extra 43–260 per cent in energy consumption. And in the USA, high input systems can consume 20–120 per cent more energy than low input systems, even though yields may be comparable. Larger farms also tend to use relatively more energy than smaller ones. In the Punjab, farms in a class 14–25 ha use three times as much direct energy per hectare as farms smaller than 6 ha (Singh and Miglani, 1976).

Comparisons within countries or even regions are much more likely to be reliable than those between countries, as so many confounding factors become important. However, a comparison of the energy consumption across systems is revealing, if only at the level of orders of magnitude (Table 3.2). Low input, resource-conserving systems of production are much more energy efficient than the high input systems typical of industrialized countries. Low input or organic rice in Bangladesh, China, and Latin America can produce 1.5–2.6 kg cereal per MJ of direct energy consumed. This is some 15–25 times more efficient than irrigated rice produced in the USA.

Table 3.1 Impact of modernization of agricultural systems on yields and direct energy consumption

Country	*Low input comparison*	*High input comparison*	*Amount of extra yield for high input*	*Amount of extra energy consumption for high input*
Philippines[1]	traditional rice	modern rice	+116%	+3000%
	rainfed rice	irrigated rice	+150%	+800%
	irrigated rice with *Azolla*	irrigated rice with N	+0–30%	+200%
India[2]*	bullock, rice	power tiller, rice	+8%	+43%
	bullock, rice	tractor, rice	+13%	+74%
	bullock, wheat	power tiller, wheat	+12%	+89%
	bullock, wheat	tractor, wheat	+6%	+266%
India[3]*	bullock, rice	mechanized, rice	+20%	+45%
	bullock, wheat	mechanized, wheat	+29%	+138%
USA[4]	low input, maize	conventional, maize	+0%	+120%
USA[5]	organic, wheat	conventional, wheat	+29%	+48%
USA[4]	low input, maize	conventional, maize	+0%	+22%

Note: for the sake of comparisons, the data in this table refer to direct energy use plus indirect energy for manufacturing fertilizers and seeds. These are not comprehensive energy accounts, in which all embodied energy is included. For those marked *, only direct energy is included.
1 Luzon 2 West Bengal 3 Uttar Pradesh 4 Midwest 5 Pennsylvania and New York
Sources: FAO, 1976; Leach, 1976; Singh and Singh, 1976; Pimental et al, 1989; Berardi, 1978; Ikerd et al, 1992; Lockeretz et al, 1981.

In industrialized countries the trend has been towards the substitution of inexpensive fuel energy for expensive human labour, so making agriculture a significant energy consumer. Since the 1940s, some 25 million draft animals and 9 million agricultural workers have been replaced in the USA; and in the UK, 340,000 jobs have been lost (Berardi, 1978; MAFF, passim). Energy consumption has increased too. According to the OECD (1993b), the absolute energy consumption per hectare has increased in OECD countries by 39 per cent from 1970 to 1989. On average, some 1734 MJ are consumed per hectare of agricultural land, rising to 46,400 MJ for the highest consumer, Japan.

One consequence of this increased substitution of energy for labour in agriculture is a growing contribution to global warming. Agriculture is a major direct source of atmospheric pollution, emitting methane, nitrous oxide, ammonia and the various products of biomass burning (Conway and Pretty, 1991; IPCC, 1990). The single main cause of global warming, however, is carbon dioxide, estimated to contribute about half of the projected warming over the next 50 years. Agriculture contributes to CO_2

Table 3.2 Amount of cereal produced (kg) per megajoule (MJ) of direct energy input in different agricultural systems

Location	*System of production*	*No. of kg cereal produced per MJ of direct energy (+ indirect energy for fertilizers and pesticides used (kg/MJ)*
Japan	Irrigated, high input rice	0.30
China	Organic rice	1.53
Philippines	High input, irrigated rice	0.22–0.36
	Low input, irrigated rice with *Azolla*	0.79
	Rainfed, upland rice	0.72–0.88
Latin America	Low input, upland rice	1.94
Bangladesh	Low input, deepwater rice	2.64
USA	High input, irrigated rice	0.09
USA	High input maize	0.25
	Low input maize, alternative rotations	0.67
UK	Very high input wheat	0.45
	Low input wheat	1.09

Sources: Adapted from IRRI, 1981; FAO, 1976; Walters, 1971; Pimentel et al, 1989; Leach, 1985

production directly through the burning of biomass and indirectly through its consumption of energy produced by fossil fuel burning. For each kilogramme of cereal from modernized high input conditions, 3–10 MJ of energy are consumed in its production; but for each kilogramme of cereal from sustainable, low input farming, only 0.5–1 MJ are consumed. A shift to low input systems could, therefore, have an impact on the process of global warming.

However, there is considerably more energy consumed between the farm and the consumer. In the USA, it is said that food travels on average 3000 km from farm to plate. In Britain, the production of a 1 kg loaf of bread consumes some 20.7 MJ (equivalent to 0.48 kg of oil), of which 80 per cent is consumed by milling, baking, transport and retailing (Leach, 1976). Making agriculture more energy efficient, by transferring to low input sustainable processes, could only decrease the energy consumed in the remaining 20 per cent. This could reduce the energy consumed in a loaf of bread to 16–17 MJ. However, for cereals processed and consumed on the farm, or those passing through fewer processing or transport stages, significant improvements in energy efficiency could be possible following a transition to a more sustainable agriculture.

Pesticide-Induced Pest Resistance and Resurgences

The reason for applying pesticides is to prevent pest damage, yet unfortunately they can cause outbreaks themselves. Pesticides can be inefficient for several reasons (Conway, 1981; Risch, 1987). They can cause resurgences by killing off the natural enemies that control pests. They can produce new pests, by killing off the natural enemies of species which hitherto were not pests. And they can induce resistance in pests to pesticides.

Resistance can develop in a pest population if some individuals possess genes which give them a behavioural, biochemical or physiological resistance mechanism to one or more pesticides. These individuals survive applications of the pesticide, passing their genes to their offspring so that with repeated applications the whole population becomes resistant. High and frequent applications of pesticides exert the greatest selection pressure on populations. Resistance has now developed in all insecticide groups and at least 480 species of insect, mite or tick have been recorded as resistant to one or more compounds (Georghiou, 1986). Resistance has also developed in weeds and pathogens. Before 1970, few weeds were resistant to herbicides but now at least 113 withstand one or more products. Some 150 fungi and bacteria are also resistant (WRI, 1994).

Unfortunately, natural enemies appear to evolve resistance to pesticides more slowly than herbivores, mainly because of the smaller size of the natural enemy populations relative to pests and their different evolutionary history (Risch, 1987). The coevolution of many herbivores with host plants that contain toxic secondary compounds means they have metabolic pathways easily adjusted to produce resistance (Croft and Strickler, 1983). In Sudan, the increasing application of pesticides to cotton over the past 50 years has steadily reduced the number of predator species. One insect, the whitefly, which was formerly kept in check by predators, is now an economically very important pest (Kiss and Meerman, 1991; PT, 1990).

Outbreaks and resurgences are more likely to occur when the landscape has been simplified to contain just a single crop. This may be of cereals, such as wheat or rice, or of plantation crops, such as bananas, cotton or coffee. In Costa Rica, 30 per cent of imported pesticides are used in the production of bananas for export. Bananas are grown in huge plantations, which are highly susceptible to pests and diseases, and there have been repeated cycles since the 1950s of heavy applications of one product, closely followed by pest outbreaks caused by the rapid development of resistance (Thrupp, 1990). When decisions were taken to stop spraying because of inefficiency and growing costs, insect pests rapidly declined: *'two years after insecticides were halted, the previous [predator] species became established again'* (C Stephens in Thrupp, 1990). Today's integrated approach to pest management requires greater technical expertise and labour of managers and operators, and now incorporates cultural methods, minimal and selective use of insecticides, and threshold monitoring. Insects rarely now present any problems, though spraying against disease, nematodes and weeds is still heavy.

In Asia, where 90 per cent of the world's rice is produced and consumed, reports of disease and insect outbreaks are numerous (Khush, 1990; Kenmore, 1991; Winarto, 1994). Brown planthopper (*Nilaparvata lugens*) outbreaks have at various times destroyed hundreds of thousands of ha of rice in countries from India in the west to the Solomon Islands in the east. In Indonesia, the first problems started occurring in 1974. Losses jumped in 1975, after the government started subsidizing pesticides and in 1977 over 1 million tonnes of rice were lost, enough to feed some 2.5 million people (Kenmore, 1991). In 1979, 750,000 ha were infested, followed by lower, but not insignificant, levels of infestation of between 20–150,000 ha per year during the 1980s. During this period, BPH was only really checked with the release of rice varieties containing genes that confer resistance, though even some of these have been attacked by new biotypes of the pest (Khush, 1990).

Studies in the Philippines and Indonesia have clearly shown that outbreaks occurred after increases in insecticide use (Kenmore et al, 1984; Litsinger, 1989; Winarto, 1993). BPH is kept under complete biological control in intensified rice fields that are not treated by insecticides. Even with over 1000 reproducing adults per square metre, the natural enemies exert such massive mortality that rice yields are unaffected. As Peter Kenmore (1991) describes *'insecticide applications disrupt that natural control, survival increases by more than ten times, and compound interest expansion then leads to hundreds of times higher densities within the duration of one rice crop. Trying to control such a population outbreak with insecticides is like pouring kerosene on a house fire'* (see Case 11, Chapter 7). Other countries in South East Asia still, however, suffer significant losses to BPH. In central Thailand, some 250,000 ha were infested in 1990, the worst year on record.

PESTICIDES AND HUMAN HEALTH IMPACTS

Mortality and Morbidity from Pesticides

There is no doubt that pesticides are hazardous. At very high dosages many are lethal both to laboratory animals and people, and can cause severe illness at sublethal levels. But just how serious is the hazard from medium to low dosages is open to question (Conway and Pretty, 1991; IARC, 1991). In the 1950s, 1960s and 1970s organochlorine insecticides were in widespread use in the industrialized countries and high levels of exposure were common in those engaged in their manufacture, in agricultural workers and, because of the presence of residues in foods, among the general public. Nevertheless, there is little evidence of serious ill-health, other than as a result of accidental exposure to high dosages. The herbicides 2,4,5-T and 2,4-D were also commonly used in that period, and were originally thought to be a cause of miscarriages. Subsequent, more thorough, studies suggest a link with increased incidence of a certain rare cancer, non-Hodgkin's lymphoma, but not with miscarriages or other reproductive effects (Hoar et al, 1986, 1988; Witt, 1980; Agresti, 1979; Conway and Pretty, 1991).

Other pesticides appear to be intrinsically less hazardous, although the organophosphates, in particular, can cause severe poisoning. These are more acutely toxic than organochlorines but since they are not stored in body tissues are probably less hazardous over the long term. Two highly hazardous pesticides are the nematocide, DBCP, which causes infertility in humans and the herbicide, paraquat, which is carcinogenic and mutagenic. However, many synthetic pyrethroids, and modern herbicides and fungicides, have very low toxicity and no known health effects.

In the industrialized countries, the major hazard lies in accidents. Even then, fatalities at work are very rare – one a decade in the UK; and eight a decade in California – and there are many other more common causes of death on the farm. There is, though, a relatively high incidence of ill-health among those engaged in applying pesticides. Farmers exposed to organophosphates during the dipping of sheep, for example, appear increasingly to be suffering a wide range of sub-acute health problems.

One problem is that the systems for recording pesticide poisoning vary within and between countries, and are difficult to compare. In the UK, there are at least four institutions collecting mortality and morbidity data, all giving different data (Conway and Pretty, 1991; HSE, 1993). These suggest some 40–80 confirmed cases each year. In California, a comprehensive system of reporting, perhaps the best in the world, records some 1200–2000 cases each year (CFDA, passim). Overall the hazard in the industrialized countries presented by pesticides appears to be not very different from that of other manufactured chemicals, such as pharmaceuticals.

By far the greatest risk, though, is from pesticides in the home and garden where children are most likely to suffer. In California alone, some 6–8000 children of less than 6 years of age are treated for pesticide poisoning each year. In Britain, some 600–1000 people need hospital treatment each year from home poisoning (Conway and Pretty, 1991). Although, in this respect, pesticides are no different from hazardous medicines, they are often not perceived as being in the same category and are less carefully guarded. None the less, there continues to be considerable public concern over the risks arising from exposure to pesticides, in particular through accidental spraying and spray drift, or from residues in foodstuffs.

Greater Hazards in Third World Countries

In Third World countries, mortality and illness due to pesticides are much more common relative to the amount of pesticide used. Lack of legislation, widespread misunderstanding of the hazards involved, poor labelling and the discomfort of wearing full protective clothing in hot climates, all greatly increase the hazard both to agricultural workers and to the general public (Conway and Pretty, 1991). Moreover, many pesticides known to be highly hazardous and either banned or severely restricted in the industrialized countries, such as parathion, mevinphos and endrin, are widely available. In 1988, the Food and Drug Administration of the USA found that 5 per cent of some 10,000 imported foods when tested were found to contain residues of products banned in the USA, indicating continued widespread export and use of such compounds (GAO, 1989).

It is very difficult to say how many people in the South are affected by pesticide poisoning. This is partly because reporting mechanisms are weak, with farmers tending not to seek medical treatment – as is also the case in the North (Dinham, 1993). Most data are gathered by doctors, researchers and activists from individual testimony and hospital records, and so are viewed as anecdotal or circumstantial. This is not to denigrate these reports; it is just that many policy makers do not accept them as sufficiently 'scientific'.

None the less, put together the data paint a picture more bleak than appeared to be the case in the 1980s (Conway and Pretty, 1991; Dinham, 1993).

- In Malaysia and Sri Lanka, for example, some 7 to 50 per cent of all farmers reported that they experienced poisoning at least once in their lives (Jeyeratnam, 1990).
- In Thailand, a survey of 250 government hospitals and health centres revealed that some 5500 people were admitted for pesticide poisoning in 1985 alone, of whom 384 died (Jonjuabsong and Hawi-khum, 1991).
- In Latin America, between 10–30 per cent of agricultural workers tested show inhibition of the blood enzyme, cholinesterase, which is a sign of organophosphate poisoning (WHO, 1990).
- In Venezuela, there were 10,300 cases of poisoning with 576 deaths, between 1980–90 (Dinham, 1993).
- In Paraguay, 75 per cent of farmers around Asunción experienced symptoms after spraying (Dinham, 1993).
- In Brazil, 28 per cent of farmers in Santa Catarina say they have been poisoned at least once; and in Parana, some 7800 people were poisoned between 1982–92 (Dinham, 1993).
- In China, a recent statement from the Agricultural Ministry in China suggested that more than 10,000 Chinese farmers died in 1993 from poisoning by pesticides (Quinn, 1994). Many were said to be victims of home-made cocktails marketed illegally and some 30 per cent of products were unlicensed by authorities. Since 1975, the value of pesticide imports into China has grown from US$76 million to $293 million

According to the latest (1990) estimates by the WHO, a minimum of 3 million and perhaps as many as 25 million agricultural workers are poisoned each year, with perhaps 20,000 deaths.

Pesticide Poisoning and Health Costs in the Philippines

Recent evidence emerging from the intensive rice-growing regions of the Philippines is confirming this picture of common mortality and illness from pesticides. These areas have greatly benefited from the Green Revolution packages and use of pesticides is still growing, with sales of pesticides increasing by 70 per cent between 1988 and 1992. Between 1980–87, the National Department of Health Statistics recorded some 4031 cases of pesticide poisoning, including 603 deaths (Casteñeda and Rola, 1990). But this may be an underestimate of the extent of poisoning, as studies are increasingly showing high incidence of poisoning symptoms that go unreported by farmers (Rola and Pingali, 1993; Marquez et al, 1992; Rola, 1989; Loevinsohn, 1987). A recent WHO report recorded 1303 cases of poisoning between January 1992 and March 1993 in one region alone (WHO, 1993).

In one study, Michael Loevinsohn (1987) examined mortality statistics in several contrasting municipalities in central Luzon for diagnosed pesticide poisoning and for other conditions that could be the result of such poisoning. Organochlorines such as endrin and HCH can cause convulsions, so that poisoning may be misdiagnosed as epilepsy, brain tumours or strokes. Similarly, poisoning by organophosphates, such as parathion, can be misdiagnosed as cardiovascular or respiratory diseases.

The study detected a 27 per cent increase between 1961–71 and 1972–84 of non-traumatic mortality rates among rural males aged 15–54 years, although in children and women it decreased. This increase closely coincided with the growth in pesticide use. When the figures were broken down, they revealed that deaths in the rural areas diagnosed as poisoning increased by 247 per cent and those from associated conditions by 41 per cent between the two periods, yet mortality from all other causes, except cancer, decreased by 34 per cent. In the case of stroke, mortality increased for all men in both the urban and rural areas, but significantly the increase was greater among young men who are generally at low risk of stroke.

Following the 1982 ban on endrin, mortality attributed to stroke decreased for all men, but the decrease was significantly greater among the younger men in rural areas. The study also revealed that mortality rates had originally peaked each year during August, the month of greatest insecticide use. But after double-cropping became widespread, a second mortality peak appeared in February, at a time when insecticides were used on the newly cultivated dry season crops. These correlations are highly suggestive of occupational exposure to pesticides.

Another important study compared the health status of farmers exposed to pesticides in Nueva Ecija with those unexposed in Quezon. In the exposed group, there were statistically significant increased eye, skin and lung problems. Some 67 per cent of farmers suffered from severe irritation of the conjunctivae (compared with 10 per cent in the unexposed group); 46 per cent suffered from eczema and nail pitting (compared with none in the unexposed); and 46 per cent suffered respiratory problems (compared with 23 per cent in the unexposed). Another study of Nueva Ecija farmers found that 50 per cent of rice farmers suffered from sickness due to pesticide use (Rola, 1989).

Agnes Rola of the University of the Philippines and Prabhu Pingali of IRRI (1993) calculated the health costs of these pesticide problems, taking into account impact on exposed farmers and the costs of restoring individuals to normal health, so as to examine the economics of various pest control strategies. The 'complete protection' strategy, in which some nine sprays are used per season, returned less per hectare than the

Table 3.3 Net benefits and health costs of four pest management strategies in lowland irrigated rice, Philippines

Pest management strategy	*Net benefits, excluding health costs (Pesos/ha)*	*Health costs (Pesos/ha)*
Complete protection: nine applications of pesticide per season	11,846	7500
Economic threshold: treatment only when threshold passed, usually no more than two applications used	12,797	1188
Natural control: pest control emphasizes predator preservation and habitat management, alternative hosts and resistant varieties	14,009	0
Current local practice: 2–3 applications of very hazardous compounds per season	13,847	720

Source: Rola and Pingali, 1993

economic threshold, farmers' practice and natural control strategies (Table 3.3). These results indicate that both farmers and the national economy at large would be better off by cutting pesticide use or eliminating it entirely by adopting more integrated and sustainable practices.

SOIL CONSERVATION AND EROSION

The Causes and Costs of Erosion

Despite the fact that indigenous systems of soil and water conservation are widespread, well adapted to local conditions, persist for long periods and are capable of supporting dense populations, soil erosion continues to be a problem throughout the world (UNEP, 1983; Reij, 1991; Kerr and Sanghi, 1992, Tato and Hurni, 1992; Hudson and Cheatle, 1993; Pretty and Shah, 1994). Indigenous systems are insufficient alone to prevent agricultural land from continuing to lose productive soil, water and nutrient resources. This is partly because not all farmland is protected by conservation measures, but also because not all erosion arises from farmland. Both roads and urban areas concentrate water flows and non-agricultural areas are also subject to erosion.

Farmers may not be conserving soil and water for a variety of reasons. They may lack the locally appropriate knowledge or skills, particularly if they have been resettled or migrated to new areas. They may be unwilling to invest in conservation measures if the economic costs of conservation are greater than the expected benefits, particularly if the future is uncertain, such as if political instability or conflict threaten the future, or if security of tenure is uncertain. Farmers may be short of labour for construction or maintenance, such as following a decline in population, outmigration in the face of better opportunities for income earning, particularly in urban centres, or simply rising labour costs. They may not be conserving because of the misguided efforts of earlier soil and water conservation programmes. Finally, they may be so responsive to policies encouraging increased food production that they simply ignore the costs.

In Britain, a major cause of soil erosion has been the shift in recent years towards the cultivation of winter cereals, driven by production-oriented policies. The high price of wheat has encouraged winter cultivation in fragile environments and this has led to a massive increase in soil erosion. It was long thought that water erosion was not a problem for British agriculture. But, since the late 1960s, the land sown to winter cereals has tripled, largely at the expense of grassland and spring cereals. Erosion can be of the order of 30–95 t/ha in fields where field boundaries and hedges have been removed from critical positions. Bob Evans estimates that some 6200 km^2 (4.4 per cent) of land in Britain is now at high risk and some 2100 km^2 (1.5 per cent) at very high risk. Erosion is greatest when there is little vegetative cover, such as during winter when winter cereals are being grown; when slopes are long, such as in big fields; and when farmers cultivate up and down slopes, rather than across the contour (Evans, 1990a, b).

On the South Downs in England, for example, erosion was uncommon until winter cereals were widely grown. In the late 1970s, only 5 per cent of these chalk downs were under winter cereals, but this increased to 65 per cent by 1992. In the past ten years, loss of soil accompanied by flooding has caused many incidents of flooding of housing and farms, causing several hundred thousand pounds worth of damage (Boardman, 1990, 1991; Boardman and Evans, 1991; Robinson and Blackman, 1990).

To farmers, erosion reduces the biological productivity of soils and the capacity to sustain productivity into the future. Although soil erosion is clearly costly to economies as well as to farmers, it is difficult to calculate reliably the precise costs (Eaton, 1993; Bishop, 1990). Studies in Mali, Malawi and Java suggest that the costs to farmers are substantial, representing 3–14 per cent of gross agricultural product (Table 3.4).

Off-site costs are also important. Soils are less able to retain water, which runs off more readily into waterways carrying sediments. These block downstream irrigation canals, reservoirs and harbours. Reduced volume means both greater maximum flows and so more likelihood of floods, and reduced minimum flows in dry seasons. The functional lifetime of reservoirs has declined in many countries. And natural systems, particularly fisheries and coral reefs, are threatened by sediments and agricultural pollutants.

In Java, sedimentation costs the economy US$25–90 million each year by shortening the life of reservoirs, reducing hydroelectric output, and increasing maintenance needs for dredging irrigation systems and harbours. In Thailand, a steadily depleting capacity of 20 reservoirs costs US$0.3 million in forgone income from reduced irrigation capacity and in lost hydroelectric capacity. Greater costs, some $18 million annually, are incurred by the need to dredge 19 million m^3 of sediment from the Chao Phraya River to keep the channel open to shipping. These off-site costs can be substantially greater than the financial losses to farmers. At one site in

Table 3.4 Selection of the on- and off-site costs of soil erosion

On-site costs	
Mali (1988)	US$ 4.6–18.7 million per year Equivalent to 3-13 per cent of agricultural GDP and 1.7 per cent total GDP
Malawi (1980s)	US$25 million per year Equivalent to 14.6 per cent of agricultural GDP and 4.8 per cent total GDP
Java, Indonesia (1980s)	US$ 320 million per year Equivalent to 3 per cent of agricultural GDP
Off-site costs	
USA (1990)	US$10,150 million of damage to freshwater and marine recreation, to water storage, navigation, flooding, fishing, water treatment, irrigation channels, roadside ditches and steam cooling
Cape Verde (1984)	US$2.6 million from a single storm
Java (1980s)	US$ 25–90 million per year for sedimentation
Thailand (1980s)	US$18 million per year for sedimentation
UK (1982–87):	
Mile Oak, Sussex	£105,000 of damage to housing, plus 150,000 on flood alleviation works
Rottingdean, Sussex	£400,000 of damage to 40 houses, gardens and roads
Breaky Bottom, Sussex	£81,000 of damage to vineyard

Sources: Ribaudo, 1989; Boardman and Evans, 1991; Bishop, 1990; Bishop and Allen, 1989; Attaviroj, 1991; Haagsma, 1990; Magrath and Arens, 1989; Faeth et al, 1991; Robinson and Blackman, 1990

the UK, where soil loss reached 250 t/ha, the estimated total loss to the farmers was just £13,000, mostly due to lost seed and fertilizer, compared with £400,000 of damage to housing (Robinson and Blackman, 1990).

More Terracing and More Erosion

Despite decades of effort, soil and water conservation programmes have had surprisingly little success in preventing erosion. The quantitative achievements of some programmes can appear impressive. In Lesotho, all the uplands were said to be protected by buffer stripping by 1960; in Malawi (then Nyasaland), 118,000 km of bunds were constructed on 416,000 ha between 1945–60; and in Zambia (then North Rhodesia), half the native land in eastern province was said to be protected by contour strips by 1950 (Stocking, 1985). In Ethiopia, during the late 1970s and 1980s, some 200,000 km of terracing were constructed and 45 million trees planted (Mitchell, 1987).

Ironically, though, many programmes have actually increased the amount of soil eroding from farms. This is because these impressive achievements have mostly been short lived. Because of a lack of consultation and participation, local people, whose land is being rehabilitated, find themselves participating for no other reason than to receive food or cash. Seldom are the structures maintained, so conservation works rapidly deteriorate, accelerating erosion instead of reducing it. If performance is measured over long periods, the results have been extraordinarily poor for the amount of effort and money expended (Shaxson et al, 1989; Hudson, 1991; Reij, 1991).

It is well established that poorly designed structures cause erosion. Yet throughout Africa, little account has been taken of how more terracing can lead to more erosion. In the early twentieth century, erosion in Lesotho was not a serious problem in cultivated fields, because grassed field boundaries were well developed and maintained (Showers, 1989). Despite this indigenous practice, contour banks were installed. Local people did not approve, because they reduced the size of fields, and either breached or the outlets developed into gullies. The administration attributed these gullies to *'unusual weather'* (Showers and Malahleha, 1990). Elsewhere in southern Africa, the first anti-erosion measures introduced in the early 1930s were large ridge terraces and bunds. But these imported measures disturbed natural patterns of drainage and permitted storm water to break through at vulnerable points. Careless construction made them susceptible to bursting and locals came to believe that *'gully erosion was caused by the government'* (Beinart, 1984).

Narrow-based terraces were introduced into Kenya from the USA in 1940 (Gichuki, 1991). For 15 years they were widely used. By 1947, some 4000 hectares were being protected each year and this rate continued until 1956–7. But these terraces filled up with sediments too quickly, were impossible to maintain and even began to aggravate erosion. And so, by 1958, the number falling into disrepair was exceeding new construction. By 1961, some 20,000 ha had fallen into disrepair. Eventually, the authorities recognized the problems and L H Brown, the chief agriculturalist, issued a memorandum in 1961 saying that *'narrow-based terraces should be abandoned as policy... we should move to strips of vegetation, preferably grass'* (in Wenner, 1992).

Bad contour ridging in the 1960s was worse than none at all in Zimbabwe (then Rhodesia), where farmers say the compulsory construction of ridges caused siltation of rivers. The ridges connected whole fields and drained in a single drainage line, so during large storms, concentrated water into powerful and fast-moving bodies that caused great damage (Wilson, 1989). The same thing has occurred with cut-off drains in Kenya. Their function is to intercept and divert storm water, but many were constructed in a way that caused erosion. *'The most severe mistakes were that cut-off drains were laid and constructed on the wrong sites. They were designed with steep gradients... The water is discharged into gullies which are deepening. The channel ridges were bare and cut-off drains were not supported by other structures below them... All these factors have made the structures more dangerous than useful. More problems were created. Gullies have widened, soil was eroded and crops destroyed'* (Hunegnaw, 1987).

Recent Project Efforts

Graded and contour bunds developed for large-scale farming in the USA are widely applied in soil and water conservation programmes in India. Even under heavy subsidies, most small farmers reject them, for very good reasons (Kerr and Sanghi, 1992). These bunds leave corners in some fields and so there is a risk of losing the piece of land to a neighbour. The central water course for drainage benefits only some farmers, damaging the land of others. Contour farming is inconvenient when farmers use multi-row implements, and so is only suitable where the holding is large and tractors are available. Contour bunding without facilities for dealing with surplus water commonly breach, again concentrating water flow that quickly forms gullies. It is, therefore, not uncommon for entire bunds to be levelled as soon as project staff shift to the next village (Sanghi, 1987).

Sometimes, successes are reversed almost immediately. In an evaluation of World Food Programme supported conservation in Ethiopia, the extent of the terracing was said to be *'impressive'*, yet monitoring in one sub-catchment found 40 per cent of the terracing broken the year after construction (SIDA, 1984). The project had expected that local people would bear all the costs of maintenance. Another example comes from the Yatenga region of Burkina Faso, where 120,000 ha of earth bunds constructed at high cost with machine graders in the early 1960s have now all but disappeared (Marchal, 1978, 1986). In the Majjia and Badéguicheri valleys of Niger, most of the 6000 ha of earth bunds constructed between 1964–80 are in an advanced state of degradation (Reij, 1988). In Sukumuland, Tanzania, where contour banks, terraces and hedges were forced upon farmers, almost no evidence remained of these conservation works by the early 1980s and now *'erosion is extremely severe'* (Stocking, 1985).

In Oaxaca, Mexico, a large-scale government soil conservation programme is also establishing contour bunds based on the US models. It is an area noted in the 1970s and 1980s by various 'expert' missions as having *'massive soil erosion'* and *'the world's worst soil erosion'*. But recent evidence is suggesting that erosion has only become serious following the imposing of terraces and bunds (Blackler, 1994). Rill erosion has been recorded within one year of their establishment and degradation has been so severe that less than 5 per cent of the bunded area is cropped.

In Cape Verde off the west coast of Africa, the state takes responsibility for erosion control by paying farmers to work on their own land. The result is that traditional practices are ignored as farmers take the money without influencing the project. *Socalco* terraces, for example, are built from top to bottom of steep slopes, with the result that foundations are often left hanging in the air (Haagsma, 1990). As Ben Haagsma has put it *'this does not stimulate... good cooperation between farmers and MDRP [the project]. It is difficult to eradicate the attitude 'MDRP knows best'*. In India, farmers have only permitted bunds to be constructed on their fields because they are attracted to subsidies. The impact on the relations between government and farmers is serious: *'in most villages farmers have become addicted to subsidies which normally come as part of development projects.'* (Sanghi, 1987).

The impact of these programmes has been to make many things worse.

A failure to involve people in design and maintenance can create considerable long-term social impact. The enforced terracing and destocking in Kenya, coupled with the use of soil conservation as a punishment for those supporting the campaign for independence, helped to focus the opposition against both authority and soil conservation (Pretty and Shah, 1994; Gichuki, 1991). This led, after independence, to the deliberate destruction of many structures because of their association with the former administration (Anderson, 1984). In neighbouring Somalia, a large FAO-funded project constructed dams during the 1970s to check gullies, but because of poor construction, many collapsed or diverted the floods, so accelerating gully erosion instead of preventing it. This induced widespread disenchantment amongst local people for all conservation projects that followed (Reij, 1988). Such attitudes are a critical constraint for many current soil conservation programmes.

THE LOSS OF BIODIVERSITY

Why Farmers Prefer Diversity

Farmers of traditional and low input agricultural systems have long favoured diversity on the farm. Today, there is still a huge variety of mixtures cultivated, including cereals, legumes, root crops, vegetables and tree crops. In Africa more than 80 per cent of all cereals are intercropped, producing in some cases highly complex patterns on the ground, with up to 20 species grown in close proximity (Vandermeer, 1989; OTA, 1988). In Latin America, about 60 per cent of maize is intercropped and 80–90 per cent of beans are grown with maize, potatoes or other crops (Francis, 1986). In one field in the Andes in Peru, Robert Rhoades recorded some 36 potato varieties growing in 13 rows (Rhoades, 1984). These were all shapes and sizes, and a variety of colours, including black, red, blue, purple, yellow and white. Altogether some 3000 traditional varieties are still grown by Andean farmers.

In very variable conditions, farmers rarely standardize their practices. They maintain diversity, develop a variety of strategies and so spread risk. Mixtures of crops and varieties clearly provide farmers with a range of outputs, and also represent logical approaches to coping with variable environments. Mixed crops can also be less variable in time and space, and combined yields are often greater, particularly if differences in root and shoot geometry allow the crops to use light, nutrients and water more efficiently (Vandermeer, 1989; Francis, 1986; Rao and Willey, 1980; Trenbath, 1974). Intercropping can reduce weed problems, so influence labour requirements; returns to labour can be increased; and erosion and run-off may be reduced by the greater ground cover given by the mixture (OTA, 1988).

Farmers themselves recognize the value of mixtures. In Indonesia, farmers in rainfed conditions plant a greater mix of crop combinations during the more risky seasons (Castillo, 1992). In the dry season, they plant 42 combinations; in the uncertain middle season, they use 25; and in the rainy season, they plant just 7. The author of a recent comprehensive review of intercropping, John Vandermeer, has put it: *'In personal conversation with [farmers] in southern Mexico, Costa Rica and Nicaragua, I have frequently... been told that two crops make a good combination because one is taller than the other and 'fits in' to the spaces where the other does not, or that the root systems go to different depths and thus use nutrients from different parts of the soil.'* Such popular knowledge is common.

It is, however, impossible to say categorically whether a mixture will result in better yields than the monocropped alternatives, except perhaps for legume–non-legume mixtures (Trenbath, 1976; Willey, 1979; Vandermeer, 1989). Much depends upon the local conditions and characteristics of the crops themselves. Pest attack is frequently reduced in intercrops, because of a variety of factors (Risch et al, 1983). Host plants are more widely spread and so harder to find; one species may trap a pest; or one species may repel the pest; and/or predators may be attracted (see Chapter 4). Weeds are also more likely to be suppressed by mixtures.

Recent surveys of non-irrigated rice systems in Cambodia, Indonesia, Laos, Madagascar, Myanmar, Nepal, Philippines and Thailand have found that farmers manage their highly diverse conditions with different land-use strategies (Fujisaka, 1990, 1991b; Fujisaka et al, 1992). Farmers described different combinations of landscape position, soil type, hydrology, and flood and drought risk, and showed how they matched these to different combinations of rice varieties and management practices. In upland Laos, for example, farmers distinguished 20 different types of soil. Each grew up to 4 varieties of rice, with 29 varieties grown in the 2 regions of Luang Prabang and Oudomasay. But these mixes are not static. Farmers are continually experimenting with new varieties or readopting old or existing ones. In Bukidnon, Philippines, farmers were cultivating 18 varieties, having dropped another 6 in recent years.

In Myanmar, 52 different varieties of rice were encountered in rainfed uplands and lowlands, and in deepwater conditions. Each farmer does not seek to maximize yield, nor do they have one preferred variety. They grow up to six varieties each, with a range of different taste, colour, pest resistance, growth pattern, duration, flood/drought tolerance, milling recovery and market price qualities (Table 3.5).

The Decline Under Modernization

It is only recently that fields monocropped to single species and varieties have become common. The introduction of modern varieties and breeds has almost always displaced traditional varieties and breeds. During the twentieth century, some 75 per cent of the genetic diversity of agricultural crops has been lost. Only about 150 plant species are now cultivated, of which just 3 supply almost 60 per cent of calories derived from plants (FAO, 1993; Fowler and Mooney, 1990). The trend has been rapidly downwards in many countries (Table 3.6). In India, once more than 30,000 rice varieties were grown, but now just 10 cover 75 per cent of the whole rice area. A comprehensive study of the decline in the USA this century has been conducted by Cary Fowler and Pat Mooney (1990), where *'the losses of fruit and vegetable varieties are staggering'*. For 65 types of vegetable, they record a consistent loss of between 80–100 per cent of the varieties of each

Table 3.5 Numbers of rice varieties and their qualities encountered in three non-irrigated rice regions of Myanmar

Region	*Number of varieties*	*Qualities of rice varieties*
Rainfed uplands	18	115 to 180-day duration red and white glutinous and sticky long and short awned and awnless drought tolerant and intolerant 100–150 cm in height for eating or for rice wine grain size length of time in stomach yields from 1.0 to 2.0 t/ha
Deepwater rice	18	elongation from water panicle in or out of water expansion when cooked eating quality yellow and white resistance to stem borer yield from 1.5 to 1.8 t/ha
Rainfed lowlands	16	125–200 day duration flood or drought tolerant glutinous or not yellow, white or black length of time in stomach fertilizer responsiveness distinctive aroma expansion when cooked yields from 1.2 to 6.2 t/ha

Source: Fujisaka et al, 1992

since the turn of the century. Of the 8207 varieties listed for these vegetables in 1903, only 607 are now held by the National Seed Storage Laboratory.

Most modern scientists have seen mixtures as a problem to be overcome. When the Rockefeller-sponsored team first visited Mexico in the 1940s to assess wheat cultivation as a precursor to establishing the national wheat breeding programme (see Chapter 2), the low-yielding traditional fields were condemned because: *'most fields were a mix of many different types, tall and short, bearded and beardless, early ripening and late ripening; fields usually ripened so unevenly that it was impossible to harvest them at one time without losing too much over-ripe grain or including too much under-ripe grain in the harvest'* (Stakman et al, 1967). These mixtures of traditional varieties may not have yielded well, but did give some insurance against pest and disease attack. Modern wheat varieties introduced in Mexico in the 1940s and 1950s were soon susceptible to new races of rust, and were quickly overcome.

Table 3.6 Decline in diversity of crops and livestock in a selection of locations

India	Once more than 30,000 rice varieties grown; now expected that just 10 rice varieties will soon cover 75 per cent of rice area.
Philippines	Before the modernization of early 1970s, 3500 varieties of rice existed; now only 3–5 are grown in irrigated areas.
Europe	Half of all the breeds of domestic animals (horses, cattle, sheep, goats, pigs and poultry) have become extinct since the beginning of the 20th century; a third of the remaining 770 breeds are in danger of disappearing by 2010.
France	71 per cent of apple production from one variety, Golden Delicious; 30 per cent of bread wheat from 2 varieties and 70 per cent from 10 varieties; In the SE, the Provençal diet contained 250 plant species at beginning of 20th century; now it comprises only 30–60.
Greece	95 per cent of local wheat varieties lost since 1920s.
The Netherlands	A single potato variety covers 80 per cent of potato land; 90 per cent of wheat planted to 3 varieties; 75 per cent of barley planted to 1 variety.
UK	68 per cent of early potatoes planted to 3 varieties; 4 wheat varieties account for 71 per cent of wheat area.
USA	Since 1900: 6121 apple varieties lost (85 per cent) 2354 pear varieties lost (88 per cent) 546 garden pea varieties lost (95 per cent) 516 cabbage varieties lost (95 per cent) 394 field maize varieties lost (91 per cent) 383 pea varieties lost (94 per cent) 329 tomato varieties lost (81 per cent 295 sweet corn varieties lost (96 per cent) Now: 71 per cent maize area planted to 6 varieties 96 per cent pea area planted to 2 varieties 65 per cent rice area planted to 4 varieties 76 per cent snap bean area planted to 3 varieties

Sources: Pimbert, 1993; FAO, 1993b; Soetomo, 1992; Fowler, Vellvé and Mooney, 1990

Similar simplification occurred during rice modernization in Asia. In Central Luzon, Philippines, for example, all 25 varieties grown by farmers in the mid-1960s were traditional varieties. By 1980, the total number of varieties grown had fallen only to 19, but just 2 were traditional (Cordova et al, 1981). Some 95 per cent of the rice area is now planted to MVs. In the USA, mixtures of wheat and oats, oats and barley, sorghum and alfalfa, maize and soybean were common in the early part of this century (Thatcher, 1925; Bussell, 1937; Bailey, 1914). All these are now mainly grown as monocrops. Mostly it has been the incentive structures provided by cheap and available inputs that has encouraged farmers to specialize. But in some countries, farmers have been prevented by law from growing traditional varieties and there have been reports of traditional crops being burned or destroyed (Soetrisno, 1982; see this chapter).

With these losses of genetic diversity could go future opportunities. Locally adapted crops or livestock can be critical for helping to deal with particular challenges brought by pests or diseases. One rice variety from India, for example, has been central to efforts to cope with a devastating virus. During the 1970s, the grassy-stunt virus devastated rice from India to Indonesia. After a 4-year search, in which over 17,000 cultivated and wild rice samples were screened, disease resistance was found. One population of the wild species, *Oryza nivara*, growing near Gonda in Uttar Pradesh, was found to have a single gene for resistance to grassy-stunt virus strain 1. Today, resistant rice hybrids containing the wild Indian gene are grown across some 110,000 km^2 of Asian rice fields (FAO, 1993b). Genetic erosion, the reduction of diversity within a species, is a global threat to agriculture.

The Value of Wild Diversity

The value of wild biodiversity to farming households has seldom been recognized by agriculturalists (Jodha, 1991; Bromley and Cernea, 1989; Scoones et al, 1992). Wild resources are often called wastes or wastelands, and represent a symbol of backwardness and underdevelopment. During the British agricultural revolution of the seventeenth to nineteenth centuries, common resources were seen by many officials as the *'trifling fruits of overstocked and ill-kept lands'* (in Humphries, 1990), and *'mere sand... and fit for nothing but rabbits'* (Burwell, 1960); and to large landowners, commons *'burdened a village with beggarly cottages and idle people. They were better enclosed [for agriculture]'* (in Thirsk, 1985). In India, common resources are still called wastelands.

It is well recognized that hunter–gathering communities, such as the !Kung San in Botswana or Indian groups in the Amazon, depend on wild resources for their complete livelihoods. What is less widely recognized is that farming households also rely heavily on wild resources (Table 3.7). If wild habitats are lost, these resources will no longer be available to rural households. And those who will suffer most are the poorest, who most often rely on wild resources as key sources of food, fuel, medicines and fodder.

Many agricultural institutions were indicating during the 1980s that agricultural area would have to expand substantially if growing world populations would be fed (FAO, 1989; TAC, 1988). More recently, these calls have been toned down, as it is being increasingly recognized that expansion on this scale will incur significant costs to rural households and national economies alike.

Recent Responses to Support Farmers

When external agencies work closely with farmers to document the variety and performance of their crops, the results can be extraordinary

Table 3.7 Use of wild plants for food and medicine by farming communities

Location	*Importance of wild resources*
Brazil[1]	Kernels of babbasu palm provide 25 per cent of household income for 300,000 families in Maranhâo State
China, West Sichuan[2]	1320 tonnes of wild pepper production; 2000 t fungi collected and sold; 500 t ferns collected and sold
Ghana[3]	16–20 per cent of food supply from wild animals and plants
India, West Bengal[4]	155 wild plants collected for food, fodder, medicine and fuel
Kenya, Bungoma[5]	100 species wild plants collected; 47 per cent of households collected plants from the wild and 49 per cent maintained wild species within their farms to domesticate certain species
Kenya, Machakos[6]	120 medicinal plants used, plus many wild foods
Nigeria, near Oban National Park[7]	150 species of wild food plants
South Africa, Natal/KwaZulu[8]	400 indigenous medicinal plants are sold in the area
Sub-saharan Africa[9]	60 wild grass species in desert, savanna and swamp lands utilized as food
Swaziland[10]	200 species collected for food
Thailand, NE[11]	50 per cent of all foods consumed are wild foods from paddy fields, including fish, snakes, insects, mushrooms, fruit and vegetables
South west of USA[12]	375 plant species used by native Indians
Zaire[13]	20 tonnes chanterelle mushrooms collected and consumed by people of Upper Shaba
Zimbabwe[14]	20 wild vegetables, 42 wild fruits, 29 insects, 4 edible grasses and one wild finger millet; tree fruits in dry season provide 25 per cent of poor people's diet

Sources: 1 Fowler and Mooney, 1990; 2 Zhaoqung and Ning, 1992; 3 Dei, 1989; 4 SPWD, 1992; 5 Juma, 1989; 6 Wanjohi, 1987; 7 Okafor, 1989; 8 Cunningham, 1990; 9 Harlan, 1989; 10 Ogle and Grivetti, 1985a; 11,12, 13 Scoones et al, 1992; 14 Wilson, 1990

(Salazar, 1992; Fowler and Mooney, 1990). Many have found varieties that perform well in low external input conditions (Box 3.1). The Phrey Phdau rice research station in Cambodia, for example, has collected 1320 local rice varieties with the help of Oxfam. One local variety, *2-Somrung 2*, yields 5 t/ha under low-external input conditions. Another, Prambei Khor, compares equally with IR42 on yield terms, and has superior straw production and grain quality. In Thailand, the NGO Technology for Rural and Ecological Enrichment (TREE) has collected 4000 accessions of rice in two years, which have been stored in the National Rice Germplasm Bank pending the development of community seed banks (Salazar, 1992; Siripatra and Lianchamroon, 1992).

Box 3.1 The role of the group 'Farmer-Scientist Participation for Development' (MASIPAG), Philippines in the conservation of biodiversity

The MASIPAG programme was established in 1986 to encourage farmers' participation in the development of improved varieties which yield well under low-external input conditions. The programme has:

- collected 210 accessions from farmers around the country;
- made cross-combinations to produce 101 selected lines by 1990; of these half showed good yield potential under low input conditions;
- distributed 40,000 kg of seed selected from 34 cultivars around the country;
- produced varieties that yield 3.7–5.7 t/ha with no applications of fertilizer or pesticides; this compares with a range for MVs needing external inputs of 3.2–5.2 t/ha.

Source: Salazar, 1992

Once again, the clear principle is that farmers, given the choice, rarely replace local varieties entirely with a single MV. They prefer to add a MV to their existing mix of options. In Indonesia, a local variety called *Rojolele* was found that compared well with IR64, a recent release from IRRI (Soetomo, 1992). IR64 yields better but, because it needs more water, weeding and fertilizer, it gives lower returns to farmers. *Rojolele* is also favoured for its distinctive taste and fragrance, as well as being fairly resistant to brown planthopper, rats and birds, unlike IR64. Farmers can also produce it locally in community seed banks, which are growing in popularity in Indonesia, despite them being contrary to existing government policy that holds that germplasm should be managed only in official genebanks.

When farmers are given the choice about new varieties, they much prefer to absorb the new technology into their existing systems. This is quite different to the way that agricultural modernization has worked so far, with farmers having to adopt the whole package or nothing at all. Recent participatory research with women farmers in Andhra Pradesh conducted by ICRISAT scientists led by Michel Pimbert has shown how important is the principle of supporting local biodiversity. Women experimented with new varieties of pigeonpea that were resistant to the pod borer, *Helicoverpa armigera*, which can devastate whole crops in bad years (Pimbert, 1991). Although some varieties were rejected because of bitter taste, interest was generated by the resistance to pod borer. Women farmers, though, said they would not replace their traditional varieties, but incorporate the new varieties of pigeonpea into their current mix of landraces. The new technology, in this case, has been adapted to suit local farming and livelihood conditions.

THE BREAKDOWN OF THE SOCIAL FABRIC OF RURAL COMMUNITIES

The social costs of modernized agriculture have been widely documented. As agriculture has increasingly substituted external inputs and resources for internal ones, so there has been a decline in the number of jobs for local people. Standardization has reduced the range of management skills needed, and many decisions have been taken out of the hands of farmers and local institutions.

Agricultural modernization has helped to transform many rural communities in both industrialized and Third World countries. The loss of jobs, the further shift of economic opportunity away from women to men, the increasing specialization of livelihoods, the increasing concentration of land in the hands of wealthy villagers and urban investors, the growing gap between the well-off and the poor, and the cooption of village institutions for the purposes of the state, have all been features of this transformation.

Social Change in Rural Britain

A period of remarkably successful agricultural growth since the mid twentieth century has brought significant social change in rural areas of Britain (see, eg, Newby, 1980). As farming has intensified its use of external inputs, so it has shed jobs, bringing poverty and deprivation to many people. Between 1945 and 1992, the number of farms in England and Wales has fallen from 363,000 to 184,000, while the total area of agricultural land has remained stable at 19 million ha. Over the same period, the number of regular hired and family workers on farms in England alone fell from 478,000 to 135,000 (MAFF, passim). In the past decade, there have been dramatic falls in the numbers of most types of people engaged in farming activities throughout Britain (Table 3.8). It is expected that the number of people engaged in agriculture will fall by a further 17–26 per cent during the 1990s (in DoW, 1992). Now, as a result of modernization, some of the worst poverty is in rural areas (Pretty and Howes, 1993).

Table 3.8 Changes in labour force (in thousands) on agricultural holdings in the UK, 1981–92

Class of worker engaged agriculture	*1981*	*1992*
Total labour force (in thousands)	709.9	621.8
Total farmers, partners, directors	293.6	280.5
Spouses of farmers, partners and directors doing farm work	74.6	76.0
Salaried managers	7.9	7.8
Regular hired whole and part-time workers	182.2	124.9
Regular family workers	54.5	46.4
Seasonal or casual workers	97.0	86.2

Source: Pretty and Howes, 1993, using MAFF Agricultural Statistics from Agricultural and Horticultural Censuses, prepared by Government Statistical Service, Guildford

In the quest for greater food production, landscapes have been homogenized, and rural livelihoods and farming systems have been progressively simplified. Where there were diverse and integrated farms employing local people, there are now operations specializing in one or two enterprises that largely rely on farm or contractor labour only. Where processing operations were local, now they are centralized and remote from rural people. The result is that few people who live in rural areas have a direct link to the process of farming. Fewer people make a living from the land and, of course, they understand it less. The lack of employment has also coincided with the steady decline in rural services, such as schools, shops, doctors and public transport.

However, the number of people in rural areas is increasing, though it appears that it is younger people migrating away, to be replaced by older, particularly retired, new entrants (DoW, 1992). More people want to move into rural areas too. Recent surveys found that 76 per cent of those who live in cities want to live in a village or country town and 37 per cent expect to move out during the next decade (Rose, 1993). The rural community, bonded in the past by a common understanding and economic interest in the land, currently appears unlikely to be brought together by close links with farming.

Various national enquiries have shown that the incidence of rural poverty is considerably greater than previously supposed (DoW, 1992; HL, 1990; ACORA, 1990). According to an unpublished government report, some 25 per cent of rural households are living on or below the official poverty line (in DoW, 1992). Farmers and farmworkers are about twice as likely to commit suicide than the rest of the population, and suicide is the second most common form of death for male farmers (in DoW, 1992). Farmers are increasingly recognized as suffering the stress and deteriorating confidence associated with lonely occupations (Martineau, 1993; Cornelius, 1993). The Duke of Westminster's report (1992) described these problems in this way: *'Hidden in the rural landscape which the British so much love, people are suffering poverty, housing problems, unemployment, deprivation of various kinds, and misery. Traditional patterns of rural life are changing fast, causing worry, shame and distress. Those most affected are often angry and bitter but feel they have little chance of being heard. The suicide rate is very high. Neither the public nor the private sector is showing any signs of caring very much about all this'.*

Small family farms have been especially vulnerable (Lobley, 1993; Moss, 1993). They rely more on diverse sources of off-farm income and so are dependent upon the wider success of the rural economy. When small farms are given up, they tend to be amalgamated into ever larger holdings, with a resulting radical change in the landscape structure (Munton and Marsden, 1991). Many successions lead to intensified land use, and the removal of woods and hedges. Continuity of farms is a goal held by many farm families. Most wish to see their heirs as successors. Yet, the evidence suggests that few will do so. Since the late 1960s, the proportion of farmers planning to pass their businesses to their heirs has fallen from about 75 per cent to 48 per cent (Ward, 1993). Su... farmers in the less prosperous areas (Marsd... declining economic fortunes seems to have erode... successors in family farming and, the prospect of a farm... to have become less attractive to farm children.

Changes in Japan and the USA

These changes are mirrored by social changes in other industrialize... countries. In Japan, similar threats to rural culture are occurring. More than half of farmers are older than 60 years of age and 75 per cent are part-time, relying on jobs in manufacturing as their main income source. Only 16 per cent of all farms have a male under 60 years devoting more than 150 days each year to farming (Ohnox, 1988). Like many other parts of the world, the next generation shows little interest in the labour-intensive work of farming: 70 per cent of farms have no successor. The number of farm households has fallen from 5.82 million in 1960 to 4.2 million in 1991. In that time, the number of people living in a farming household has fallen from some 34 million (30 per cent of the population) to just under 17 million (14 per cent of population) (Iwamoto, 1994; MAFF, passim).

One woman farmer of a 0.2 ha plot on the outskirts of Tokyo says her son, who works in an electronics factory, wants her and her 70-year-old husband to retire and sell the land to property developers: *'the young are not interested in the old ways and the old values. We have always owned land, it is the foundation and strength of our family. Our son says the land should be sold for building, or as a car park, but we believe everyone benefits from having farmers in the heart of the city'* (in Davies, 1992). However, many say that looking down on their tiny plots of land from their tower blocks has helped them protect their sanity. Farmers fear that their rice culture is under terminal threat: *'people are lamenting the coming extinction of the two thousand year rice culture of Japan'* (Furusawa, 1988). None the less, it is also true that this culture has been heavily protected by national policies.

Just as in Britain and Japan, there have been huge changes in rural culture during this century in the USA. Since 1900, the proportion of the national population who are farm residents has fallen from 40 per cent to just 1.9 per cent, or 4.6 million people (AAN, 1993a). Family farms have been consolidated into larger farms; labour opportunities have fallen; and farm enterprises have been concentrated in fewer hands. This modernization has been most visible in the declining number of farms and the replacement of family farming by modern large-scale farming. But it has also had significant impacts on social systems.

A classic study conducted in 1946 by Walter Goldschmidt showed what happens when the social structure in the countryside changes during modernization (Goldschmidt, 1978) . He studied the two rural Californian communities of Arvin and Dinuba in the San Joaquin Valley. These were matched for climate, value of agricultural sales, enterprises, reliance on irrigation and distance from urban areas. The differences were in farm scale: Dinuba was characterized by small family farms, and Arvin by large, commercialized farms. There were striking differences between the two communities. In Dinuba, there was a better quality of life, superior public services and facilities, more parks, more shops and retail trade, more diverse businesses, twice the number of organizations for civic improvement and social recreation, and better participation by the public. The small farm community was a better place to live *'perhaps because the small farm offered the opportunity for 'attachment' to local culture and care for the surrounding land'* (Perelman, 1976). A study of the same communities in the late 1970s reaffirmed these findings (Small Farm Viability Project, 1977).

Recent years have brought severe financial crises for family farmers. They were squeezed by debt and low product prices. Many thousands lost their businesses. Many others did not see this as a problem, but as desirable. It was widely perceived to be the way to agricultural prosperity. Michael Perelman (1976) quotes a Bank of America official who said in 1969 *'what is needed is a program that will enable the small and uneconomic farmer – the one who is unwilling or unable to bring his farm to the commercial level by expansion or merger – to take his land out of production with dignity'.*

Small farmers were widely taken to be economically inefficient. But their loss has been a severe loss to rural society. Linda Lobao's study (1990) of rural inequality shows the importance of the locality that Goldschmidt illustrated. The changing structure of farming has brought a decline in rural population, increased poverty and income inequality, lower numbers of community services, less democratic participation, decreased retail trade, environmental pollution and greater unemployment. The decline of family farming does not just harm farmers. It hurts the quality of life in the whole of society. Corporate farms are good for productivity, but not much else: *'this type of farming is very limited in what it can do for a community... we need farms that will be viable in the future, correspond to local needs and remain wedded to the community'* (Lobao, 1990).

Wendell Berry, the influential poet and farmer, has long drawn attention to what happens during modernization. Agricultural crisis is a crisis of culture: *'A healthy farm culture can be based only upon familiarity and can grow only among people soundly established on the land; it nourishes and safeguards a human intelligence of the earth that no amount of technology can satisfactorily replace. The growth of such a culture was once a strong possibility in the farm communities of this country. We now have only the sad remnants of those communities. If we allow another generation to pass without doing what is necessary to enhance and embolden the possibility now perishing with them, we will lose it altogether.'* (Berry, 1977)

One indicator of the crisis is the suicide rate among farmers. In the mid-West, suicide rates among male farmers were twice the national average during the 1980s (Gunderson, in FW, 1991a). Some 913 took their own lives between 1980–8, producing annual rates higher than for any other documented occupation.

Mexican Indians and Modern Agriculture

After centuries of avoiding incorporation into the wider culture of what is now Mexico, the Yaqui Indians of Sonora, situated in the north of Mexico and bordering the Gulf of California, have been entirely changed by rapid integration into the modern way of farming (Hewitt de Alcantara, 1976). They had been known for their tradition of strong social cohesion, cultivation of a great diversity of crops, and the use and management of wild foods, including wild fruits and oysters. They relied on small-scale water harvesting and irrigation structures to irrigate their crops from the seasonal rivers. But this has all changed.

In the 1940s, modern water control projects at Potam in the Yaqui Valley began to divert water to commercial farms. As more water was removed, the river ran low, making many families' plots unusable. Yaqui agriculture began to be undermined as they were coerced into joining the government scheme so as to have access to water. At the time, the state was trying to assure the permanent tranquillity of the tribe, and it was felt that commercializing Yaqui agriculture through water control and the formation of collective credit societies would help. During the 1950s, farmers were grouped into 40 credit societies, each containing some 30 members and their plots were joined to form the common land for each society. But external banks were given complete control of all farming decisions. Farmers had to grow wheat and cotton, and soon only 10 per cent of land was left under maize-beans-squash, with 90 per cent under wheat and cotton. The varied production of fruit and vegetables noted by visitors 20 years earlier had disappeared.

The local people had lost control over their own land, and the social and economic changes were significant. *'Instead of preparing and working their family plots by hand, using seeds from previous crops and silt from periodic river floodings, Yaqui cultivators found themselves observing the march of tractors and combines driven by bank employees across common land planted with high-yielding seeds and fertilized with chemical products. Most Yaqui indians intervened themselves only occasionally, when some menial task like cleaning irrigation canals or picking cotton demanded unskilled labour.'* (Hewitt de Alcantara, 1976)

As a result of complete control by federal employees and no local participation or involvement, feeder canals soon became blocked, land badly levelled, fertilizer applied incorrectly and planting dates missed. Yields of wheat were too low to repay the loans which local credit societies had forced them to take on. All local households were heavily in debt in the early 1960s. Despite the Yaqui's fear of water, the government tried to set up a fishing cooperative. This failed. They tried a cattle cooperative, but this was run by a local bank and so lands were quickly overgrazed.

By the early 1970s, the whole process had thoroughly undermined traditional institutions, not least the family. It has ceased to be a productive unit, and its disintegration has led to clashes between the old and young, and a weakening of traditional religious and cultural components of life. Ceremonies previously good at maintaining a relatively equal distribution of wealth had broken down. A group of landless had appeared. Those made wealthy were mainly the shopkeepers, who were *mestizos* from outside the community. A few large farmers had appeared, utilizing hired labour only during peak seasons and occasionally renting additional land. About the only communal activity that remained was the cooperation for *fiesta* celebrations. As Cynthia Hewitt de Alcantara (1976) noted, such integration with the surrounding *mestizo* culture was no doubt inevitable, but '*the way in which it was enforced seriously damaged a tradition of economic and social democracy, local self-government, and community service which should have been valued at least as highly as material progress. More to the point, the modernization of Potam undermined these attributes without really bringing material prosperity at all – except to a very few.*'

Social Change in Indonesia

One of the earliest technological changes during rice modernization was the replacement of the traditional *ani-ani* knife with sickles and scythes for harvesting (Collier et al, 1973; KEPAS, 1984). By tradition, Javanese and Sundanese rice farmers did not restrict anyone wishing to participate in the rice harvest. The harvesters were mostly women from their own and neighbouring villages. They used the *ani-ani*, a small hand knife, to cut each stalk of rice separately. The rice sheaves were carried to the owner's house, where the harvester would receive a share of the harvest. In this *bawon* system, the owner kept seven, eight or nine shares to one for the harvester.

With the adoption of MVs with their short straw and simultaneity of maturation, rice could be harvested much more quickly by sickle or scythe. Owners increasingly adopted the new *tebasan* system of cash-and-carry for harvesting, in which the standing rice crop was sold to a trader, who then arranged for harvesting. With these changes, bands of men increasingly became itinerant harvesters and opportunities for income generation for women fell. Many were entirely excluded from the process of harvesting. In some parts of Java, there were 200 or more women harvesting each hectare of rice in 1970. By 1990, they had been replaced by 10–20 men (Salazar, 1992). At the Agro-Economic Survey at Bogor, Collier and his colleagues calculated that women's share of the harvest fell from 65 per cent in the 1920s to 37 per cent in the late 1970s (Collier et al, 1982).

But modern technologies have affected more than just women alone. The two-wheeled tractor is now used extensively in land preparation, and water pumps and tube wells have been introduced for irrigating rice. The potential impact of this mechanization can be gauged by the calculation that if all these modern mechanization techniques were introduced to Java, then over 3 billion person-hours of labour would be lost (Collier et al, 1982). This is equivalent to 2 million full-time employees or many more part-time workers.

In the early 1980s, Loekman Soetrisno and colleagues (1982) interviewed landless and nearly landless farmers in a well-irrigated and apparently prosperous village in Central Java. They were asked how agricultural development had affected their lives: 60 per cent answered that *'development has made us in a difficult position'*. They said they were confused by the various government regulations, saying in particular that two were unfair. These were that all farmers in the village had to plant modern varieties on their rice paddies and that they had to follow a strict cropping schedule. Farmers appreciated the rationale, but could not abide by the regulations. They could not afford the modern inputs. They also *'did not have the courage not to abide by the regulations, as this would mean direct confrontation with the village bureaucracy.'* They reported to Soetrisno that in 1978 one small farmer had to burn his rice when the local officials found out that he was planting a local variety.

The order to all farmers to plant rice in a fixed schedule had a major impact on social structure too. Unlike the richer farmers, the poorest did not have buffaloes or cows to help them plough the land. Neither could they hire extra labourers to work on their land. It was also the custom that before small farmers prepared their land, they would work first for richer farmers for the additional income. The new regulation prevented them from earning this income. As a result, they had to rent their land to richer farmers and become tenants on their own land, with the resulting loss of social status in the community. One January, this village was featured on national television news as 60 people had been found suffering from severe malnutrition. According to the village head, most of these were landless and small farmers (Soetrisno, 1982).

The decline of the traditional *sawah* system of rice production in Bali is an example of what can happen when a sustainable system is changed (Poffenberger and Zurbuchen, 1980). It was self-sufficient within the boundaries of a single watercourse. Complex social, ecological and economic linkages made the system sustainable and resilient for at least 1100 years. But, over just a few years, rice modernization broke apart these local relationships by substituting external processes. Pesticides replaced predators, and fertilizers replaced cattle and traditional land management. Government officials made decisions rather than local institutions, and local labour groups were replaced with specialist workers and tractors (Box 3.2).

Wheat and Pastoralists in Tanzania

Another example of the social damage caused by modern farming comes from Tanzania, where millions of dollars of Canadian aid were spent between 1969 and 1993 in developing wheat farms on the dry Basotu Plains. Yields were comparable with those on the Canadian plains and the farms came to supply nearly half of the national wheat demand. But the plains are also the homeland of some 30–50,000 Barabaig pastoralists. The impact on their lives of these wheat farms has been recorded at first hand and documented in depth by Charles Lane (Lane, 1990, 1993, 1994).

The Barabaig economy is based on livestock production. Their herds of cattle, sheep and goats utilize the forage, water and salt licks found scattered throughout their territory. They have a complex grazing rotation system in which they move among eight different forage regimes. This can mean that some land is free of people and animals for long periods, which allows it to be preserved from overuse. All members of the

Box 3.2 The impact of modernization on the traditional wet rice cultivation system of Bali

Wet rice has been cultivated in Bali since at least AD882. Irrigation cooperatives, the *subaks*, are responsible for the allocation of water and the maintenance of irrigation networks, as wet rice is too complex for one farmer to practice alone. Each *subak* member has one vote regardless of the size of landholding. Soil fertility is maintained by the use of ash, organic matter and manures. Rotations and staggered planting of dry and wet crops control pests and diseases. Bamboo poles, wind-driven noise-makers, flags and streamers scare off birds. And rice is harvested in groups, stored in barns and traded only as needs arise. Rice yields are typically 1–2 t/ha, and sometimes as high as 3t/ha.

Modernization depends on the adoption of the whole new package. The major impacts have been as follows.

- Yields could be 50 per cent greater than under the traditional system, but only under optimum conditions, as the new rice was more susceptible to climatic and water variation.
- Pests and diseases increased as a result of the continuous cropping and the killing of predators and frogs by pesticides.
- Farmers sold cattle, as no they were longer needed for ploughing and manures.
- Mechanised rice mills displaced groups of women who used to thresh and mill the rice.
- Harvest teams replaced the communal *banjar* activities at harvest.
- As the new rice could not be stored for long periods (it had a thinner, looser husk and softer kernel), so it had to be sold immediately after harvest when the prices were lowest. This meant men received large sums of cash, and the women could no longer plan for the year's food security by monitoring the rice barn. '*Wisdom lies in keeping the family's capital in the rice barn, where they can regulate and dispense it, rather than in the form of risky, hard-to-manage cash lump sums*'.
- The *subak* organisations, once in complete control, lost many decisions to higher level institutions – now the government decides cropping patterns, planting dates and irrigation investments.
- The *subak* also organized redistribution through local religious and ritual culture, as the better off were expected to give more goods and services to community ceremonies.
- Reduced labour and employment in rice cultivation forced rural people to seek work elsewhere.

Source: Poffenberger and Zurbuchen, 1980

community have access to communal land. But this access is not uncontrolled: certain areas and resources are protected by rights and obligations for individuals, clans and local groups. In the past, the customary rules and institutions had been effective in both maximizing production and conserving resources. The Barabaig, like many people who live in variable environments, have a tradition of respect for the land they rely on for their survival. Their elders recently said *'We value and respect the land. We want to preserve it for all time'* (in Paavo, 1989).

But in order for wheat to be grown on the Basotu Plain, about 40,000 ha of land was taken from the Barabaig. This was their most fertile prime grazing land. Some of them were forcibly removed and their homes burned. They were prevented from following traditional routes across the farms to reach pasture, water or salt resources. Many of their sacred graves were ploughed up and are no longer recognizable. There are also ecological problems, as the soil is left bare soon after the July harvest until the time of planting in February. This makes the soil susceptible to rain-induced erosion, and deep gullies have been created and the sacred Lake Basotu is being silted up.

Although the farms cover only 12 per cent of the total land area of the district, the loss of this area is crucial for pastoralist production (Lane and Pretty, 1990). By losing access to these fertile areas, the whole rotational grazing system has been disrupted, so reducing the pastoral productive capacity beyond the direct impact of the wheat farms. This loss has resulted in a drastic reduction of livestock numbers and a decline in production which the Barabaig say has caused them *'great suffering'* (in Paavo, 1989).

Part of the problem is that outsiders misunderstand pastoralists and their production systems. Rangeland is common land to the Barabaig and individual herders move about in response to their assessments of range productivity or social needs. People who fail to understand this can be misled into thinking land is vacant or poorly managed by the pastoralists, so justifying their dispossession. One study of Canadian aid to Tanzania said: *'The project has many of the characteristics of a frontier development effort. Traditional pastoralists... are being displaced and absorbed into the project as labourers. Previously idle land is being brought under cultivation'* (Young, 1983).

From the viewpoint of the farms, wheat production is financially profitable. A project evaluation conducted in 1980 arrived at a benefit/cost ratio of 1.59. The return to the capital of nearly 40 per cent also indicated that it was a *'very profitable investment for the Tanzanian economy'* (Stone, 1982). But if the wider social and environmental impacts are accounted for, then the picture changes dramatically. The costs actually far exceed the benefits, and there would appear to be many better ways to use aid and scarce foreign exchange. As one economic assessment put it: *'The results of this study indicate that wheat production on the Hanang farms is profitable from the viewpoint of the farms... However, from the standpoint of contributions to, and resources used within, the Tanzanian economy the Project is shown to be uneconomic. In strict economic terms, the costs have exceeded the benefits'* (Prairie Horizons Ltd, 1986). The financial resources spent developing a high input system in a remote and dryland region would have been more efficiently used for buying wheat on the world market.

Social Change in India

Despite a wealth of studies on the impact of modern agricultural technology on the rural economy in India, there is no clear consensus as to whether labour opportunities for men and/or women have increased or decreased (Palmer, 1981; Agarwal, 1985; Chand et al, 1985; Whitehead, 1985; Sardamon, 1991; Chaudhri, 1992; Kaul Shah, 1993).

One recent study is now widely cited as evidence that the Green Revolution can lead to large, across-the-board gains in income, nutrition, and standard of living for small and large-scale farmers, and even for the landless poor (Hazell and Ramasamy, 1991). In North Arcot, Tamil Nadu, between 1974 and 1984, regional paddy output increased by 57 per cent, and this growth has had significant economic impact on the region's villages and towns. In addition to an increase in the wage rate, the distribution of income improved and absolute poverty declined. Small paddy farmers and landless labourers, who were initially among the poorest households, gained the largest proportional increases in family income, virtually doubling their real income during the decade. Non-paddy farmers and non-agricultural households increased their real family incomes by 20–50 per cent. As the authors put it: *'none of the predictions of the critics – that smaller farmers would be either unaffected or made worse off by the green revolution and that unnecessary mechanization would significantly reduce rural employment, thus worsening absolute poverty – came true'*.

There is, however, considerable evidence that large-scale technological innovations tend to be followed by mechanization of some women's work (Whitehead, 1985; Palmer, 1981; Billings and Singh, 1970). This change displaces landless women workers and many of the tasks become 'male'. But there is also evidence to indicate that the demand for female agricultural labour can increase as a result of modernization (Agarwal, 1984, 1985; Chand et al, 1985).

One feature common to most studies on the impact of agricultural modernization is that they focus solely on employment, production and income distribution. Very few have considered the wider changes in livelihood strategies and the quality of life (Kaul Shah, 1993). But the introduction of technologies does not necessarily have to lead to social disruption. Where external institutions work closely with local people, then these new technologies can lead to improved welfare for men and women. A recent review of the activities of AKRSP in Gujarat has shown that since the introduction of modern varieties, fertilizers and plant protection measures, coupled with soil and water conservation measures organized on a watershed management basis, incomes of all households have increased since 1986 as a result of intensification (Kaul Shah, 1993). Rising agricultural productivity has increased the demand for local labour and so increased the opportunities for local work. This has made an enormous difference to the quality of people's lives. Both men and women have experienced an increase in workload on their own farms, but it is the qualitative shifts in their livelihoods that people themselves say have been most significant (Table 3.9).

These qualitative changes were particularly important for women: their

Table 3.9 Changes in livelihoods of men and women of Samarpada village, Bharuch District, Gujarat, following the adoption of modern varieties, some fertilizers, and soil and water conservation measures as part of a participatory watershed management supported by the Aga Khan Rural Support Programme

Criteria	*Before programme (1987)*	*After programme (1992)*
No. households migrating to Surat for 2–10 months of every year	42	6
No. women migrating without men	14	0
No. households not migrating	13	34
Agricultural wage labour (Rs/day)		
men	2	13–15
women	3	13–15
Number of village children in school	8	56
Regularity of full meals during summer	1 per 3 days	2 per day
Vegetable consumption	rainy season only	every day
No. bullock carts owned	0	19
Sources of livelihoods:		
agriculture	20%	60%
migrating labour	80%	20%
non-farm income/employment in village	0	20%
Women's involvement in decisions on purchase of vegetable seeds, milch animals, tree saplings, clothing	none	regular
Ownership of clothing by women in landless and poorest group	1 set	3 sets

Sources: Kaul Shah, 1993; Shah, 1992

former work in labouring gangs in Surat was *'at the unbearable cost of insecurity, poor health, overwork, shame and loss of social respect'* (Kaul Shah, 1993). The decrease in migration through substitution from local income-earning opportunities represents a significant improvement in welfare. Asked how they would describe being 'happy', women replied *'when we don't have to migrate to Surat and have enough to feed ourselves and our children'*. This is despite the fact that wages are two to three times greater in Surat. Some of the most interesting changes recorded have been shifts in the work burden of men and women. Bullock carts are now used for gathering fuelwood and carting harvested rice from the fields, both formerly time consuming and heavy work for women. And where agricultural activities, such as weeding and paddy transplantation, were formerly segregated, they are now carried out by both women and men.

As indicated in Chapter 7, this is in fact a successful case of agricultural regeneration based on a judicial mix of local and external resources. Where it differs from the bulk of agricultural modernization efforts is in the formation of local institutions necessary for sustaining the changes. Once the migration cycle had been broken, people could stay in the village and were able to be fully involved in local decision making. These local institutions are a critical part of any effort for sustainable and self-reliant agriculture.

SUMMARY

The pursuit of increased productivity and conserved natural resources in the course of rural modernization has produced benefits in the form of improved food production and some improvements in resource conservation. But these improvements look so good that it is easy to forget there have been losers as well as winners. All sectors of economies have been affected by modernization. The drive for agricultural efficiency has drastically cut the numbers of people engaged in agriculture in industrialized countries. External inputs of machines, fossil fuels, pesticides and fertilizers have displaced workers in Green Revolution lands. Rural cultures have been put under pressure, as more and more people have been forced to migrate in search of work. Local institutions, once strong, have become coopted by the state or have simply withered away.

Many environmental and health impacts have increased in recent years; others have continued to persist despite all efforts to reduce them. These costs of environmental damage are growing, and are dispersed throughout many environments and sectors of national economies. A largely hidden cost of modern agriculture is the fossil fuel it must consume to keep outputs high. Modern agriculture has tended to substitute external energy sources for locally available ones. For each kilogramme of cereal from modernized high input conditions, 3–10 MJ of energy are consumed in its production; but for each kilogramme of cereal from sustainable, low input farming, only 0.5–1 MJ are consumed. A shift to low input systems could, therefore, have an impact on the process of global warming.

Pesticides have caused problems by inducing resistance in pests and damaging the health of farmers, farmworkers and consumers. The hazards are greater in Third World countries and emerging evidence is producing a bleaker picture that appeared to be the case in the 1980s. According to the latest estimates from the WHO, a minimum of 3 million and perhaps as many as 25 million agricultural workers are poisoned each year, with perhaps 20,000 deaths. Studies in the Philippines, in particular, are showing how costly these problems are to national economies as well as to the affected individuals.

Despite the fact that indigenous systems of soil and water conservation are widespread, well adapted to local conditions, persist for long periods and are capable of supporting dense populations, soil erosion continues to be a problem throughout the world. To farmers, erosion reduces the biological productivity of soils and the capacity to sustain productivity into the future. Although soil erosion is clearly costly to economies as well as to farmers, it is difficult to calculate reliably the precise costs, though studies in Mali, Malawi and Java suggest that the costs to farmers are substantial, representing 3–14 per cent of gross agricultural product.

One surprising cause of soil erosion is bad soil conservation programmes. There are many examples throughout the world of impressive terracing and bunding disappearing when local people have not been involved in planning and implementation. Poor terracing results in worse erosion. The impact of these programmes has been to make many things worse. A failure to involve people in design and maintenance can create considerable long-term social impact, inducing widespread disenchantment among local people for all conservation projects that followed.

Biodiversity has fallen under modern agriculture. Farmers of traditional and low input agricultural systems have long favoured diversity on the farm, and it is only recently that fields monocropped to single species and varieties have become common. The introduction of modern varieties and breeds has almost always displaced traditional varieties and breeds. During the twentieth century, some 75 per cent of the genetic diversity of agricultural crops has been lost. Only about 150 plant species are now cultivated, of which just 3 supply almost 60 per cent of calories derived from plants.

Agricultural modernization has helped to transform many rural communities in both industrialized and Third World countries. The loss of jobs, the further shift of economic opportunity away from women to men, the increasing specialization of livelihoods, the increasing concentration of land in the hands of wealthy villagers and urban investors, the growing gap between the well-off and the poor, and the cooption of village institutions for the purposes of the state, have all been features of this transformation. Cases are described of social change in Britain, Japan, the USA, Mexico, Indonesia, Tanzania and India. In all of these the social costs have been substantial. Modern agriculture, though not the sole cause, has clearly been a contributor to these changes.

4

RESOURCE-CONSERVING TECHNOLOGIES AND PROCESSES

'And the soil said to man: take good care of me or else, when I get hold of you, I will never let your soul go.'
Kipsigis proverb, as told by Mr arap Keoch, Chemorir, Kenya, 1990

ADOPTING RESOURCE-CONSERVING TECHNOLOGIES

The Multifunctionality of Technologies

Sustainable agriculture involves the integrated use of a variety of pest, nutrient, soil and water management technologies and practices. These are usually combined on farms to give practices finely tuned to the local biophysical and socioeconomic conditions of individual farmers. Most represent low-external input options. Most such farms are diverse rather than specialized enterprises. Natural processes are favoured over external inputs and by-products or wastes from one component of the farm become inputs to another. In this way, farms remain productive as well as reducing the impact on the environment.

This chapter gives details of both proven and promising resource-conserving technologies. These draw on a range of experiences from both farms and research stations, where the impacts of pests, diseases and weeds have been reduced; the viability of natural predators enhanced; the efficiency of pesticide and fertilizer use improved; and nutrients, water and soil conserved. Many of these are examples of farmers already taking steps to reduce costs and the adverse environmental effects of their operations. Some have done so by improving conventional practices; others by adopting alternatives. Most have tried to take greater advantage of natural processes and beneficial on-farm interactions, so reducing off-farm input use and improving the efficiency of their operations.

These technologies basically do two important things. They conserve existing on-farm resources, such as nutrients, predators, water or soil. Or they introduce new elements into the farming system that add more of these resources, such as nitrogen-fixing crops, water harvesting structures or new predators, and so substitute for some or all external resources.

Many of the individual technologies are multi-functional. This mostly implies that their adoption will mean favourable changes in several components of the farming system at the same time. For example, hedgerows encourage predators and act as windbreaks, so reducing soil erosion. Legumes introduced into rotations fix nitrogen, and also act as a break crop to prevent carry-over of pests and diseases. Grass contour strips slow surface runoff of water, encourage percolation to groundwater and are a source of fodder for livestock. Catch crops prevent soil erosion and leaching during critical periods, and can also be ploughed in as a green manure. The incorporation of green manures not only provides a readily available source of nutrients for the growing crop, but also increases soil organic matter and hence water retentive capacity, further reducing susceptibility to erosion.

The principles of integrated farming, a key element of sustainable agriculture, focus on increasing the number of technologies and practices, and the positive, reinforcing linkages between them. But this multi-functionality also makes classification of the technologies problematic. In this chapter, technologies and practices are presented in sections for pest and predator management, integrated plant nutrition, soil conservation, and water management systems. Some of this has to be arbitrary. A green manure can, for example, act as a break crop so preventing pest carry-over; add nitrogen and organic matter to the soil; and prevent soil and water loss by providing ground cover.

The best evidence for the effectiveness of resource-conserving technologies must come from farms and communities themselves. If a technology, such as a nitrogen-fixing legume is taken by farmers and adapted to fit their own cropping systems, and this leads to substantial increases in crop yields, then this is the strongest evidence of success. Wherever possible, the evidence for this chapter is drawn from the field. Some of these are 'traditional' practices that have been in existence for generations. Others are of recently introduced technologies, transferred from other farmers and communities or from research efforts.

As indicated in Chapter 2, it is possible to develop any number of productive and sustainable systems on research stations. The ultimate test for these, though, is whether different types of farmers find them useful and whether they can adapt them to their own conditions. A sign of sustainability, therefore, is the degree to which the skills and knowledge of farmers are enhanced, and whether they become involved in their own experimentation with technologies (see Chapter 6).

Transition Costs for Farmers

Although many resource-conserving technologies and practices are currently being used, the total number of farmers using them is still small. This is because their adoption is not a costless process for farmers. They cannot simply cut their existing use of fertilizer or pesticides and hope to maintain outputs, so making their operations more profitable. They will need to substitute something in return. They cannot simply introduce a new productive element into their farming systems and hope it succeeds. They will need to invest labour, management skills and knowledge. But these costs do not necessarily go on for ever.

These transition costs arise for several reasons. Farmers must first invest in learning. As recent and current policies have tended to promote specialized, non-adaptive systems with a lower innovation capacity, so farmers will have to spend time learning about a greater diversity of practices and measures. Lack of information and management skills is, therefore, a major barrier to the adoption of sustainable agriculture. During the transition period, farmers must experiment more and so incur the costs of making mistakes, as well as of acquiring new knowledge and information.

Another problem is that we know much less about the resource-conserving technologies than we do about the use of external inputs in modernized systems. As external resources and practices have substituted for internal and traditional ones, knowledge about the latter has been lost. Much less research on resource-conserving technologies is conducted by conventional research institutions. In India, for example, postgraduate research on modernized farming greatly exceeds that on sustainable or low input systems (see Table 8.4).

The on-farm biological processes that make sustainable agriculture productive also take time to become established. These include the rebuilding of depleted natural buffers of predator stocks and wild host plants; increasing the levels of nutrients; developing and exploiting micro-environments and positive interactions between them; and the establishment and growth of trees. These higher variable and capital investment costs must be incurred before returns increase. Examples include for labour in construction of soil and water conservation measures; for planting of trees and hedgerows; for pest and predator monitoring and management; for fencing of paddocks; for the establishment of zero-grazing units; and for purchase of new technologies, such as manure storage equipment or global positioning systems for tractors.

For these reasons, it is not uncommon for resource-conserving returns to be lower than conventional options for the first few years. One remarkable set of data from 44 farms in Baden Würtemburg, Germany, has shown that wheat, oats and rye yields steadily increased over a 17-year period following transition to a strictly organic regime (Dabbert, 1990).

It has been argued that farmers adopting a more integrated and sustainable system of farming are internalizing many of the agricultural externalities associated with intensive farming, and so could be compensated for effectively providing environmental goods and services. Providing such compensation or incentives would be likely to increase the adoption of resource-conserving technologies. None the less, these periods of lower yields seem to be more apparent in conversions of industrialized agriculture. Current evidence appears to suggest that most low input and Green Revolution farming systems can make rapid transitions to both sustainable and productive farming (see Chapter 7).

THE MANAGEMENT AND CONTROL OF PESTS AND DISEASES

Why Pesticides are Not Ideal

Although agricultural pests, weeds and pathogens are thought to destroy some 10–40 per cent of the world's gross agricultural production, pesticides are not the perfect answer to controlling pests and pathogens (Conway and Pretty, 1991). Here the term pesticide is used to refer to products that control insects, mites, snails, nematodes and rodents (insecticides, acaricides, molluscicides, nematicides and rodenticides), diseases (fungicides and bactericides) and weeds (herbicides). Throughout this book, the term pest is also used to refer to all these harmful organisms. Pesticides can be dangerous to human health and damage natural resources but, more importantly to the farmer, pesticides are often inefficient at controlling pests. They can cause pest resurgences by killing off the natural enemies of the target pests. They can produce new pests, by killing off the natural enemies of species which hitherto were not pests. Pests and weeds can also become resistant to pesticides, so encouraging further applications. And, lastly, pesticides provide no lasting control and so, at best, have to be repeatedly applied.

Ideally, pesticides should not lead to pollution, interfere with natural enemy control, or result in pests evolving resistance. Needless to say, this is unlikely. Many of the newer pesticide compounds are more selective, less damaging to natural enemies and less persistent in the environment. One consequence of greater regulation is the development of a number of chemicals that are highly targeted in their effect. But one problem is that many of these are more expensive to farmers than broad spectrum products. What farmers need is a wide range of possible technologies that can make use of the agroecological processes of predation, competition and parasitism to control pests more effectively than pesticides alone.

Most pest species are naturally regulated by a variety of ecological processes, such as by competition for food or by predation and parasitism by natural enemies. Their numbers are more or less stable and the damage caused is relatively insignificant in most cases. High input farms, though, are very different from natural ecosystems. Fields are planted with monocultures of uniform varieties, are well watered and provided with nutrients. Not surprisingly, these are ideal conditions for pest attacks, and frequently the scale and speed of attack means that farmers can only resort to pesticides.

Integrated pest management (IPM) is the integrated use of a range of pest (insect, weed or disease) control strategies in a way that not only reduces pest populations to satisfactory levels but is sustainable and non-polluting. IPM as an external intervention was first applied in 1954 for the control of alfalfa pests in California by making use of alternative strip cropping and selective pesticide use (Conway, 1971). Also in the 1950s, cooperative cotton growers in the Canete Valley of Peru developed IPM in the face of massive breakdown of control due to excessive use of pesticides (Smith and van den Bosch, 1967).

Inevitably IPM is a more complex process than, say, relying on regular calendar spraying of pesticides. It requires a level of analytical skill and certain basic training in crop monitoring and ecological principles. Where farmers have been trained as experts, such as in Honduras (Bentley et al, 1993) and in the rice-IPM programmes of South-East Asia (Kenmore, 1991), then there are substantial impacts. But where extension continues to use the conventional top-down approach of preformed packages, then few farmers adopt the practices, let alone learn the principles. As Patricia Matteson (1992) put it: *'few IPM programmes have made a lasting impact on farmer knowledge, attitudes or practice'*. The large-scale IPM for rice programmes are demonstrating that ordinary farmers are capable of rapidly acquiring and applying the principles and approaches (see Case 11, Chapter 7). These programmes are not necessarily teaching farmers new technologies and knowledge as this can become outdated very rapidly; rather they are concerned with developing farmers' own capacity to think for themselves and develop their own solutions. These are producing substantial reductions in insecticide use, while maintaining yields and increasing profits (Table 4.1).

Table 4.1 Impact of IPM programmes on pesticide use, crop yields and annual savings

Country and crop	*Average changes in pesticide use (as % of conventional treatments)*	*Changes in yields (as % conventional treatments)*	*Annual savings of programme (US $)*
Togo, cotton[1]	50%	90–108%	11–13,000
Burkina Faso, rice[1]	50%	103%	nd
Thailand, rice[2]	50%	nd	5–10 million
Philippines, rice[2]	62%	110%	5–10 million
Indonesia, rice[2]	34–42%	105%	50–100 million
Nicaragua, maize[3]	25%	93%*	nd
USA, nine commodities[4]	no. of applications up, volume applied down	110–130%	578 million
Bangladesh, rice[5]	0–25%	113–124%	nd
India, groundnuts[6]	0%	100%	34,000
China, rice[2]	46–80%	110%	400,000
Vietnam, rice[2]	57%	107%	54,000
India,rice[2]	33%	108%	790,000
Sri Lanka, rice[2]	26%	135%	1 million

* even though yields are lower, net returns are much higher
nd = no data
Sources: 1 Kiss and Meerman, 1991; 2 Kenmore, 1991; Winarto, 1993; van der Fliert, 1993; Matteson et al, 1992; FAO, 1994; 3 Hruska, 1993; 4 NRC, 1989; 5 Kamp et al, 1993; Kenmore, 1991; 6 ICRISAT, 1993

Cutting pesticide use by at least half produces substantial savings for governments. In Togo and Burkina Faso, farmer monitoring of pests in cotton and rice cultivation has also led to large cuts in pesticide use with no loss to farmers (Kiss and Meerman, 1991). In Madagascar, an IPM programme in the Aloatra basin is showing that the aerial pesticide applications on 60,000 ha of rice during the 1980s to control African stem borer were completely unnecessary (von Hildebrand, 1993). Pests here are well controlled by natural enemies and never actually cause economic losses. Integrated measures in Madagascar focus on cultural control, plant resistance, moderate herbicide use and a surveillance system. Substantial savings in foreign currency have also been made. Similar savings are being made in Nicaragua, where CARE is training farmers to use pesticides on maize more appropriately. Not only are net returns better, but farmers who received IPM training did not suffer decreased levels of the blood enzyme, cholinesterase. By contrast, farmers receiving no training had a 17 per cent reduction, indicating chronic exposure to organophosphate pesticides (Hruska, 1993).

In recent years IPM has become widely adopted in the USA, focusing mainly on better scouting for pests, rotations and other cultural practices. On many crops IPM is employed on more than 15 per cent of total acreage; for some, such as apple, citrus and tomato, it is now the preferred approach. A wide range of studies have shown that farmers can maintain or improve yields following adoption of IPM, as well as maintain or increase profits (Allen et al, 1987; NRC, 1989). In general, more farmers have increased the number of pesticide applications as a result, though the volume of pesticides has declined because of precise timing and the use of more specific products.

Using Resistant Varieties and Breeds

A major line of defence is to have crops and animals that are resistant to the likely pests and diseases. During selection and breeding to produce high yielding crop varieties and livestock breeds, many natural defence mechanisms are lost. This may be deliberate since bitter compounds reduce the palatability of plants to humans as well as wild animals. But often the loss is inadvertent. The breeders' primary aim is increased yield and, by focusing selection on the genes that govern yield characteristics, the genes that confer protection may not be retained. High yielding, modern varieties of rice in the Philippines, for example, suffer proportionately higher yield losses, on average 20 per cent of yield, compared to 13 per cent for traditional varieties, although of course the yields of the former are in absolute terms larger (Litsinger et al, 1987).

Modern livestock breeds are also less resistant to diseases. One of the most economically important of these is trypanosomiasis,which is transmitted by tsetse fly and affects some 10 million ha of Africa. Annual losses of meat production alone are estimated to be some $5 billion (FAO, 1993). But some African cattle are trypanotolerant, having developed resistance to the parasite over thousands of years. One such breed is the N'Dama, which have long been kept by West African farmers in marginal areas. They are less productive than modern cattle, though thrive on low-quality forage, and have better survival and longevity. Embryo transfer techniques have been used to enhance these N'Dama cattle, and they have also been crossed with the Red Poll, a rare British breed, to produce the Senepol breed. This has been introduced into the Caribbean and southern USA (FAO, 1993).

Much of the success of modern agriculture has centred upon breeding varieties resistant to known pests and diseases. During the 1940s, wheat in Mexico suffered several destructive epidemics of wheat rust and the first task of the improvement programme was to breed varieties with stem rust resistance (see Chapter 2). By 1949 four pioneer hybrids with high levels of resistance were available to farmers and, by 1956, national average yields had risen from 650 to 1100 kg/ha. Rice, too, has benefited from the incorporation of resistant genes drawn from a variety of Asian sources. The first modern varieties had narrow genetic resistance as breeders had selected for a limited number of desired characteristics, including short straw, high tillering ratio, insensitivity to photoperiod and early maturity. Subsequently, however, protection has been built in, year by year, so that modern rice varieties are resistant to a much wider range of pests and pathogens (Khush, 1990).

Evolution, though, also works to counter the breeders' selections. New species of pests, weeds and pathogens appear, and, more important, new strains of existing pests and pathogens that overcome the hard-won resistance may develop. One example is the brown planthopper, a serious pest of rice, of which at least three strains have appeared in recent years. Each new strain results in a major outbreak and the hurried distribution of new resistant rice varieties. Another is the sorghum greenbug in the USA. In 1968, the greenbug caused US$100 million loss to the sorghum crop and farmers spent some $50 million in the following year to control the pest. By 1976, however, resistance to the greenbug was found and the new hybrids were being grown on 1.5 million hectares. A new biotype of greenbug capable of attacking this hybrid then emerged in 1980, but again researchers were successful in developing another resistant variety (NRC, 1989).

For low external input farmers, resistant crops and livestock represent an important alternative to pesticides in controlling pests and pathogens. The 'treadmill' nature of breeding for resistance does, however, mean that farmers must rely on regular supplies of new seed. Most of these treadmill problems occur because modern varieties are not planted in mixtures and, if palatable, present pest and diseases with unchecked opportunities for population growth. However, planting a diversity of varieties or genotypes in a field can help to harness the inherent variability in pest and pathogen resistance. One option is to create multilines by mixing seeds from similar lines of a crop variety. The lines are very similar in most of their characteristics, but have different genes for resistance. In theory, when new strains of a disease appear only one or two of the lines will prove susceptible. Build up of the disease is slow, an epidemic is prevented and most of the crop escapes damage.

Alternative 'Natural' Pesticides

Many farmers know which locally available plants have insecticidal or disease-controlling properties, and there is a wide range of locally available compounds to repel, deter or poison pests of their crops and animals (Table 4.2). Many of these are both selective in their action, killing pests and not predators, and degrade rapidly so do not contaminate the environment. Of those that are repellents, the effectiveness is short lived and usually considerably reduced by rain. Increasingly, scientists are identifying the mechanisms behind these 'traditional' approaches to control. Some, though, are toxic to people and broad spectrum in their action, and thus are not so different to many synthetic products (Conway and Pretty, 1991). Non-plant products are also widely used, such as solutions of cattle manure and animal urine to repel insects and animals; soil added to leaves to abrade the cuticle of insect pests; and sand or ash added to stored grain to stop the movement of weevils.

The most widely used natural plant compounds are the antifeedants that render plants unattractive and unpalatable to pests. The most common is neem (*Azardirachta indica*), which occurs over wide areas of Asia and Africa. Almost every part of the tree is bitter, although the seed kernel possesses the maximum deterrent value. The derivatives are known to control more than 200 species of insects, mites and nematodes (Saxena, 1987; FAO, 1993). Yet neem does not harm birds, mammals (including people) and beneficial insects such as bees. The seed is most commonly formulated in an oil or cake: in parts of India neem cake has been applied to rice for centuries. In the USA, neem extract controls Colorado potato beetle sufficiently well to give 27–47 per cent better yields than unsprayed potatoes (Zehnder and Warthen, 1988). Neem is also less toxic to beneficial predators.

But there is a disadvantage, since neem degrades fairly rapidly in sunlight and as a consequence the most successful applications have been in the control of stored grain, rather than field pests. However, a multinational corporation from the USA has recently synthesized a product which stabilizes azardirachtin, the active ingredient of neem. Although this is potentially good news for farmers worldwide, it has instead become a threat to the traditional technology. The company has now been granted a sole patent on both the stabilizer and azardirachtin. This means that it can, in theory, charge farmers for using their own traditional technology or prevent them from using it. In practice, this will of course be difficult. What is more likely to happen is that local supplies of neem will be bought up, formulated with the stabilizer, and then sold back to farmers. In 1993, the world's first commercial-scale facility, capable of processing 20 tonnes of neem seed each day, was opened (FAO, 1993).

Farmers also rely on many local plants for the control of livestock diseases. In industrialized countries, much traditional veterinary knowledge has died out with the decline of horses and the mechanization of farming since the 1950s. For centuries, British horsemen made use of many hundreds of herbs and wild plants, including agrimony to control fever, burdock for conditioning, feverfew for curing colds, horehound for keeping

Table 4.2 Selection of locally available compounds used for pest control in a range of countries

Plant	*Country*
Chili pepper (*Capsicum frutescens*)	*Kenya:* ground, stirred in water, left to stand and sprayed against aphids or fed to chickens to treat diarrhoea *Papua New Guinea:* ground, stirred in water with soap, sprayed to repel aphids *Benin:* milled with earth and mixed with beans during storage *Philippines:* pulverized and burnt monthly beneath food stores *Honduras:* mixed with garlic in water, left to stand, diluted and applied next day to repel insects
Custard apple, sweetsop (*Annona* spp)	*China and Philippines:* pulverized seeds used against human lice *West Africa*: water suspension of seeds controls insect pests
Turmeric (*Curcuma domestica*)	*Sri Lanka:* root shredded and added to cow urine, sprayed against insect pests; threads dipped in grated turmeric and stretched across fields to repel insects *Various locations:* dried, pulverized root added to stored produce to repel weevils and borers
Neem (*Azardirachta indica*)	Neem effective as aqueous solution, oil, kernel powder and press cake for insect pests and fungal control *India:* used on vegetables, citrus, cereal and bean crops *Ghana:* leaves burned, ashes mixed with water and spread on crops
Muna (*Minthustachys* spp)	*Peru:* muna twigs used to line inside of potato stores and pits
Croton oil tree (*Croton tiglium*)	*Thailand:* water extract made from pulverized seeds and used against aphids
Mexican marigold	*Kenya:* cut and laid around livestock bomas to repel safari ants
Simson weed (*Datura stranonium*)	*Cameroon:* leaves, stems, flowers and seeds shredded and soaked in water, soap and kerosene solution, and sprayed against leaf-eating caterpillars and aphids
Gliricidia spp	*Sri Lanka:* fresh leaves applied as a mulch to control transmitter of mosaic virus
Castor oil (*Ricinus communis*)	*Cameroon:* seeds mashed and heated in water, soap and kerosene solution, mixture sifted, diluted and sprayed immediately *Ecuador:* leaves placed in maize fields to attract beetles, which are paralyzed by the castor *India:* castor widely grown as intercrop with cereals and cotton to repel insects
Ryania (*Ryania speciosa*)	*South America*: powder or spray used against maize and fruit pests
Daluk (*Euphorbia antiquorum*)	*Sri Lanka:* chips of Euphorbia placed in water at point of impounding of irrigation water to control thrips

Sources: Stoll, 1987; Matthias-Mundy, 1989; Schrimpf and Dziekan, 1989; Pretty, 1990a; Upaswansa, 1989; Fre, 1993; Catrin Meir, personal communication

horses on their food and celandine for clearing worms (Evans, 1960). In Peru, shepherds use wild tobacco to control skin parasites of sheep and feed cattle with artichoke to prevent liver flukes (Matthias-Mundy, 1989). Pastoralists in Africa have been widely recorded as using many herbs for pest and disease control, as well as for setting broken bones and treating wounds (Schwabe and Kuojok, 1981; McCorkle, 1989; Fre, 1993).

Bacterial and Viral Pesticides

Pesticides based on bacteria and viruses are also promising in terms of selectivity and reduced potential for pollution. The greatest successes so far have been preparations of *Bacillus thuringiensis* (*Bt*). The bacillus produces a crystalline compound, which dissolves when ingested by insects producing toxic proteins that paralyse the gut and mouthparts. Strains of *Bt* have been used against moth pests for some 25 years, though new strains have been shown to be active against a range of other pests including nematodes, mites and beetles.

The crystal toxins of *Bt* are produced by a single gene, which has now been cloned and inserted into non-pathogenic bacteria that colonize plant roots, and also directly into crop plants such as tobacco and tomato. The potential for engineering plants to contain their own defensive compounds in this way is considerable. According to the OECD, field releases with transgenic tobacco, cotton, maize and tomato have taken place in the USA, Israel and Spain. As yet restrictions on genetically engineered micro-organisms have not permitted extensive field trials. Sales of *Bt* are now worth US$100 million per year. However, resistance to *Bt* has been reported in Australia, the USA, Japan, Philippines, Thailand and Taiwan, particularly where *Bt* products have been repeatedly applied as a spray or incorporated into crops (PT, 1993).

Some strains of bacteria are also effective at controlling crop diseases, such as *Agrobacterium*, which produces an antibiotic that controls crown gall tumours of orchard trees and ornamental plants (NRC, 1989). Antibiotic substances produced by the bacterium *Streptomyces* have been formulated into a biofungicide, which inhibits growth of *Rhizoctonia* in oil seed rape, *Fusarium* on cereals, *Pythium* on sugar beet and *Alternaria* on cauliflowers. Other new biopesticides include products based on fungi and toadstools. Although the demand for these 'biopesticides' is growing rapidly, the market share still remains small in relation to the remaining pesticide market.

A successful example of the use of viruses has been the release of live coconut rhinoceros beetles infected with a baculovirus in islands of the South Pacific. The virus spread at 3 km/month and within 18 months of release beetle populations had declined at some locations by 60–80 per cent (Bedford, 1980; Young, 1974). Some naturally occurring viruses can also give good control. In Brazil, a major pest of cassava, the cassava hornworm, is being controlled by spraying with extracts of hornworm infected with a virus, *Baculovirus erinnyis* (CIAT, 1987). Mixtures made from recently infected larvae result in 90–100 per cent mortality within seven days of application, though virus from four-year-old frozen larvae still kill some 70 per cent of hornworms. Frozen virus is now available on a semi-commercial basis in Brazil, and farmers have been shown how to collect, prepare, store and apply infected larvae in newspaper, radio and TV campaigns. Farmers in the USA have also successfully made their own preparations (Box 4.1)

Box 4.1 Farmer's own technology for pest management, Florida

Eugene Alford and his neighbours use a naturally occurring fungal disease as part of their IPM programme to control velvet bean caterpillars on their soybeans. He collects dead dried caterpillars infected with the mould *Nomurea rileyi* and grinds them into a powder which he freezes for application the following summer. He applies this powder, which contains fungal spores, to parts of his fields most affected by caterpillars a few weeks before spraying time and monitors its effects. When conditions are optimum the disease will wipe out all caterpillars within 4–5 days. Alford estimates annual pesticide savings of up to US$10,000 on his 200 ha farm.

Source: International Ag-Sieve, 1988

Cutting Input Use in Industrialized Systems

The alternative to seeking safer compounds is to rely on more efficient and careful application of existing pesticides. Most damage arises today, not so much because of the intrinsic characteristics of the pesticide compounds but because of the way they are used. There is increasing evidence that farmers can reduce their pesticide applications through the precise targeting of pests and weeds in crops without suffering any reduction in profitability (Pretty and Howes, 1993).

In Britain, evidence is emerging to show that if farmers get the timing of applications of fungicide on cereals right, they can cut rates by 50–75 per cent and still maintain yields (see Table 7.2). Farmers regularly have to examine crops and then apply a quarter-rate mix when 75 per cent of plants are showing at least one active mildew spot (Wale, 1993). Careful monitoring and sequential sampling for pests on brassicas has reduced the need for pesticides by 85 per cent, while maintaining yields. In the fruit sector, many farmers have been able to cut their use of fungicides to 12–25 per cent of former levels following the adoption of a range of IPM technologies (Doubleday, 1992). But these low dose approaches do place extra management demands on farmers. As Stuart Wale put it: *'the use of low dose mixtures is not appropriate for all growers. It is primarily intended for those who can inspect crops regularly and make a timely application of [pesticide]'* (FW, 1993a).

An important new approach is patch spraying. This needs a combination of modern technology, regular field monitoring and modified spray systems that allow application exactly where there are known problems. A field map showing the location of weeds or pests is first constructed by using a combination of aerial photography, image analysis of maps and field walking. This information is then stored in a tractor-mounted computer which also controls the sprayer. In the field the operator enters the location of the tractor and a distance/speed monitor tracks its position as it moves. The position is compared with the pest or weed maps and herbicides or pesticides dispensed only when they are needed. The impact on cost reduction can be considerable, with some farmers cutting herbicide bills substantially with no impact on cereal yields. John Morrison (FW, 1993b), who farms in Derbyshire, has cut herbicide bills by 95 per cent in some fields. In one field he saved £1700 by patch spraying: *'we'd have had to spray the whole lot if it hadn't been for the modified spray system'*. In effect, farmers are substituting labour and knowledge for the former dependence on external measures for pest control.

Patch spraying can be more efficient with the use of global positioning systems (GPS). A GPS utilizes signals from satellites to fix the precise position of a tractor or combine harvester within a field. The system can produce yield maps for fields by combining data from existing yield monitors on combines with the exact position in the field. Even in modern agriculture, yield variations within a single field can be up to 4 t/ha. Seed, pesticide, herbicide and nitrogen rates can be matched to the variations within a field. One farmer cut nitrogen rates by 30 per cent, which reduced the amount of nitrogen leaching out in the field drains by 60 per cent (FW, 1991b). As another farmer, John Fenton, put it: *'it must make sense to tailor input levels to as small an individual area as possible. Blanket rates over a large area are wasteful'* (FW, 1993c). And because many factors affect crop yield, the technology is best used in combination with soil analysis, regular field walks and monitoring.

Pheromones for Disrupting Pest Reproduction

Some pest populations can be controlled by disrupting their reproduction. Synthetic chemicals that mimic pheromones, which are hormones released by female insects to attract males, will greatly reduce the chances of mating by confusing male insects, while the release of larger numbers of pre-sterilized males will ensure that most matings are sterile. Both of these, though, are high-input options and require intervention at a very large scale, involving cooperation among large numbers of individual farmers. So far they have only been effective on large enterprises, or as part of government or co-operative run schemes.

Slow-release formulations of synthetic pheromones that confuse males are very effective in disrupting mating in a variety of pests if applied on a large scale. Control of the pink bollworm (*Pectinophora gossypiella*) on cotton has been successful in Egypt, Pakistan and the USA, and of grape oriental fruit moth on peaches, tomato pinkworm and some pests of stored products (Campion et al, 1987; NRC, 1989). In Egypt pheromone formulations are sold at the same price as conventional insecticides and two to three applications are as effective as four to five sprays of conventional pesticide. Predators are more numerous in pheromone-treated than in insecticide-treated fields, and also more bees survive, leading to bumper crops of honey (Campion and Hosny, 1987). In Pakistan, only 7 per cent of cotton bolls are infested by pink bollworm in pheromone-treated fields, while 25 per cent are infested in conventionally treated fields (NRI, 1994). In the USA, the pink bollworm disruptant is applied on some 40,000 ha annually, holding infestations down to 1 per cent or less with a single application, so permitting a reduction of insecticide applications have by nearly 90 per cent (NRC, 1989). In the UK, the use of pheromone traps to monitor codling and tortrix moths has become standard practice in apple orchards (Doubleday and Wise, 1993).

Another approach, the release of sterile males, requires an even larger scale of operation. Massive rearing facilities are needed to raise large numbers of pests that are then sterilized by irradiation and released in a sufficient quantity to swamp the natural population. The first pest eradicated by this technique was the screwworm, *Cochliomya hominivorae*, a serious pest of cattle in the south-western USA (Knipling, 1960; Conway, 1971). From early 1958, over 50 million sterile flies were released each week, eventually eradicating the pest over large areas within a year. Although there have been other successes, the technique is only likely to succeed where the pest populations are relatively small and isolated.

Similar to pheromones are juvenile hormones, which kill or prevent insects from reaching a mature stage for reproduction. As metamorphosis is prevented, the insects are biologically dead, and the population eventually ceases to exist. These compounds offer the possibility of being active only in certain insects, with no biological activity in other organisms. They have not yet been used in field crops.

MANAGING NATURAL ENEMIES

Releasing Predators and Parasites

The natural enemies of pests include a great variety of parasitic wasps and flies, predators such as ladybird beetles, spiders, hoverflies, wasps, ants and assassin bugs, and larger animals such as lizards, birds and fish. Pests are also attacked by a range of pathogenic bacteria, fungi, viruses and nematodes. In natural conditions many such natural enemies may be present, acting together to regulate pest numbers. One problem with pesticide use is that if all the pests are killed, the predators may have nothing to feed on and so also die. This may mean pest numbers increase (see Chapter 3). Ideally the pest population needs to be brought down to a desired level and maintained there, hopefully permanently. This implies that the pest population is not eradicated and, indeed, is tolerated to an extent.

The use of natural enemies is commonly referred to as biological control. The term sometimes implies any form of non-pesticidal control, but it is less confusing if restricted to the use of natural enemies. It is also important to distinguish between what is called *classical biological control* which involves the release of new or exotic natural enemies and *augmentation* which relies on improving the degree of existing control. There has been considerable effort over recent years to develop effective biological control programmes (Waage and Greathead, 1988; Jutsum, 1988).

Occasionally the results are spectacular. One such success was the control of the prickly pear, *Opuntia*, a cactus that was introduced into Australia as a garden plant from Mexico at the end of the last century. It soon spread to pasture land and by the 1920s some 25 million ha were infested. But eventually *Cactoblastis cactorum*, the larvae of which tunnel into and destroy the cactus, was discovered in Argentina and taken to Australia. The cactus now only occurs as individual plants or in small patches (Conway, 1971).

Equally successful is the use of *Trichogramma*, an egg parasitoid of moth pests, which is used on 15 million ha worldwide. To be effective, though, the releases usually have to be carefully managed. A single, carefully timed release of the egg parasitoid *Trichogramma* early in the growing season gives as good control of moth pests on maize and sugar cane in China as weekly mass releases later in the season (Waage and Greathead, 1988).

A recent success story has been the parasitic wasp, *Epidinocarsis lopezi*, introduced from South America to control the cassava mealybug (*Phenacoccus manihoti*) in Africa (Kiss and Meerman, 1991; Neuenschwander and Herren, 1988). The mealybug first appeared in Congo and Zaire in 1973, and is now found in a wide belt from Mozambique through Zaire and across to Senegal. Severe attacks can cause up to 80 per cent reductions in cassava yields. Widespread searches were conducted throughout South America, the original home of cassava, and the wasp discovered in Paraguay was released in Nigeria in 1981. It is now present in 25 African countries. After the release of *E. lopezi*, cassava mealybug populations decline by up to 50 per cent, often leading to yield gains of some 2.5 t/ha (Hammond et al, 1987; Neuenschwander et al, 1989).

Although there have been some attempts to calculate benefit:cost ratios, these have been controversial as data is poor and many generalizations have had to be made. Norgaard concluded returns of $2.25 billion for an investment of $14.8 million, a ratio of 149:1, and IITA researchers concluded returns of $3 billion for a ratio of 178:1 (Norgaard, 1988; IITA/ABCP, 1988). The most controversial aspect of the programme, however, has been that farmers have not had to be involved, unlike other similar programmes (see Box 2.3).

Some of the most successful biological control programmes have been against pests of glasshouse crops, such as tomatoes, cucumbers and ornamentals. Pests in glasshouses rapidly multiply in the controlled and favourable environment, often with devastating effects. However, the high degree of environmental control can also favour the planned release of natural enemies. There has been rapid growth in the use of natural enemies in recent years and some expect more growth in this market than for conventional pesticides. One private company that supplies biological control agents to farmers, recently indicated that in the future the *'conventional crop protection market is expected to stagnate, while the natural biological products sector expands by 10–15 per cent per year, accounting for 5–10 per cent of the total market within the decade'* (FW, 1993d).

None the less, there have been many more failures with biological control than successes (Jutsum, 1988). Spectacular failures include the introduction of cane toads to sugar cane crops in Australia, that not only failed to control cane beetles but also became pests themselves; and the attempts to control cedar scale in Bermuda involving the introduction of over 50 species of natural enemies between 1946 and 1951. The problem is that a great deal needs to be known about the detailed population dynamics of pests and their natural enemies and, in some cases, where growth rates of pest populations exceed natural mortality, then biological control is physically impossible.

Improving the Habitat for Natural Enemies

Natural enemy populations can also be encouraged by increasing the diversity of farms and their neighbouring environments. Many natural enemies need food sources in the form of pollen or nectar, which can often be provided by wild vegetation near or in the crops. There are usually more natural enemies in fields bordered by diverse hedgerows and in orchards adjacent to woodlands (Lewis, 1969; Altieri and Schmidt, 1986; Herzog and Funderbank, 1986; El Titi and Landes, 1990). These non-crop plants that harbour natural enemies of pests often give good control (Table 4.3). The predators, however, often need some encouragement to invade crops and fields. Perennial stinging nettle, for example, is a source of predators of aphids and psyllids, and, as predator numbers increase in the spring, so their dispersal to crop fields can be encouraged by cutting the nettles (Herzog and Funderbank, 1986).

One traditional technique in orchards has been to encourage populations of predatory ants. In China, bamboo bridges have for some 1700 years been placed between branches to encourage movement of citrus ants (*Oecophylla smaragdina*) from tree to tree (Huang and Pei Yang, 1987). These feed on various insects that attack orange, tangerine, lemon and pomelo trees. Whole orchards can be colonized by securing a nest on one tree and then connecting this to others with the bamboo strips.

The practice is also common in Indonesia. Raymon Wibisana (1987), a Javanese farmer, described his practice in this way: *'I have encouraged big red ants to breed in my orchard for they stop worms and mites from pestering the trees. The ants do not seem to harm the trees, and if their numbers become too great they can be used for chicken feed. The ants can also be used against the hopping insect. Two of my trees that were without red ants suffered, but a third, which was inhabited by the ants, grew well and bore fruit. So I ran a rope from the healthy tree to the other two so that the ants could move across. It took them two weeks to do so, but after that the ants ousted the pests and the trees had new leaves'.*

Birds and fish can also be important. Farmers in Sri Lanka encourage birds to come to their rice fields by putting food on unstable discs attached to stakes so that when birds attempt to perch, the food falls into the rice and in following it down they see the pests; and by inserting coconut fronds in the fields as owl perches to aid rat control. Farmers also preserve large trees and pockets of woodland close to the paddy to provide nesting

Table 4.3 Selection of cropping systems from the USA and UK in which non-crop plants enhance the biological control of certain pests

Crop	*Non-crop plants*	*Pest regulated*	*Mechanism*
Apple	*Nasturtium*	Aphids	Enhancement of hoverfly populations
Brassicas	*Amaranthus and Chenopodium*	Green peach aphid	Increased abundance of predatory beetles
Cotton	Ragweed (*Senecio*)	Boll weevil	Alternative host for parasite *Eurytoma*
Peach	Ragweed	Oriental fruit moth	Alternative host for parasite *Macrocentrus*
Sugar cane	*Euphorbia*	Sugar cane weevil	Provision of nectar and pollen for *Lixophagus* parasite
Grape vines	Wild blackberry (*Rubus* spp)	Grape leafhopper	Alternative host for parasitic wasp, *Anagrus*
Wheat, barley, oats	Timothy grass *(Phleum pratense*) and Yorkshire fog (*Holcus lanatus*)	Cereal aphids	Grass beetle bank provides habitat for predatory ground beetles
Wheat, barley, oats	*Phacelia*	Cereal aphids	Flowering *Phaecelia* encourages hoverfly populations

Sources: Altieri and Liebman, 1986; Herzog and Funderbank, 1986; Pretty and Howes, 1993

and perching places for predatory birds (Upaswansa, 1989). A similar approach has been taken by groundnut farmers in Andhra Pradesh, India, who have stuck small tree branches into the soil (ICRISAT, 1993). These perches attract birds, which eat pod-borer caterpillars. Farmers on 400 ha of Guntar District stopped using pesticides, saving themselves some 1 million rupees (US$34,000). They also had higher yields than their neighbours still using pesticides, because natural enemies could survive in their fields.

Rice pests can also be controlled by cultivating fish in the paddy water. The fish move with the flood water in the rainy season from refuges, and help to keep down the incidence of insect pests and pathogens (see Table 4.6).

Beetle Banks and Flowering Strips in Industrialized Systems

A recent development in temperate agriculture to encourage predators while reducing pesticide applications has been to use beetle banks, flowering strips and conservation headlands. In Britain, several hundred potentially beneficial species of predators and parasites may live in or by cereal crops. Most of these are killed when the crops are sprayed to control pests. But if the field habitat is manipulated to increase plant diversity, then the need for spraying pesticides can be greatly reduced. When grass strips are constructed across large fields, then predatory beetles proliferate and can get to the field centres, the regions where aphid populations are greatest (Wratten, 1992; UoS/GC, 1992). The cost of establishing a 400 m bank in a 20 ha field is about £90, including cultivation, grass seed and loss of crop. In succeeding years the cost of land taken out of production is £30 for the same field. One aphid spray costs £300 across the same field, plus the cost of yield reduction due to aphid infestation. One farmer, Michael Malyon, recently created five beetle banks, indicating that *'we never get good yields in our large fields... The cost of putting in the banks in a field is negligible compared with the potential benefits'* (FW, 1993e).

Wild flowers also encourage predators. Hoverfly larvae are voracious predators of aphids, and because the adults need pollen and nectar to lay eggs, they thrive on farms rich in wild flowers. Headlands left unsprayed with herbicides support many more predators than those where flowering plants are removed; the weeds attract non-pest herbivorous insects, which encourage hoverflies and other predators of cereal aphids, such as the beetles *Agonum dorsale* and *Bembidion lampros*. The survival rate of partridge and pheasant chicks, which feed on the herbivorous insects, is also greater in these conservation headlands (Game Conservancy, 1993).

These practices are growing in popularity. Some 1800 km of conservation headlands were recorded in England and Wales in 1992. Recently, farmers have been experimenting with *Phacelia tanacetifolia*, a blue-flowering ornamental introduced from the USA. This has a long flowering period and again attracts hoverflies. Where it has been planted in strips, the number of eggs laid per aphid is twice as great as in fields with no flowering strips (Wratten, 1992). It is ironic, of course, that farmers who are putting in beetle banks and flowering strips may well have been encouraged to remove hedgerows in the recent past.

Rotations and Multiple Cropping

Crop rotations are a central component in the development of resource-conserving farming, with the maximum use made of crops that contribute to soil fertility and reduce pest damage. The approach is to rotate non-host crops with susceptible crops in sequence. While the non-host crop is present, the pest populations decline so that they are very low or even absent when the susceptible crop is grown again. The non-host crop provides a 'break', disrupting the relationship between a pest or pathogen and its host. It is a practice that rarely has ecological or economic drawbacks and many farmers regard rotation as an essential component of prudent management.

The retention of spatial and structural diversity through multiple cropping practices is as important as rotations. Many small farmers still rely on multiple cropping. In Latin America, some 60 per cent of maize is grown with beans (Francis, 1986; Altieri, 1990). Rice, cotton, beans and cassava are widely grown in mixtures. Generally the more diverse an agroecosystem, the less abundant are herbivore pests though, in some mixtures, herbivores do prevail (Conway and Pretty, 1991; Risch et al, 1983). Different crops can be grown row by row, or in alternate strips each consisting of several rows of the same crop, or they may be grown in a more complicated spatial pattern or, indeed, at random. Mixtures of spring barley varieties, for example, provide good control of powdery mildew: even though pure stands treated with fungicides yield slightly better than untreated mixtures, the untreated mixtures provided better economic returns (Wolfe, 1981; Wolfe and Barratt, 1986).

There are various factors in these crop mosaics that help constrain pest attack. A host plant may be protected from insect pests by the physical presence of other plants that may provide a camouflage or a physical barrier. Mixtures of cabbage and tomato reduce colonization by the diamond-back moth, while mixtures of maize, beans and squash have the same effect on chrysomelid beetles. The odours of some plants can also disrupt the searching behaviour of pests. Grass borders repel leafhoppers from beans and the chemical stimuli from onions prevents carrot fly from finding carrots.

Alternatively one crop in the mosaic may act as a trap or decoy – the 'fly-paper effect'. Strips of alfalfa interspersed in cotton fields in California attract and trap *Lygus* bugs. There is a loss of alfalfa yield but this represents less than the cost of alternative control methods for the cotton. Similarly crucifiers interplanted with beans, grass, clover or spinach are damaged less by cabbage maggot and cabbage aphid. There is less egg-laying on the crucifiers and the pests are subject to increased predation. Interplanting can also be combined with selective use of pesticides, applying them at the appropriate time solely to the trap crop.

INTEGRATED PLANT NUTRITION

Just as with integrated pest and predator management, there is a wide range of nutrient management measures that can both maintain soil fertility and sustain productivity. These are increasingly being known as integrated plant nutrition systems (FAO, 1991). These focus on improving the efficiency of inorganic fertilizers, introducing new crops into rotations that fix nitrogen or utilizing organic sources of nutrients.

Improving the Efficiency of Fertilizers

It is virtually impossible to maintain crop production without adding nutrients. When crops are harvested, nutrients are invariably removed and so have to be replaced. There are a variety of sources: the mobilization of existing nutrients in the soil and parent rocks; the fixing of nitrogen from the atmosphere; or the supply of organic or inorganic fertilizer. The application of fertilizer, ideally, should closely match the needs of plants but often farmers, for reasons of cost, will apply fertilizer in fewer and larger doses. Commonly, fertilizer is applied in excess of need, so some nutrients are lost from the farm as nitrates to surface or ground water, or as ammonia or nitrous oxide to the atmosphere. On average, some 30–60 per cent of applied nitrogen is lost in non-irrigated farming, rising to 60–70 per cent from paddy cultivation (Conway and Pretty, 1991). This represents a substantial loss to farmers.

Crops vary in the efficiency with which they take up nutrients and so breeding for efficiency of nitrogen use is a potentially productive approach. For instance, the widely cultivated rice variety IR36 was superseded by the more nitrogen-efficient IR42. The soil type and sources of nitrogen other than the inorganic fertilizers are also important factors. If reserves in the soil are known then it is possible to make fertilizer recommendations tailored for the specific requirements of each field and each crop.

Fairly precise recommendations are now available for farmers in Europe and North America, though not generally for farmers in the South (Conway and Pretty, 1991). In the UK, these are largely based on the previously grown crop (MAFF/ADAS, 1988). Cereals are assumed fully to deplete reserves, for instance, whereas pasture leaves high reserves for the next crop. The outcome is a set of recommendations for nitrogen fertilizer application rates dependent on both reserves and soil type. For instance, it is recommended that winter wheat likely to yield less than 7 t/ha when grown on sandy soil with low reserves should receive 175 kg N/ha. But if the reserves are high and the soil a clay, then no fertilizer needs to be applied.

Nutrient uptake and absorption can also be improved by using foliar sprays, slow-release products or by incorporating, with the fertilizer, certain compounds that inhibit the bacterial conversion of nitrogen compounds. Foliar fertilization is efficient because of rapid absorption and translocation into the plant (Alexander A, 1993). In some vegetable crops, it is possible to reduce fertilizer applications by 25 per cent by substituting foliar sprays.

Low input farmers are likely to be the greatest beneficiaries of deep placement fertilizers such as urea briquettes, urea marbles or urea supergranules (USG), as a small quantity of fertilizer is now capable of going further. In Taiwan, for example, USG increases rice yields by 20 per cent on farms in marginal areas, but has no impact in the already high yielding zone (De Datta, 1986). Nutrient uptake and absorption can also be improved by using slow-release products coated with sulphur. Sulphur-coated urea reduces the need for split applications and helps to fulfil sulphur requirements of the crop, with economic returns of the order of US$4–7 for every dollar spent (De Datta, 1986). Urea can also be combined with various aldehydes to make insoluble products, with methylene-urea the most common and also the best at preventing nitrate leaching. Polymer or resin-coated fertilizers can be tailored in such a way that the release period extends up to 12 months or more. Granules of fertilizer are coated with a diffusion barrier through which nutrients slowly pass (Alexander A, 1993).

Nitrification inhibitors, such as dicyandiamide, prevent the conversion of ammonium to the more mobile nitrate form. These products can improve yields as well as cut losses to the environment. But they are 10–20 per cent more costly than conventional fertilizers and are only available to farmers in industrialized countries. Inhibitors that reduce gaseous ammonia losses from broadcast applications of nitrogen to rice paddies delay the build up of ammonia in the water, but it is not clear whether there is a positive impact on yields too.

Livestock Manures and Composts

Farmers who can neither afford nor rely on a regular supply of inorganic fertilizers must find alternative organic sources of nutrients. These sources are often cheaper, more efficient than inorganic compounds and focus on recycling of nutrients. Livestock are therefore a critical component of sustainable agricultural systems. The nutrient value of manures largely depends on how they are handled, stored and applied. Losses of nitrogen tend to be highest when liquid systems of storage are used and when the manure is broadcast without incorporation. Livestock manures from cattle, pigs and chickens are important, as they positively affect soil structure and water retention, and benefit soil organisms. Soils under integrated farms, for example, have more earthworms than those under conventional management (El Titi and Landes, 1990; Edwards and Lofty, 1977).

It is becoming more common for farming households with only small farms to keep their animals permanently penned in zero-grazing or stall-feeding units rather than permit them to graze freely. In Kenya, zero-grazing units are a central part of efforts to improve soil and water conservation. Fodder grown on the farm in the form of improved grasses, tree fodder and the residues of cultivated crops are cut and carried to the animals. Because of the proximity to the crops, manures can be returned directly to the land, so improving nutrient supply and soil structure. In dryland regions migrating pastoralists and their livestock are frequently welcomed by farmers during the dry season. The livestock are kept in pens overnight on the crop fields and in some areas farmers are even willing to pay herdsmen for this overnight kraaling (McCown et al, 1979; Scoones and Toulmin, 1993).

Where manures are in short supply, farmers are often willing to pay for them to be imported (Wilken, 1987). In Mexico, farmers are willing to pay US$8–12 for a truckload of chicken manure and vegetable growers in Quetzaltenango in Guatemala buy chicken wastes that are transported 100 km from Guatemala City. In Oaxaca, Mexico the highest value organic material for fertilizing crops is the nutrient-rich debris from the nests of ants. The material is collected in bags and carefully applied to individual plants of high value crops, such as tomatoes, chilis and onions. The decline in use of ant refuse in recent years is said to be a result of substitution by commercial fertilizers (Wilken, 1987).

Composting is a technique of long standing that combines the use of animal manures, green material and household wastes. The materials are heaped or placed in a pit in such a fashion that anaerobic decomposition occurs. Harmful substances and toxic products of metabolism are broken down, while pathogens, and the seeds and roots of weeds are destroyed by the heat generated within the compost heap.

Composting is particularly valuable in the tropics since organic matter stores nutrients and protects them against leaching. It also makes the soil more friable and easier to plough, improves moisture retention and aeration, and remedies the problems caused by inorganic fertilizers. Farmers in Tanzania make compost from stall litter which includes crop residues, leafy tree branches and old roofing grass; in Rwanda farmers mix household wastes, crop residues, weeds, dried leaves and twigs of trees; and in Nepal farmers use a combination of up to 25 wild plants mixed with animal manures (Kotschi et al, 1989; Tamang, 1993). Wood ash is also commonly used, being carried from burnt bushland to compost heaps. But composting is demanding of labour, both in the building of heaps and in spreading on the fields. Its use is thus likely to be limited to kitchen gardens, though whole farms can benefit (see Case 13, Chapter 7).

Legumes and Green Manures

The impact of legumes grown together with or before a cereal crop can reduce, and sometimes eliminate, the need for nitrogen fertilizers. Symbiotic bacteria present in specialized nodules that develop on the roots of legumes can fix nitrogen directly from the atmosphere. The cultivation of cereals and legumes crops together can improve both total yields and stability of production. Bushes and trees with nitrogen-fixing capacity also have beneficial effects on plants growing with or after them.

In the Americas, the interplanting of maize, beans and squash, often the seeds being placed in the same planting hole, is a practice of great antiquity, probably dating to soon after agriculture began in the valleys of Mexico (Gleissman, 1990). In such situations, with soils of low inherent fertility, the cultivation of cereals and legumes crops together can improve both total yields and stability of production. In maize and cowpea mixtures, some 30 per cent of the nitrogen taken up by the maize is obtained from the legume (Aggarwal and Garrity, 1987). Cowpea and lablab are particularly useful legumes for inter-cropping with cereals, the former because it is adapted to acid, infertile soils, and the latter because it is drought-tolerant, produces good fodder and can regrow well after clipping. Here, legumes contribute not only through nitrogen fixation, but also because the green matter can be used as a mulch or green manure.

Undersowing is a once-common practice used now by only a few farmers in industrialized countries. Cereals are sown with a legume and/or grass, and these are already established at harvest. This can help control pests and diseases, provide ground cover and supply nitrogen. Undersowing cereals and brassica with trefoil and clover increases the number of insect predators, reduces the numbers of pests and gives better crop yields than monocrops (Potts, 1977; Dempster and Coaker, 1974; El Titi and Landes, 1990).

Legumes have long been used in milk production systems. However, the advent of readily available and cheap inorganic fertilizers has led to a decline in the reliance on legumes to maintain soil fertility. Mixed grass-clover swards gave way to high nitrogen input grass pastures as producers attempted to maximize yields in response to modern price incentives. Adding nitrogen reduces the content and production of clover, leading to monocultures of grass. But with the 1984 introduction of milk quotas in Europe, there has been renewed interest in the use of legumes in dairy production as a means of reducing unit costs. Grass-clover swards with no application of inorganic nitrogen can successfully support long term-dairy cattle grazing and intensive silage making under commercial farm conditions. These clover-rich swards can fix 80–280 kg N/ha/yr. The financial returns from high nitrogen input systems are no greater, and are often substantially lower, than the grass-clover system (Bax and Fisher, 1993; MMB/SAC, 1992; Younie, 1992; Pretty and Howes, 1993) (see Table 7.3).

Nutrients are also supplied when vegetation is incorporated in the soil as a 'green manure'. Green manures increase nutrient levels as well as improve the physical properties of the soil. This has long been practised; the Romans grew lupins and ploughed them in before sowing cereals more than 2000 years ago. Quick-growing legumes are valuable green manures for many low input systems, and have the potential to meet much, if not all, of the nitrogen requirements of succeeding non-legume crops. The equivalent amount of nitrogen fertilizer required to match the green manures can be 80–200 kg/ha. Many green manures can also add large amounts of organic matter, up to 30 tonnes/ha (Flores, 1988).

One of the most remarkable is the velvetbean (*Mucuna pruriens*). This has been widely promoted as part of the work of World Neighbors in Central America, though its effectiveness is attested by its spontaneous spread from village to village without outside intervention. It grows rapidly, is palatable to animals and people, fixes large amounts of nitrogen and can produce as much as 60 t/ha of organic matter (CIDICCO, *passim*). It can grow on most soils and its spreading habit suppresses weed growth. This compares with an average for the country of just 0.6 t/ha. Incorporating such green manures into cropping systems can substantially increase yields (Table 4.4). Honduran farmers are able to harvest some 2.5–3.2 t/ha of maize when grown after velvetbean (see Case 4, Chapter 7).

Sesbania rostrata, though, is probably the fastest nitrogen-fixing plant, accumulating 110 kgN/ha in only 45 days (Lathwell, 1990). In Rwanda, the shrub *Tephrosia vogelii* grows to a height of 3 m in 10 months and produces 14 tonnes/ha of above-ground biomass which, when worked into the soil, can increase cereal yields by as much as four fold to some 2800 kg/ha, a response equivalent to 120 kg/ha of inorganic fertilizer (Kotschi et al, 1989). In Nepal, some green manures can produce rice yields that outperform those produced by as much as 100:30:30 kg of NPK/ha (Joshy, 1991). In dryland North-East Thailand, short duration legumes can be grown following the first rainfall peak, which is insufficient for rice transplanting. Cowpeas, grown for 45–60 days before the rice, have increased rice yields by 5–20 per cent for poor farmers compared with conventional fallowing (Craig, 1987). The benefits are such that inorganic fertilizers are no longer necessary.

In Bhutan, *Sesbania aculeata* substitutes for external inputs, but the best performance occurs when farmers have access to some inorganic nitrogen (Table 4.5). *Sesbania* with no fertilizers produces the same rice yields as 40:40:30 kg NPK/ha; but if fertilizers are added to rice after the *Sesbania* then yields increase to 5.4–5.5 t/ha, levels that can be achieved only if 120

Table 4.4 The impacts of green manuring of legumes on cereal yields

Country	*Green manure*	*Cereal*	*Impact on yields (as % of conventional)*	*New yields (kg/ha)*
Rwanda[1]	*Tephrosia vogelii*	Maize	400%	2800
NE Thailand[2]	*Vigna* spp (Cowpea)	Rice	105–120%	2875
Honduras[3]	*Mucuna pruriens* (velvetbean)	Maize	295%	2500
Brazil[4]	*Mucuna aterrima*	Maize	nd	6800
	Crotolaria striata	Maize		5800
	Zornia latifolis	Maize		
Bhutan[5]	*Lupine mutubilis*	Potato	133%	21500
	Sesbania aculeata	Rice	131%	4560
Vietnam[6]	*Tephrosia candida*	Rice	136%	2160
	Stylosanthes spp	Rice	145%	2000
	Vigna spp	Rice	145%	2100
Nepal (mid-hill region)[7]	*Sesbania cannabeana*	Rice	116%	5845
	S rostrata	Rice	118%	6030
	Vigna radiata	Rice	145%	6600
Nepal (terai region)[7]	*Sesbania cannabeana*	Rice	194%	3340
	S rostrata	Rice	218%	3690
	Vigna radiata	Rice	200%	3380
Brazil (Santa Catarina)[8]	*Mucuna pruriens* (velvetbean)	Maize	nd	3000–5000
	Canavalia ensiformis (jackbean)	Maize		3000–5000
	Dolichos lablab (lablab)	Maize		3000–5000
	Vigna spp	Maize		3000–5000
	Melilotus albus (sweet clover)	Maize		3000–5000

nd = no data
Sources: 1 Kotschi et al, 1989; 2 Craig and Pisone, 1988; 3 Bunch, 1990; Flores, 1991; 4 Lathwell, 1990; 5 Norbu, 1991; 6 Thai and Loan (1991); 7 Joshy, 1991; 8 Bunch, 1993

kg N/ha are added. Use of *Sesbania* as a green manure can save the use of between 40–120 kg N/ha (Norbu, 1991). The key lesson would appear to be that green manures increase crop yields significantly by providing nitrogen. But if farmers are able to get hold of small amounts of inorganic nitrogen, then they will benefit still further.

There are 19 million hectares of upland rice worldwide and average yields are only about 1 tonne/ha. Yet if cowpea or lablab are intercropped with rice, and then allowed to continue growing through the dry season, the biomass can be incorporated as a green manure before the next rice crop (Aggarwal and Garrity, 1987). Rice yields increase to 1.4–1.9 t/ha as a result and there is the added bonus of the legume grain yield of 0.5–1 t/ha.

Table 4.5 Impact of the green manure *Sesbania aculeata* on rice yields in Bhutan

Presence or absence of Sesbania green manure	*Presence or absence of NPK fertilizer (kg/ha)*	*Rice yields (t/ha)*
Zero	Zero	3.5
Green manure	Zero	4.6
Zero	40:40:30 of NPK	4.7
Green manure	40:40:30 of NPK	5.6
Zero	120:40:30 of NPK	5.4

Source: Norbu, 1991

Recent research in semi-arid India has shown that some legumes, such as chickpea and pigeonpea, have a unique mechanism that allows them to access phosphate in phosphate poor soils (Johansen, 1993). They release acids from their roots, which react with calcium-bound and iron-bound phosphate to release phosphate for plant uptake. As their deep rooting also helps water infiltration, they have a positive residual affect on subsequent crops, as both phosphate and water availability are increased.

Azolla and *Anabaena*

Blue-green algae are another important source of nitrogen, the most widely exploited being the alga *Anabaena azollae*. This fixes atmospheric nitrogen while living in cavities in the leaves of a small fern, *Azolla*, that grows on the water of rice fields in both tropical and temperate regions. *Azolla* quickly covers the water surface in the ricefield, but does not interfere with the normal cultivation of the rice crop.

Very high nitrogen production is possible following *Azolla* inoculation in rice fields. In the Philippines, 57 tonnes of freshweight *Azolla* can be harvested after 100 days yielding more than 120 kg/ha of nitrogen (Watanabe et al, 1977; Kolhe and Mitra, 1987). Over the whole year, *Azolla* can fix more than 400 kg N/ha, a rate in excess of most tropical and subtropical legumes. This nitrogen is only available to the rice crop after *Azolla* has decomposed and so exploitation consists of incorporating the ferns into the soil while wet as a green manure or removing them for drying and then reapplying them to the ricefields.

The results of at least 1500 studies in China, Philippines, Vietnam, India, Thailand and USA have shown that when *Azolla* is grown in paddy fields, rice yields increase by on average 700 kg/ha, with a range of 400 to 1500 kg/ha (Liu and Weng, 1991; San Valentin, 1991; Kikuchi et al, 1984). In India, wheat crops following rice with *Azolla* have also been shown to produce improved yields (Kolhe and Mitra, 1987). Like most resource-conserving technologies, the incorporation of *Azolla* is labour and knowledge intensive. The timing of incorporation is also critical, since a sufficient period has to elapse for the green manure to decompose.

For most farmers, *Azolla* offers the opportunity of substituting for inorganic fertilizers. The incorporation of *Azolla* as a green manure in parts of Brazil has permitted for a 30–50 per cent reduction in the use of nitrogen fertilizers (Kopke, 1984). In the Philippines, recent studies have shown that incorporation of *Azolla* would allow nitrogen applications to be reduced by at least half (San Valentin, 1991). *Azolla* with 30:30:20 kg NPK fertilizer per hectare yields 34 per cent better than rice with the same NPK and no *Azolla*, producing yields of some 5.9 t/ha. This is 74 per cent better than rice with neither (3.4 t/ha). Some studies have, however, shown an important transitional period during which yields in the two years after *Azolla* drop from 4.95 to 3.6 and then to 3.8 t/ha (Castillo, 1992). The decreased variable costs did not offset this fall, but by the third year yields had recovered and net returns were higher.

Although the benefits of *Azolla* would appear obvious, many farmers are not using it. The National Azolla Action Programme was established in the Philippines to reduce the burden of high costs to farmers. A programme of working closely with farmers has established that *Azolla* combined with 30 kg/ha of nitrogen will sustain current yields, saving the country some US$23 million each year in foreign exchange (Box 4.2).

Box 4.2 The National Azolla Action Programme, Philippines

The National Azolla Action Programme was established in 1982 to reduce the burden of high costs to farmers in the Philippines. The objective was to replace half of the fertilizer nitrogen requirement for rice production with internal resources. The programme aims to cover 300,000 ha of irrigated lowland rice areas. The process has been as follows:

- establishment of a National Inoculum Center (NIC), with a network of regional sub-centres in agricultural universities and colleges, which screen and test local *Azolla* varieties;
- establishment of propagation centres to provide materials to municipalities and villages;
- preparation of information and materials on the culture and utilization of *Azolla* for extension workers and farmers;
- conduct of training, demonstrations and on-farm trials.

At the end of extensive on-farm trials, the results indicated that *Azolla* plus a small amount of nitrogen fertilizer (30 kg/ha) would give equivalent grain yields. The NAAP has estimated that if *Azolla* substitutes for half of the nitrogen requirement for rice in this way, this would generate annual savings of at least US$23 million.

Source: San Valentin, 1991

Agroforestry

There is a huge diversity of agroforestry systems throughout the world, in which the bushes and trees have many benefits. Those with nitrogen-fixing capacity have beneficial effects on plants growing with or after them. Some of this is a result of the fixed nitrogen, but significant quantities can also be supplied in the leaf litter or from deliberate pruning. Trees also improve the microclimate by acting as windbreaks, by improving the water-holding capacity of the soil and by acting as shade trees for livestock – so focusing the deposition of manure (Young, 1989). In the Majjia Valley, Niger, windbreaks of neem trees help to conserve moisture and soil, raising the yields of cereals grown between by some 20 per cent (Kerkhof, 1990).

On the southern coast of China, there are some 140,000 ha of coastal fields protected by windbreaks and shelter belts (Luo and Han, 1990; Zhaohua, 1988; Beckjord, 1991). The trees are species of mainly *Casuarina, Metasequoia, Leucaena, Acacia, Paulownia* and various bamboos. These protect crops from typhoon damage in the rainy season and cold spells in early spring and late autumn. As a result, wheat and rice yields can be 10–25 per cent higher than in unprotected zones. *Paulownia* is successfully intercropped with cotton, maize, beans, groundnut, sweet potato, rape, garlic, watermelon and vegetables. *Paulownia* is well suited to agroforestry systems, as its deep tap root does not compete with shallow rooted crops for nutrients and water. A tree can grow 2.5 m in one year, reaching 10–20 m after ten years, when it can supply 400 kg of young branches and 30 kg of leaves for fodder or soil amendment.

Woody shrubs and trees planted on the contour can protect the soil and provide fodder, fuelwood and timber. It has long been the practice in the countries of the Mediterranean to plant rows of trees such as olives between rows of cereals or vines. Most recently in tropical countries there has been a considerable research on alley cropping, in which trees of various kinds are planted in contour rows with, usually, subsistence crops in between. Often the trees are fast-growing legumes, which fix nitrogen into the soil. They also provide fodder for animals, green manures and fuelwood. However, much of this research has been conducted on research stations, where the constraints experienced by farmers are not replicated. As a result, very few alley cropping systems have been adopted as designed and many projects have failed because of the desire to stick to the rigid technical model (see Chapter 2).

The sloping agricultural and technology (SALT) model is one such alley cropping technology being promoted on Mindinao Island in the Philippines (Palmer, 1992; Tacio, 1991, 1992). Over the past 20 years, it has been developed on demonstration farms as a highly productive and potentially sustainable system. Contour hedgerows of *Leucaena* are mixed with maize, which yields three to four times as much as non-SALT farms and net returns are better. However, farmers have not as yet been willing to adopt the package and there is little evidence of widespread farmer interest (Garrity et al, 1993). Some farmers have, however, taken components and adapted them into their own systems (Palmer, 1992).

Many agroforestry systems also combine livestock, so increasing the number of internal linkages. Rubber monocrops can be transformed with the introduction of animals (Nair, 1989). Although intercrops are cultivated in immature rubber plantations, when the canopy closes only weeds survive. These are costly to control. In Malaysia, sheep rearing in rubber plantations can keep down weeds as well as give added returns from the sheep. Bees and chickens are other animals that will also survive in plantations.

SOIL CONSERVATION

Conservation Tillage

The way in which the soil is tilled can have a significant influence on how well soil, nutrients and water are retained. In conventional tillage, the topsoil is inverted and mixed by means of a mouldboard plough or disc, or a handtool such as a hoe. This incorporates most of the crop residues or stubble and the nutrients they contain. However, there is a lag period from the time the seed is sown to when there is sufficient vegetative cover to prevent soil erosion by wind or water. An alternative approach is to use conservation tillage in which the soil surface is disturbed as little as possible. Significant amounts of residue then remain on the soil surface, so helping to reduce runoff, sediment loss and loss of nutrients. The seed is directly drilled through the layer of residues. In no-till farming soil preparation and planting are done in one operation; in reduced-till farming there is limited preparation with disc or chisel plough.

These conservation tillage systems are widely promoted by the Soil Conservation Service in the USA. Between 1980 and 1993, the area devoted to conservation tillage grew from 16 to 40 million ha and so now covers some 35 per cent of all harvested land (AAN, 1993b). The main focus is on reduced tillage with chisel ploughing, using crop residues to provide a mulch cover. In practice this reduces soil erosion by up to 50 per cent. Other conservation tillage practices, such as no tillage, strip tillage and ridge tillage systems, reduce erosion by 75 per cent or more, but are less widely adopted. By 1992, no-till farming covered 11 million ha. The fastest growth was in Iowa, where some 1 million ha are under no-till systems. One problem with no-till, however, is that many farmers have to use more herbicides as weeds are no longer controlled by ploughing operations (NRC, 1989).

Contour Farming

Another approach for conserving soil nutrients is to resort to physical structures, such as terraces or bunds, of varying scale. These are common to many indigenous agricultural systems throughout the world (see Reij, 1991; Kerr and Sanghi, 1992; Kassogue et al, 1990). Most of these are designed to check the surface flow of water, and thus perform the dual role of water harvesting and retention. The simplest approach is to construct earth banks across the slope to act as a barrier to runoff. These are suitable on shallow slopes and are frequently used in conjunction with contour planting. Sometimes the earth bunds are reinforced with vegetation such as crop stalks, or planted with grass or trees, to create greater stability. As such vegetative bunds are partly permeable, crops planted in front of the bund also benefit from water runoff. These are not quickly damaged by runoff, and thus maintenance costs are low.

Simple walls may also be constructed along the bunds and these are quickly strengthened by natural processes. Elsewhere, rocks may be most appropriate substances for the construction of contour bunds or walls. After the first heavy rains, fine soil, branches and leaves begin to fill in the walls, making them more impermeable.

More costly to construct are various forms of terrace. Diversion and retention terraces are appropriate for shallow slopes and bench terraces are effective on steeper ones, but not on thin soils where the parent rock is close to the surface. All can raise crop yields by some 30–50 per cent over those on non-terraced slopes. But construction costs for bench terraces are usually very high. Many soil conservation projects have expended huge sums on food for work during the course of terracing (see Chapters 2 and 3).

Rather than construct physical structures that generally require a large labour input, a lower input alternative is to plant crops along contours. As water flows across the surface so it meets with rows of plants growing perpendicular to the flow, which slows it down and improves infiltration. In strip cropping the main row crop is grown along the contour in wide strips alternating with strips of protective crop, such as grass or a legume. If the protective strips are of grass they can be effective at filtering out particulate matter and nutrients from surface flow of water. Contour grass strips not only reduce loss of soil but help in the process of establishing terraces.

There is widespread evidence for improved crop yields and reduced erosion following terracing of fields in many countries (Pretty and Shah, 1994; Tato and Hurni, 1992; Reij, 1991). In Ethiopia, for example, one study of the impact of the *fanya juu* terrace (which involves throwing soil uphill to make a bund) found improved yields over non-terraced fields of some 30–40 per cent (Michael, 1992). The variability of yield on the terraced fields was also lower. In neighbouring Kenya, *fanya juu* terracing has improved yields of maize and beans by some 50–60 per cent (Pretty et al, 1994; SWCB, 1994; Tjernstrom, 1992; Figueiredo, 1986; Grönvall, 1987; Hunegnaw, 1987).

But much of this evidence is small scale and localized. As we discuss in Chapter 5, for the full benefits of conservation to accrue to farmers, it is necessary to consider the impacts at a wider, community scale. It is, however, increasingly being well established that whole communities are capable of adopting and adapting soil conservation practices and principles. A recent comprehensive study of Machakos District in Kenya has shown that even though there has been a three-fold increase in population since 1945, net imports of maize to the district have fallen from 17.4 to 7.6 kg per capita (Tiffen et al, 1993). More conservation has led to increases in agricultural yields and the diversity of crops grown. Land that was severely degraded in colonial times is now intensively and sustainably managed.

Other studies are illustrating that the Ministry of Agriculture's catchment approach to soil conservation in Kenya is leading to substantial local improvements (see Case 12, Chapter 7). Where there is mobilization of the community, support to local groups and committed local staff, there is also increased agricultural productivity, diversification into new enterprises, reduction in resource degradation and independent replication to neighbouring communities.

For further details of similar community-led soil conservation initiatives in Burkina Faso, India, Lesotho, Mali and the Philippines, see Chapter 7 (Cases 2, 3, 6–10, 14, 15 and 19).

Mulches and Cover Crops

Soil, water and nutrient conservation is also improved with the use of mulches or cover crops. Organic or inorganic material is spread on the soil surface to provide a protective physical cover, the mulch, for the topsoil. Mulches protect the soil from erosion, desiccation and excessive heating, thus promoting good conditions for the decomposition and mineralization of organic matter. Mulches can also help to reduce the spread of soil-borne diseases, as they reduce the splashing of lower leaves with soil during rainfall. The cheapest approach is to use plant residues from previous crops, from nearby perennials or from wild areas, such as reeds from swamps. But equally as useful as organic materials are non-degradable mulches like plastic film. Black plastic, for example, excludes light and thus prevents weed growth. Other types of non-crop mulches include newspaper, cardboard, sawdust, woodchips, leafmould and forest bark.

No-till and reduced tillage systems both result in good cover with residues, reinforcing the conservation value of an undisturbed soil. In Guesselbodi, Niger, mulching with twigs and branches permits cultivation on hitherto abandoned soils, producing some 450 kg cereals per hectare. In drought years, the yields on mulched soils are some five times better than on non-mulched (OTA, 1988; Heermans, 1988). In China, wheat or rice straw mulches can increase tea, fruit and legume yields by 6–16 per cent, as well as reduce splash erosion (Jin, 1991). And in the hot savannah region of northern Ghana, straw mulches minimize erosion as well as increase yields. Combined with livestock manures, these mulches produce double the maize and sorghum yields than the equivalent amount of nitrogen added as inorganic fertilizer (Bonsu, 1983). Many farming communities use or have used wild plants for mulches and green manuring, such as in Nepal (Tamang, 1993), India (Poffenberger, 1990), Britain (Pretty, 1990b), Canada (Omohundro, 1985) and Guatemala (Wilken, 1987). In Guatemala, farmers of Quetzaltenango collect from the mixed pine-oak forests up to 20–30 tonnes of leaf litter for each hectare of cropland. This is incorporated into the soil to improve moisture retention and soil tilth.

Cover crops consist of vegetation that is deliberately established after or intercropped with a main crop, not necessarily with a view to harvest but more to serve various regenerative and conserving functions. The best example of the effectiveness of cover crops comes from the work of

EPAGRI in the State of Santa Catarina in Brazil (see Case 1, Chapter 7). EPAGRI are working intensively with some 60 species of cover crops, which are intercropped with subsistence crops or planted during fallow periods, mostly in the winter months. These plants act as both a green manure and mulch: some fix nitrogen, and all are cut and left on the soil surface. Several thousand farmers have now benefited from these green manure/mulch/cover crops. What they are showing is that providing ground cover is more important than constructing physical structures to prevent erosion.

Silt Traps and Gully Fields

Silt traps and gully fields are one particularly effective soil and water conservation measure used widely by farmers. Stones are placed across gullies or valleys, so as to capture nutrients, silt and moisture. Stones are often bedded into the upper surface of spillway aprons and walls to provide support for the next layer. The principle is to capture runoff from a broad catchment area and concentrate it in a reduced area, so transforming meagre rainfall into utilizable soil moisture. As water slows, any suspended debris is deposited, helping to form organic-rich soils. These gully or deposition fields have been recorded in India (Chambers, 1991; Shah et al, 1991); Pakistan (personal observation in Punjab and north-west frontier provinces); Ethiopia (ERCS/IIED, 1988); Mexico, known as *atajadizos, trincheras* and *trancas* (Johnson, 1979; Blackler, 1994); Nepal (Tamang, 1993); and Burkina Faso (Reij, 1988).

A well maintained silt trap creates a flat, fertile and moist field with a micro-environment quite unlike the surrounding area (UNEP, 1983; Chambers, 1991). Crops can thus be grown which may be of higher value than field crops on nearby drylands, such as rice in India, wheat and rapeseed in Pakistan, sorghum and rice in Burkina Faso, and chat and coffee in Ethiopia. Agriculture in these gully fields is often more productive and dependable. In Mexico, they permit earlier planting and in Gujarat they are the most stable component of a household's food supply (Griffin and Dennis, 1969; Shah et al, 1991). In Burkina Faso, sorghum yields can range between 970–2670 kg/ha, and in some fields rice can be grown (Reij, 1988). Farmers additionally benefit from these traps as groundwater levels are raised and damage to crops on the downstream side is reduced (Johnson, 1979; Reij et al, 1988).

In Gujarat, India, farmers have been plugging nullahs with earth embankments and stone pitching for at least 20 years. These are labour intensive for construction, but require relatively little maintenance. The yields of paddy in these fields are higher than in irrigated fields, and farmers are also able to take a residual crop after the rice and raise mango trees on the embankment. These structures are still not part of any official watershed management programme in the area (Shah et al, 1991).

Check dams are intrinsically incremental systems, in which farmers add to the height of their structures year by year. They do this to keep the wall above the level of the accumulating alluvium. Gene Wilken (1987) reports the narrative account of an Otomí farmer of Hidalgo, Mexico: '*An atajadizo isn't built all at once. Usually a farmer starts with a low wall across the path of an arroyo (gully). It takes a few years until the water has brought down enough debris and soil to level with the top of the wall. Then, the farmer will build up the wall a bit more, and so on, little by little until s(he) has built up a tall strong wall and a large level field. A well-made atajadizo is level so that the trapped water will cover all parts of the field evenly. It may be necessary to level the field by hand and, also, to tear down parts of the gully in order to enlarge the field. A well-made atajadizo always has a wall that is higher than the field behind it. This is necessary because water must be trapped so that it can soak into the field. But if the field is at the same level as the wall, the water will just flow over it and waste. There is no need to fertilize an atajadizo because every rainy season the water brings down new debris and soil'.*

It should not be surprising that these fields have been overlooked by professionals. Most soil conservation programmes focus on a restricted number of technologies, ignoring the diversity that already exists. In Niger, traditional stone lines in the Ader Doutchi Maggia can be observed by anyone driving on the main road from Konni to Tahoua (Reij, 1991). Despite the presence of conservation projects in the region since the early 1960s and visits by many soil conservation experts, no reports contain reference to these stone lines. In both Niger and Burkina Faso, farmers prefer stone lines and bunds, yet all major projects have constructed only earth bunds, which have of course not been maintained by local 'beneficiaries' (Reij, 1991). In India, John Kerr and N K Sanghi (1992) reflected on their own survey of soil conservation practices *'the fields which were neglected badly could be spotted easily from a distance (even while driving on the road). The indigenous technology (where soil and water conservation was successful) could not be appreciated until the specific fields were visited individually'.*

WATER MANAGEMENT SYSTEMS

Water Conservation and Harvesting

Where rainfall is unreliable and inadequate, water shortages often severely limit crop production. Water conservation and harvesting can carry crops over an otherwise disastrous dry period, can stabilize and increase production, and can even make agricultural production possible for the first time. Water harvesting systems commonly include a runoff-producing and a runoff-using area. Water harvesting systems can be found in many parts of the world, including the Middle East, south Asia, China, North America and sub-Saharan Africa (UNEP, 1983; Reij et al, 1988; Reij, 1991).

Water harvesting systems from short slopes are simple and cheap, and have a relatively high efficiency, because water is not transported over long distances. One very old system of micro-catchment use is *meskat* in Tunisia, where fruit trees, mainly olive, are fed by runoff from upper slopes in a 200–400 mm rainfall area. The *zai* system in Burkina Faso is another example. *Zai* involves the digging of small pits, local application of manure and the construction of stone bunds to catch runoff. The concentration of both water and nutrients have made *zai* a popular method to rehabilitate degraded land. Yields in the Yatenga region can be 1000 kg or more per ha, in areas where average yields are only 400–500 kg/ha (see Case 2, Chapter 7). In Kenya, semi-circular earth bunds with stone spillways collect water and increase sorghum yields from virtually nothing to some 2000 kg/ha (UoN/SIDA, 1989).

For water harvesting from long slopes, semi-permeable stone contour lines and bunds are necessary. Water runoff is slowed down, rather than concentrated and so has more time to infiltrate below the stones. Half-moon shaped bunds are used to concentrate water, almost always for forest or fodder trees. In the Tarija Basin of Bolivia, *media lunas* dug on the contours of very degraded hills are planted with trees for fodder and fuel, and legumes on the earth ridge (Bastian and Gräfe, 1989). The ground is rapidly colonized by wild grasses, and as a result soil erosion is halted, infiltration increased, and the vegetation period prolonged into the dry season.

Floodwater harvesting in the streambed, whether a valley bottom or floodplain, blocks the water which flows intermittently and often in flash floods. In North Africa and the Middle East, *wadi* floors are blocked, and fill with water from the adjacent slopes and the main water course. Many local variations of this basic principle have been documented, including from Mexico, India, Pakistan and Burkina Faso. In Burkina Faso on the Central Plateau, gully formation in the valley bottom results in concentration of runoff rather than the preferable even spread over the floodplain. Low semi-permeable dams of loose rock are constructed in the gullies to slow the water flow and push the water out of the gullies on to the floodplain. Soil is also conserved in the process, with rapid formation of terraces between the dams. Sorghum yields are 200–300 per cent higher on fields connected to the dams than unimproved fields (Scoones, 1991; Reij et al, 1988; Critchley, 1991).

Water harvesting systems do not only use water locally but can manipulate the direction of water flows to reach areas suitable for crop, tree or pasture production. By running into and around a series of obstacles water is forced to spread to parts that would otherwise not benefit from runoff. In China, such warping systems combine water manipulation with increased nutrient efficiency. Storm and floodwater is diverted at moments when nutrient and organic matter content of the water is high, so making use of water resources more efficiently and decreasing soil erosion. One warping area is the Zhaolao Gully, in Shanxi Province, over 2300 years old and feeding water to 2260 hectares. Soil moisture of warped farmland is 10 per cent higher compared to unwarped land, increasing to almost 80 per cent during dry periods. Both organic matter and nitrogen content of the soil increases, with obvious benefits for agricultural production. Warping increases the yields of maize, millet and wheat by some two to four fold over unwarped land (UNEP, 1983). Warping was also common in seventeenth- and eighteenth-century Britain, in some areas being the principal technology on which diversified and productive agriculture was based (Pretty, 1991).

Despite the apparent benefits of water harvesting systems, they are not widely used. One constraint may lie in labour requirements for both initial investment and maintenance. Where stones are transported over long distances or soil movement high on steep slopes, labour inputs rise greatly. Usually investments in water harvesting systems are high in the first year, after which labour inputs can drop by almost half. However, if there is no maintenance of structures as is likely to occur if farmers are forced to adopt the measures (see Chapters 2 and 3), the yields quickly fall. Many water harvesting systems require collective action on a large area for them to be effective. Financial costs of water harvesting can range between US$100/ha with simple, low-cost techniques and US$1000/ha with sophisticated systems.

The main question as to whether the investment in water harvesting systems will be worthwhile is determined by soil fertility. Where nutrient levels greatly limit agricultural production, an increase in water availability will have only temporary impact. Yield increases will be a passing phenomenon, and where water was once the main limiting factor, soil fertility takes over. The key lies in combining water harvesting with integrated nutrient management. By slowing water flows, water harvesting effectively controls soil erosion, and nutrients and water are harvested and conserved. In Burkina Faso, for example, the effect of *digue filtrantes* (permeable rock dams) on the quick decomposition of deposited organic matter has been signalled by farmers as of particular importance, reducing the need for manure application (Reij et al, 1988).

When the supply of water becomes more regular, then water harvesting becomes small-scale irrigation. Although this is not the place to discuss adequately the aspects of irrigation, it is none the less important to note the importance of irrigation relying on relatively local sources of water (Guijt and Thompson, 1994).

Land Drainage for Saline and Waterlogged Soils

Overuse of water in agriculture has led directly to the rapid increase in recent years of land lost to waterlogging and salinity. Although precise data are hard to come by, it is thought that something of the order of 1.5 million ha are lost annually (WCED,1987). Curing saline and waterlogged soils requires lowering the water table below the root zone of crops, followed by leaching to remove the excess salts. These salts then have to be removed from the soil by sub-surface drainage systems. The drainage technologies have been widely proven, but the implementation is far more difficult. Collective action is essential, as drainage technology is indivisble and cannot be implemented in parts (see Chapter 5).

In India, where the area of saline and waterlogged soils is 5–13 million ha, drainage technology costs some US$325–500 per ha. Returns, though, make this investment worthwhile, as immediate improvements occur in the form of increased cropping intensity, changed cropping patterns, higher yields and lower costs (Datta and Joshi, 1993; Datta and de Jong, 1991). But as Datta and Joshi make clear, technology alone is insufficient.

Participation by local people is essential for long-term success: *'planning and executing the drainage systems to manage saline and waterlogged soils by government agencies may not yield the desired results unless there is a positive attitude and strong will of the beneficiaries to participate in the programme'* (Datta and Joshi, 1993).

The problem is that most state action has been to suppress this very action needed at local level. Just as in rehabilitation of irrigation systems, it is the attention to participation and local institutional strengthening or building that is critical (Uphoff, 1992a).

Raised Beds and Chinampas

Where there is too much water, raised beds are technologies that make effective use of available resources. The basic principle is that crops are cultivated on raised fields, which are surrounded by water channels. The channels are used for transport, provide additional food in the form of frogs, fish and ducks, and are a source of aquatic plants for composts and green manures. Nutrients are cycled between the two systems. Such raised beds are traditional in China, known as high-bed, low-ditch systems; in Mexico, known as chinampas; in Kashmir, known as 'floating gardens'; and in the high Andes of Peru, known as *waru-waru*.

In Mexico, chinampas have been under continuous cultivation for at least two and perhaps three thousand years (Gómez-Pompa and Jimenez-Orsonio, 1989; Gleissman, 1990; Wilken, 1987; Gómez-Pompa et al, 1982). A wide variety of crops are grown, the most common being maize, beans, chili, amaranth and squash. Willow and alder trees grow on the margins of the fields to provide shade, windbreaks and organic matter. They also are a good habitat for birds, as well as helping to protect crops from heavy frosts and rains. The canals acquire deposits of eroded soils, decomposed plants, and wastes from villages and farmhouses, and runoff from fields. Much of this is returned as farmers dredge the muck from the canals and replace it on the fields. Even though no external inputs are used, crop yields are high.

In the Lake Titicaca basin in Peru, *waru-waru* were used widely by the pre-hispanic farmers to cope with poor soils and frequent frosts, but had fallen into disuse. Efforts have been made in recent years to redevelop this ancient technology, leading to improved agricultural production in as many as 30 altoplano communities (see Case 17, Chapter 7)

In the Pearl River delta of China, much of the land is close to or below sea level. Farmers raise soil from ditches to form beds of width 1–10 m depending on the type of crops. Narrow beds are used for sugar cane and vegetables, while systems for longer duration crops, such as banana, citrus and lychees have wider beds and ditches. In the ditches rice, fish and edible snails are cultivated, and mud is excavated to put on the beds. These high-bed low-ditch systems have helped to lower water tables, reduce soil erosion and nutrient loss, preserve organic matter in ditches and increase the internal cycling of nutrients (Luo and Lin, 1991; Zhu and Luo, 1992).

Fish Production in Irrigation Water

One of the best examples of integrated farming is when fish production is combined with rice cultivation. For at least 2000 years, farmers of South and South-East Asia have combined rice–fish culture. With the advent of the Green Revolution technologies, however, many systems have been abandoned because of the toxicity to fish of the pesticides used.

The basic principle is that fish live in the water of the paddy fields, retreating to specially constructed refuges or ponds during the dry season. The fish are beneficial because they eat weeds, algae and insect pests, and help to keep disease carriers in check. Their manures help to fertilize the rice crops. When *Azolla* is present, they eat *Azolla*, converting it into forms of nitrogen readily available to the rice. Not only are the fish a source of protein for farming families, but rice yields are usually improved too. In recent years, there have been coordinated efforts to increase rice–fish culture in the Philippines (Bimbao et al, 1992; de la Cruz et al, 1992); in Thailand (Jonjuabsong and Hawi-Khen, 1991; Boonkerd et al, 1991); Bangladesh (Kamp et al, 1993); Indonesia (Fagi, 1993); and Taiwan (Chen and Yenpin, 1986).

Although rice–fish culture in Thailand was first important in the central region, this fell away with the advent of the Green Revolution technologies. Recent spread has been in the rainfed regions of the north-east, where a wide range of government agencies and NGOs are working with farmers to improve fish yields. Fish farming can be technically difficult to get right. Bunds must be raised around fields to keep the fish in and predators out. A nursery pond has to be constructed to hold the fry until they reach fingerling size and a refuge has to be dug for the dry season. There then needs to be careful choice of fish, and control of stocking rates and supplementary feeding. In addition, farmers themselves have to reduce or eliminate pesticide use, and ensure they are not affected by neighbours.

In Bangladesh, a recent programme coordinated by CARE combines rice–integrated pest management with fish culture (Kamp et al, 1993). It is demonstrating that farmers can eliminate pesticides entirely, improve rice yields and get a harvest of carp (Table 4.6). Farmers in the programme monitor their insect populations on a regular basis and they soon see that their fields are not more infested with pests than their neighbours who have sprayed with pesticides. Reduced pesticide use could have further beneficial impacts, on human health and on duck and wild fish populations.

In the Philippines, the government rice–fish culture programme was launched in 1979, but was hampered by the modern varieties' need for heavy use of fertilizers and pesticides (Bimbao et al, 1992; de la Cruz et al, 1992). Since then, the area of rice–fish has slowly increased, as the shift from rice monoculture to rice–fish culture increases net returns by up to 40 per cent. Rice production also improves, by some 4 per cent, and farmers also benefit from vegetables grown on the banks of the raised bunds. Some 200–300 kg fish per ha are also harvested. However, there are technological constraints, such as pesticide applications, and availability of fingerlings and water that are holding back spread. In addition, most farmers are now accustomed to rice monoculture, and a shift to rice–fish means at first more work and higher costs.

Sometimes these rice–fish systems are developed into more complex polyculture farms. In Taiwan, pigs and ducks are also common elements of farms, with tilapia the most common fish (Chen and Yenpin, 1986; Lightfoot and Noble, 1992).

Table 4.6 The impact of combinations of IPM training and fish culture on rice farming in Bangladesh

Farmer trained in:	*Change in pesticide use (as % of normal practice)*	*Rice yields (as % of normal practice)*	*Food and income from fish*
Normal practice	100%	100%	No
IPM only	24%	110%	No
Rice–fish	0%	117%	Yes
IPM and rice–fish	0%	124%	Yes

Source: Kamp et al, 1993

SUMMARY

There are many proven and promising resource-conserving technologies that can be integrated to produce a more sustainable agriculture. These technologies do two important things: they conserve existing on-farm resources, such as nutrients, predators, water or soil; and/or they introduce new elements into the farming system that add more of these resources, such as nitrogen-fixing crops, water harvesting structures or new predators, and so substitute for some or all external resources.

Many of the individual technologies are also multifunctional, implying that their adoption will mean favourable changes in several components of farming systems at the same time. But their adoption by farmers is not a costless process. Farmers cannot simply cut their existing use of external inputs and hope to maintain or even improve outputs. They need to substitute labour, management skills and knowledge in return. Farmers must, therefore, invest in learning. As recent and current policies have tended to promote specialized systems, so farmers will have to spend time learning about a greater diversity of practices and technological options. Lack of information and management skills is a major barrier to the adoption of resource-conserving agriculture.

IPM is the combined use of a range of pest control strategies in a way that not only reduces pest populations to satisfactory levels but is sustainable and non-polluting. It is a more complex process than relying on spraying of pesticides, and makes use of resistant varieties and breeds, alternative 'natural' pesticides, bacterial and viral products, and pheromones for reducing the impact of pests. Predators and parasites are encouraged by direct releases, improving their physical habitat, increasing farm diversity, and adopting rotations and multiple cropping.

Integrated plant nutrition involves a combination of a more efficient use of fertilizers with the adoption of alternative sources of nutrients, such as livestock manures, composts, legumes, green manures, *Azolla* and agroforestry. Soil conservation can be enhanced through the use of conservation tillage, contour farming and physical structures, mulches and cover crops, and silt traps and gully fields. Many of these technologies, in one form or another, have been in existence in traditional agricultural systems for centuries. There are a range of water management systems that ensure the efficient use of available water. Water conservation and harvesting can improve agricultural yields in dry areas. Where too much water has been used, leading to waterlogging and salinization, then land drainage technologies making use of collective action can be used. Where environments are very wet, then thoroughly integrated systems making use of aquaculture, livestock, trees and crop production, can be remarkably efficient and productive.

For these resource-conserving technologies to be fully effective, however, they need to be adopted by whole groups or communities of farmers or land managers.

5

LOCAL GROUPS AND INSTITUTIONS FOR SUSTAINABLE AGRICULTURE

> *'One resurrected rural community would be more convincing and more encouraging than all the government and university programmes of the past 50 years. Renewal of our farm communities could be the beginning of the renewal of our country and ultimately the renewal or urban communities. But to be authentic, a true encouragement and a true beginning, this would have to be a resurrection accomplished mainly by the community itself.'*
>
> Wendell Berry, in Enshayan, 1991

COLLECTIVE ACTION AT LOCAL LEVEL

Individual Actions Only Provide Partial Protection

The widespread and growing evidence for the economic and environmental viability of resource-conserving technologies (see Chapter 4) appears to suggest that a more sustainable agriculture is a likely outcome. Once farmers get to hear of the potential benefits, of increased yields or reduced costs, then they will adopt widely and the transition will be under way.

But without attention to local institutions, this is far from likely. Sustainable agriculture cannot succeed without the full participation and collective action of rural people and land managers. This is for two reasons. First, the external costs of resource degradation are often transferred from one farmer to another, and second, the attempts of one farmer alone to conserve scarce resources may be threatened if they are situated in a landscape of resource-degrading farms.

This need for coordinated resource management applies to most aspects of resource conservation, including pest and predator management; nutrient management; controlling the contamination of aquifers and surface water courses; maintaining landscape value; and conserving soil and water resources. There are many examples of individual initiatives that are unlikely to succeed in the long term because of lack of collective support. These include the following scenarios.

- One farmer encourages predators through farm habitat management, but on neighbouring farms non-selective pesticides which kill predators are used, so local predator populations do not reach a viable size.
- One farmer uses crop rotations and mosaic patterns as part of IPM to keep pest populations below threshold values, with occasional use of pesticides, but neighbours' pesticide overuse leads to the development of localized resistance to pesticides.
- One farmer maintains a diverse farm of high landscape value, but neighbouring farms reduce the overall value by removing trees, hedges and ponds.
- One farmer opens up land for access to the public, but neighbours do not provide similar access.
- One farmer adopts practices that reduce nitrate leaching to groundwater, but other farmers on land overlying the same aquifer continue to apply large amounts of nitrogen or manures, or use practices which permit leaching.
- One farmer reduces livestock waste losses to surface water, but farmers upstream continue to pollute and so the water quality continues to be poor.
- One farmer attempts to save traditional seed, but does not receive sufficient support for viable multiplication.

There are fewer cases where farmers adopt regenerative technologies which cause damage on neighbouring land. One case might involve the adoption by a farmer of soil and water conservation terraces on a steep farm. These would capture and channel water along the contours, so slowing water flow and increasing percolation, but could also lead in heavy rainstorms to channelling of water on to unprotected neighbouring land. This would lead to the formation of gullies, so causing more erosion than if the whole hillside had remained unprotected. In most cases, however, the adoption by an individual of more sustainable practices produces benefits for the wider environment and society – either by not polluting the environment or by actively improving resource value.

Indigenous Collective Management Systems

For as long as people have engaged in agriculture, farming has been at least a partially collective business. Farmers and farming households have worked together on resource management, labour sharing, marketing and a host of other activities that would be too costly, or even impossible, if done alone. Local groups and indigenous institutions have, therefore, long been important in rural and agricultural development.

These may be formal or informal groups, such as traditional leadership structures, water management committees, water users groups, neighbourhood groups, youth or women's groups, housing societies, informal beer-brewing groups, farmer experimentation groups, burial societies, church groups, mothers' groups, pastoral and grazing management groups, tree-growing associations, labour-exchange societies and so on. These have been effective in many ecosystems and cultures, including collective water management in the irrigation systems of Egypt, Mesopotamia and Indonesia; collective herding in the Andes and pastoral systems of Africa; water harvesting and management societies in Roman north Africa, India, and south-west North America; and forest management in shifting agriculture systems. Many of these societies were sustainable over periods of hundreds to thousands of years.

The manorial system of medieval Britain, a classic example of integrated farming, was sustained for some 700 years by a high degree of cooperation between farmers (Pretty, 1990b). Local groups established detailed management measures for sustainable use of village resources; they provided support and mutual help through sharing arrangements; and they took communal decisions against individuals who attempted to overconsume or under-invest in common resources. Some of the earliest pollution control measures were established at this time, controlling, for example, contamination of water courses by wastes. Local regulations, or by-laws, covered a wide range of activities and potential resource users, and provided for controlled and sustainable use of resources (Table 5.1).

Table 5.1 Medieval agriculture in Britain: selection of by-laws established between AD1150–1400 at local level and designed to prevent long-term damage to village resources

Activity	*Management measure*
All hunting, gathering and collecting activities	Licences required
Pig feeding	Nose-rings to discourage deep-rooting
	Fines for owners of destructive pigs
	Elected swineherd responsible for any damage
Cattle grazing	Stocking rates limited
Trees	Regulation of cutting and selling
	All villagers permitted only to carry own firewood
	Heavy fines for possession of woodcutting tools without licence
	Prohibition of lopping of oak and beech trees as key source of food for pigs
	Replacement trees planted every year
Hedges	Require regular repairs
Fencing and gates	Compulsory around gardens to prevent livestock escaping and causing damage
Rushes and reeds	Mowing controlled
	Gathering permitted for own use only, not for sale outside village
Manures	Not to be sold out of village
Fishing	Permitted during daylight hours only
Watercourses	Pollution by human wastes, animal offal and hemp or flax residues prohibited

Source: Pretty, 1990b, from Ault, 1965

Later, during the agricultural revolution of the eighteenth and nineteenth centuries, farmers' groups were central in the spreading of knowledge about the new technologies (Pretty, 1991). At a time when there was no ministry of agriculture, no research stations and no extension institutions, farmers were extremely effective at organizing their own experiments and extending the results to others through tours, open days, farmer groups and publications. Farmer groups and societies were central to the diffusion of new technologies. The first were established in the 1720s and increased in number to over 500 by 1840. These groups offered prizes for new and/or high quality livestock, crops and machines; encouraged experimentation with new rotational patterns; held regular shows and open days; bought land for experimental farms run by the group; arranged tours to visit well-known innovators; and articulated farmers' needs to national agencies and government.

But indigenous local groups can have shortcomings. Some groups may institutionalize unequal but secure access to natural resources, such as in tank management and water allocation in southern India during times of water scarcity (Mosse, 1992), common property management in the open field system in Britain (Pretty, 1990b) and access to forest resources in Nepal. In highly stratified societies, it cannot be assumed that existing institutional arrangements are equitable. The persistence of an indigenous system does not necessarily indicate it has the support of all the community.

Forms and Functions of Local Institutions

The success of sustainable agriculture depends not just on the motivations, skills and knowledge of individual farmers, but on action taken by local groups or communities as a whole. This makes the task facing agriculture today exceptionally challenging. Simply letting farmers know that sustainable agriculture can be as profitable to them as conventional agriculture, as well as producing extra benefits for society as a whole, will not suffice. What is also required will be increased attention to community-based action through local institutions. Local institutions are effective because *'they permit us to carry on our daily lives with a minimum of repetition and costly negotiation'* (Bromley, 1993). Local organizations do this in a variety of ways (Box 5.1).

The problem with the term 'local' is that it can mean anything that is not national. But 'local' does have its own special characteristics. It provides the basis for collective action, for building consensus, for undertaking coordination of responsibilities, and for collecting, analysing and evaluating information (Uphoff, 1992a). It does not happen automatically. It requires the presence of institutions at these local levels.

The uniting factor is that these have in common the prevalence of face-to-face interpersonal relationships, which are more frequent and intense within small groups (Uphoff, 1992a,b). The fact that people know each other creates opportunities for collective action and mutual assistance, and for mobilizing resources on a self-sustaining basis. People feel more mutual rapport and a sense of obligation at these levels than at district or sub-district levels, which are really political constructions. At the household or individual levels, decisions and actions oriented towards

Box 5.1 Functions of local organizations and institutions

The functions of local organizations and institutions are to:

- organize labour resources for producing more;
- mobilize material resources to help produce more (credit, savings, marketing);
- assist some groups to gain new access to productive resources;
- secure sustainability in natural resource use;
- provide social infrastructure at village level;
- influence policy institutions that affect them;
- provide a link between farmers and research and extension services;
- improve access of rural populations to information;
- improve flow of information to government and NGOs;
- improve social cohesion;
- provide a framework for cooperative action;
- help organize people to generate and use their own knowledge and research to advocate their own rights;
- mediate access to resources for a select group of people.

Sources: Uphoff, 1992c; Cernea, 1991, 1993; Curtis, 1991; Norton 1992; IFAP, 1992

sustainable development are not likely to be long lasting unless they are coordinated with what other households are also doing.

It is also important to distinguish between the terms institution and organization. Here the conventions of Norman Uphoff (1992c), Alan Fowler (1992) and others are followed. There are many types of institutions, some of which are also organizations, such as banks or local governments and others which are not, such as the law or taxation. An institution is a complex of norms and behaviours that persists over time by serving some socially valued purpose, while an organization is a structure of recognized and accepted roles.

Institutions can be organizations and vice versa. Marriage is an institution that is not an organization, while a particular family is an organization with roles but not an institution, which has longevity and legitimacy. The 'family', on the other hand, is both an institution and an organization. In this chapter the concern is with institutions that have an organizational basis.

THE PERILS OF IGNORING LOCAL INSTITUTIONS

Throughout the history of agricultural development, it has been rare for the importance of local groups and institutions to be recognized. Development professionals have tended to be preoccupied with the individual, assuming that the most important decisions affecting behaviour are made at this level. As a result, the effectiveness of local groups and institutions has been widely undermined. Some have struggled on. Many others have disappeared entirely.

The Suffocation of Local Institutions

Without realizing it, governments have routinely suffocated local institutions during agricultural modernization. Local management has been substituted for by the state, leading to increased dependence of local people on formal state institutions. Local information networks have been replaced by research and extension activities; banks and cooperatives have substituted for local credit arrangements; cooperatives and marketing boards have replaced by input and product markets; and water users' associations have replaced local water control.

In South-East Asia, the Green Revolution technologies forced through social and institutional changes that were neither planned nor controlled (Palmer, 1976, 1977; see Chapter 3). In Malaysia, farmers had to be members of farmers' associations if they were to get access to credit and inputs. These were established and controlled by extension workers, and could contain 1000 to 2000 members. Extension workers conducted credit assessments for each member and had the power to reject or accept requests. In the Philippines, farmers were obliged to join smaller *Samahang Nayon* groups and on joining had to accept the whole technical package, including inputs and the guidance of the extension workers. If so, they received coupons to go the rural banks for a loan to buy subsidized fertilizer. As fulfilment of all these conditions was a prerequisite for giving land title, this created strong hostility among farmers. Such hostility is common if farmers and rural people are forced or coerced into forming groups.

In their study of 30 years of government coordinated rural development in India, Jain et al (1985) show how, initiative by initiative, the state has systematically undermined the efforts of local people. Their analysis of local administrative regions in five states containing some one million people has shown that poverty-alleviation programmes had not reached the poor. The scale of self-deception is extraordinary. The state believes it is having an impact, but the field evidence says otherwise (Box 5.2) State activities have substituted for local initiatives; they have concentrated on infrastructural investments rather than people; they have favoured the use of direct subsidies; and they have pursued schemes that are harmful to local people. Jain and his colleagues indicate that *'the foremost reason for this unfortunate state of affairs is that the people themselves have no place in rural development, as every available space is occupied by the bureaucracy. The community, which was once central to the rural development strategy, is now peripheral to it'*.

As local institutions decline, so cultures change and become less resilient. In western Kenya, the *kokwet* resource management groups of the Marakwet formerly had responses laid down for every contingency. They had a regular rota for checking irrigation structures and making small running repairs, imposed fines for illegal use, and called occasional groups of young men together for large repairs, and held ceremonies with dancing, beer drinking and ox-roasting to celebrate the upkeep. Elspeth Huxley (1960) described the decline of this management mechanism in the 1950s. At one breach of a canal where a landslide had occurred, one old man said they now waited for the government to come and mend it, as the

Box 5.2 A selection of indicators showing the widespread bias against the most needy of poverty alleviation efforts in India, 1950–85

In selected areas (blocks, districts or states):

- 75% of households helped under poverty-alleviation programmes were above the poverty line, and so should have been ineligible;
- 74% of total loans by a bank in one very poor state was for tractors, which did not help any small farmers;
- 80% of credit extended to a large number of farmers was to those above the poverty line;
- 75% of training opportunities were taken up by those above the poverty line;
- support to bonded families was too little to eliminate any bondage;
- only 12–15 per cent of children in child development and nutritional programmes were from farming families with no animals or land;
- almost no immunized children are from families below the poverty line;
- 80% of drinking water wells dug in one block were concentrated in just the larger and wealthier villages.

Source: Jain et al, 1985

'young men are tired and no longer interested'. She recognized then that: *'the end has begun; and with that old, traditional way of mending furrows will go the songs and laughter, the roasted oxen and all-night dance, the tests of skill and courage for young men. Progress will make them into clerks and storekeepers, messengers and teachers, houseboys and politicians, instead of masters of the rivers high above the plain'*.

Increased Degradation in India

When traditional social institutions collapse or disappear, it is common for natural resources to degrade. In India, the loss of local institutions for the management of common property resources has been a critical factor in the increased over-exploitation, poor upkeep and physical degradation over the past 40 years.

N S Jodha (1990) studied the importance of common property resources to rural people in 82 villages scattered over 7 states of India. Almost all rural poor households depend for their fuel, fodder and food items on common resources. Income from these resources also accounts for 14–25 per cent of total household income for these groups. The rural rich, by contrast, depend on them much less (Table 5.2). But since the 1950s, the area of common property resources has declined by at least 30 per cent, and in some villages by more than 50 per cent. Coupled with this is the dramatic increase in population pressure on the remaining resources. Most villages have seen at least a three-fold increase in the number of

Table 5.2 Extent of dependence of poor and wealthy households on common property resources (CPRs) in dryland regions of India

State	*Household category*	*CPR contribution to income (%)*	*CPR contribution to fuel supplies (%)*	*CPR contribution to animal grazing(%)*	*Days of CPR employment per household*
Andhra	Poor	17	84	–	139
Pradesh	Wealthy	1	13	–	35
Gujarat	Poor	18	66	82	196
	Wealthy	1	8	14	80
Karnataka	Poor	20	–	82	185
	Wealthy	3	–	29	34
Madhya	Poor	22	74	79	183
Pradesh	Wealthy	2	32	34	52
Maharashtra	Poor	14	75	69	128
	Wealthy	1	12	27	43
Rajasthan	Poor	23	71	84	165
	Wealthy	2	23	38	61
Tamil Nadu	Poor	22	–	–	137
	Wealthy	2	–	–	31

Source: Jodha, 1990

people per hectare of common land, resulting in a dramatic decline in the number of products that local people can gather from the commons. Species diversity has declined and species mix has changed. The number of trees and shrubs has also fallen, and so people must spend more time collecting to get the same amounts of products.

The physical degradation is a product of both over-exploitation and poor upkeep. It is the inability to enforce local regulations that has led to poor upkeep. These failures have come about because of the abolition or complete collapse of traditional formal or informal management practices (Jodha, 1990; Chambers et al, 1989b). Compared with the 1950s, only 10 per cent of the villages still regulate grazing or provide watchmen; none levy grazing taxes or have penalties for violation of common regulations; and only 16 per cent still oblige users to maintain and repair common resources.

The future is bleak in the absence of these disappearing institutional structures. There is considerable evidence that when collective management is replaced by private operation, then resource degradation occurs. This is nowhere more apparent than when it comes to groundwater issues (T Shah, 1990; Datta and Joshi, 1993). Both over-extraction leading to groundwater depletion and over-irrigation leading to increased waterlogging and salinity have occurred because collective local management has been replaced by unfettered private operation. In just Haryana and Gujarat alone, some 2.1 million ha are seriously degraded by salinity and waterlogging. Farmers cannot solve these problems alone, as *'drainage technology in indivisible and cannot be executed in parts'* (Datta and Joshi, 1993). Individuals investing in isolation will not improve the lands. Yet when farmers are organized in groups, substantial yield improvements for wheat, mustard and millet have occurred.

In the coastal region of Gujarat, large areas are experiencing saltwater intrusion into depleted aquifers (T Shah, 1990). Saline water has made irrigation impossible in many areas, with farmers' incomes falling rapidly in recent years. As Tushaar Shah put it: *'the conditions for farmers located in the saline zones, both large and small, is desperate'*, yet in the neighbouring areas not yet affected farmers continue to pump because *'they are certain that wells will soon become saline regardless of how much they themselves restrain pumping if others do not restrain pumping as well'*. Clearly, destruction will continue unless farmers stop operating privately. In Tamil Nadu, private water extraction, favoured by state policies and cheap or free electricity, has undermined collective management systems. Farmers have opted out of local organizations and so overuse common goods (Mosse, 1992; Reddy, 1990). In Tamil Nadu alone, there are some 36–39,000 tanks of various sizes. Many are now in a state of disrepair, silted up and encroached upon. Falling water tables and degraded tanks are a result of the state's progressively increasing control over irrigation systems, the trend to private water use and management, and the systematic undermining of local institutions.

State-Imposed Institutions

Just as bad as ignoring existing local institutions is the practice of imposing new ones without consideration of the likely impacts on local people. Outside interventions are liable to warp and weaken local institutions. There are dangers that the state will suffocate local initiative and responsibility, or capture and harness local initiatives and resources for other purposes. Local politicians may seek to take over local successes or gain reflected glory from them. As has been indicated above, not all initiatives are seen by local people as legitimate.

In West Africa, governments have tended to restrict the freedom of local, self-help organizations, suppressing them by favouring state-created groups. Peter Gubbels (1993) says that: *'the historical trend in many West African countries has been to deny the establishment of community-based indigenous organization, and to suppress those already in existence by restricting their freedom and autonomy... Looking at these experiences over more than 30 years of agricultural development, it is difficult to argue that the failure of agricultural development for the mass of peasant farmers has primarily been due to technical or financial shortcomings. Indeed, there is evidence to suggest that this failure is partially due to government control over indigenous processes of agricultural development'*.

It has been argued that the great success of Kofyar farmers in Nigeria has occurred precisely because they were ignored by development programmes, and so were free to develop and adapt new cropping systems according to changing needs and demands (Netting et al, 1990).

In India, the presence of panchayats has been one reason why voluntary agencies have tended to support the formation of new institutions in recent years (Agarwal and Narain, 1989). The panchayat poses two major problems for resource management at local level. They tend to be both the product of village factionalism and dominated by the more powerful groups in the village. This raises fears among the rest that the benefits will be expropriated for the privileged few. Another problem is that panchayats are too far removed from the grassroots to be effective agents for good resource management. A village often consists of several hamlets and a panchayat usually covers several villages. As a result, *'these panchayats are just too big to become an effective forum for village-level environmental management'*. Well-intentioned development efforts focusing on panchayats as the appropriate local body can cause many problems. In Maharashtra, the social forestry directorate has tried to involve panchayats in the management of village woodlots, but even though Agarwal and Narain visited many villages, in none did they find that *'the panchayat leaders cared to explain to the villagers that these village woodlots were a community resource. Most villagers were shocked to hear this. They believed these trees were all government trees'*.

Some Comparisons of Individual and Group Approaches

Studies of agricultural development initiatives are increasingly showing that when people who are already well organized or who are encouraged to form groups, and whose knowledge is sought and incorporated during planning and implementation, they are more likely to continue activities after project completion (de los Reyes and Jopillo, 1986; Cernea, 1987, 1991, 1993; Kottak, 1991; USAID, 1987; Finsterbusch and van Wicklen, 1989; Bossert, 1990; Uphoff, 1992a; Bunch and Lopez, 1994; Pretty et al, 1994a). If people have responsibility, feel ownership and are committed, then there is likely to be sustained change.

A study 4–10 years after the completion of 25 World Bank-financed agricultural projects found that continued success was associated clearly with local institution building (Cernea, 1987). Twelve of the projects achieved long-term sustainability and it was in these that local institutions were strong. In the others, the rates of return had all declined markedly, contrary to expectations at the time of project completion. At project completion, staff had estimated rates of return between 15 per cent and 30 per cent; in reality they had disastrously fallen to 2.7 per cent on average.

This clearly indicated that projects were not sustainable where there had been no attention to institutional development and farmer participation. Michael Cernea (1987) commented: *'such a high number of unsustainable projects was certainly not expected... I have often been struck by how little interest there has been in learning the true reason for failure... Not only does the failure to consider the cultural context of a project undercut the technical package promoted by the investment, but it leads to projects that at best are less effective than they could be or, at worst, outright failures'*.

In the Muda Irrigation Project in Malaysia, water users' associations were *'established carefully, patiently and successfully, taking into account farmers' resource needs, their willingness to cooperate, the physical location of their plots'* (Cernea, 1987). The endurance of these associations after the project completed was the single most important factor in ensuring the continued benefits to farmers. In contrast, the negative rate of return of the Hinvi Agricultural Development Project in Benin was caused by the disintegration of the cooperatives developed for the cultivation of oil palm. These had been imposed on the farmers and run by a parastatal with no self-management delegated to farmers. The technical and agricultural package financed by the project failed to account for traditional land tenure systems. The farmers also opposed the organizational arrangements imposed on them, so when these collapsed, the technical innovation (growing of oil palm) collapsed too. Seven years after completion, more than 75 per cent of members had opted out, refusing to work in cooperative blocks and had returned to cultivating food crops.

In India, Anil Agarwal and Sunita Narain (1989) have indicated that *'all good cases of environmental regeneration... are invariably those cases where voluntary agencies have set up an effective institution at the village level... It is the creation of a village level institution which brings people together, spurs them into action and ensures the protection and development of the natural resource base'*.

NEW CHALLENGES DURING INTERVENTION

Establishing Self-Reliant Groups

The process of establishing self-reliant groups at local level must be an organic one and so should not be forced or done too quickly (Orstrom, 1990; Röling, 1994). The International Federation of Agricultural Producers (1992) describes four essential elements of any self-supporting farmers' organization. They should have developed a financing capacity with resources of their own, the major part of which are obtained directly or indirectly from the membership. The should have developed a structure for electing farmer representatives. They should have obtained recognition as a legitimate voice of farmers. They should have developed self-reliance for planning, for management and for the provision of effective services.

The Gal Oya irrigation scheme in Sri Lanka provides some of the best evidence for the success of local groups and how they can best be established. Over the years, dramatic and lasting changes in the efficiency and equity of water use have been made (Uphoff, 1992a, 1994). At least 13,000 farmers are now involved in an area exceeding 10,000 ha. Despite many difficulties, including ethnic conflict, budget cutbacks, massive turnover of trained organizers, and bureaucratic interference, farmers' associations have maintained themselves and progressed institutionally. Water use efficiency almost doubled and yields were raised about 50 per cent over a larger service area. Norman Uphoff (1994) indicates this is because of the particular process of group formation and development itself. This has eight distinct elements and is characterized throughout by attention to the development of learning processes.

1. Use of catalysts. If people are not already organized, then organizers, animators or motivators will be needed.
2. Starting with informal organization. Begin by focusing on a particular problem and bring a small group together to solve it. The sequence 'work first and organize later' brings forward better leadership and more support among members. This is in contrast to an alternative approach of calling a meeting and forming an organization before doing work.
3. Evolve a formal structure. Let groups evolve from informal to formal status at their own pace.
4. Mobilizing a new kind of leadership. Farmer representatives and village extensionists are chosen by their groups not by election but by consensus. If the representatives must be acceptable to all members, then factional leaders are less likely to come forward. Those selected feel accountable to every member, as all had assented to their selection. Their terms of reference are prepared by the members of the group.
5. Importance of small groups at base. It is easier to create and maintain a better sense of solidarity and mutual responsibility in small groups. The wider impact on rural development occurs through the federation of these groups within higher level associations that offer benefits of scale.
6. Problem solving process. Groups regularly follow a process taking them from prioritizing problems through action and self-evaluation of progress. The philosophy is embodied by the catalysts being told there is no disgrace in making a mistake, only in not identifying them, learning from them, and avoiding repeating them.
7. Start with limited number of tasks. Groups start with one or two tasks, then expand when they wish. Groups starting with too many tasks tend to do them poorly and so cease to function. In Gal Oya, organizations proceed to deal with crop protection, credit, bulk input purchasing, mortgage releases, settlement of domestic disputes, land consolidation and even dealing with drunkenness.
8. Make provision for horizontal diffusion. It is important that farmer-to-farmer elements of communication and learning be established and sustained. This gets away from the more common vertical communication style.

Dealing with Inequity and Distorting Existing Groups

It is sometimes assumed that groups are easy to establish. This is not the case, particularly if they are to be concerned with equitable decision making and improvement of the livelihoods of poorer groups. This is partly because traditional institutions often institutionalize inequitable access to productive resources. Some of the already established institutions are full of local biases and so may not be the best representatives of local people (Matose and Mukamuri, 1992).

In Tamil Nadu, tanks have been the traditional form of irrigation for at least 2000 years (Mosse, 1992). Cultivators hold rights to a share of water from the tank and these are part of a wider system that also defines rights to shares in crop produce, to artisanal and ritual services, and to worship local deities during festivals. These systems tended to give privileged access to dominant groups, though they did ensure security to the poorest groups in times of scarcity. Today, a wide variety of situations exist. In some villages, informal rights and responsibilities still exist. In others, existing rights are overlooked by the powerful, who protect water by force. But, says David Mosse (1992), *'where they exist, institutions of community management are likely to be based upon and protect the interests of a dominant caste group'*.

The Centre for Water Resources of Anna University, Madras is working with communities on the rehabilitation of irrigation tanks, where it has had to initiate social change so as to improve agricultural performance. Existing systems of water allocation are often the critical constraint. The first task has been to establish water users' associations by first reconciling some of the differences between interest groups. In one village, where there were three types of water distribution in operation, various meetings first led to agreements to change the local rates of labour payment. The project team then arranged several practical collective tasks, such as repairing sluices, cleaning channels and agreeing a site for a community well. These led to the formation of a village water users' society, the structure of which was agreed over the course of a further 12 farmers' meetings. This group has clear rules for protection of resources and distribution of benefits. They have undertaken long neglected maintenance, so increasing tank capacity; imposed fines for cattle trespass; and most significantly shifted the local balance of power. In creating new rights, this group has provided the context for changing social relations.

It is impossible to say, without knowledge of the particular local circumstances, whether existing local groups should be built upon or entirely new ones formed. Sometimes there are no existing farmers' organizations, such as in parts of West Africa (Gubbels, 1993). If they do exist, they are more likely to be based on export-oriented cash crop production, rather than organized for farmers in the poorer, more remote areas. Sometimes existing organizations institutionalize unequal access to resources or opportunities. In both these cases, it would be necessary to build new organizations.

One problem is that external institutions are usually neither sufficiently patient nor capable of only spending small amounts of money when it comes to supporting local groups. As a result, they tended to distort and undermine local efforts (IFAP, 1992). Funders face specific circumstances and constraints that are very different to receiving organizations. Quick results are important to show success and that funds were spent well. In addition, money once committed is preferably spent in large sums, as this cuts administrative costs. The result of this is first a confidence upswing rapidly followed by a downswing: *'the confidence upswing usually starts with a success story. It attracts considerable publicity. The organization is over-rewarded as funders become extremely favourable towards the organization. This leads inevitably to unrealistic expectations about what the organization can do next. Disappointment follows. Confidence downswing usually starts with dissatisfaction by one funder. News spreads fast, and funders become distrustful. They pull out, as no funder wants to be associated with a failure'* (IFAP, 1992).

One example of where an external institution wanted to move too quickly, and so skip the complexities of local social changes, was the Hill Resource Management Program of Haryana, India. There users' groups for natural resource management were established to fill the gap left by the decline and near disappearance of indigenous management systems (Poffenberger, 1990; Misra and Sarin, 1988). The project began in Sukhomajri, where Gujar herders agreed to stop grazing the severely degraded hills if a dam were built to supply irrigation water. This 'social fencing' initiative established water users' associations in four communities during the pilot phase, who managed irrigation water and cutting of fibrous and fodder grasses from the regenerating hills. The impact on agriculture was remarkable: yields rose by 100–400 per cent, diversification increased, livestock were stall fed and fodder grass yields on the hills rose by 400–600 per cent. For the expansion phase the Haryana Forest Department became the lead agency, building 57 dams in 39 communities. But only in 30 per cent of these communities has the department successfully established management societies. In the long run the whole effort may be jeopardized as local people become less involved in participatory planning and management.

All this can be avoided if external agencies move slowly, do not expect immediate results, build up leadership, and measure success in terms of developing social relations and institutional strength. When this is done, local groups become stronger and more self-reliant, so improving the livelihoods of their members and the environments of their communities.

Evolving Roles

Groups commonly form to take charge of a new activity and/or manage a new resource, such as water users' associations for irrigation, credit groups for loans' access, water point committees to manage pumps or farmers of a common micro-catchment to control soil erosion. But such local groups do adapt and change their roles and responsibilities as internal and external conditions change. It is common for them to pass through several phases, growing increasingly strong. Local people themselves recognize these as being stages on the route to sustained action. In one self-evaluation in Sri Lanka, farmers identified the health of their own groups by referring to the fullness of the moon. The full moon signifies fulfilment and achievement of the highest order, and is represented by the indicators hardest to achieve (Box 5.3).

In the early stages, groups focus on establishing agreed rules for management and decision making. These can then be used by members as a vehicle to channel information or loans to individual members. Once small homogenous groups have successfully achieved initial goals and confidence has grown, it is common for members to turn their attention to development activities that will benefit themselves as well as the community at large. This may involve the nominating of individuals to receive specialized training, such as in soil and water conservation, pest control, veterinary practice, horticulture or book-keeping, so that they will be able to pass knowledge back to the whole group in their new role as

Box 5.3 Local indicators of health in farmers' organizations, Sri Lanka

In a self-evaluation of farmer organizations supported by the National Development Foundation, an NGO in Sri Lanka, farmers were encouraged to produce their own indicators which would identify successful or healthy farmer groups. They indicated that groups pass through three phases before they reach full unity, which they visualized in terms of the moon.

New moon groups have regular attendance at meetings by more than 90% of members; there is punctuality by all who come; and more than 75% of members participate in common activities.

Half moon groups regularly clear and maintain tank bunds. They also help others in need, including non-members, by offering their labour and not drawing on the group fund.

Three-quarter moon groups implement common decisions; have common property and use it for the benefit of all members, eg a sprayer that is rented out to members at lower than market rate. The group take over a member's share of common work when she or he is unable to do it for some valid reason; and share benefits among members, eg watered land for vegetable cultivation, in disregard of ownership.

Full moon groups help poorer members with loans from the group fund, eg for buying the decided variety of seed paddy; help redeem the mortgaged land of members; and have the strength to face external forces.

Source: Mallika Semanarayake, personal communication; Harder, 1991

paraprofessional or extension volunteer (Shah et al, 1991; Pretty et al, 1992).

As confidence further grows with success and resource bases expand, group activity can evolve to an entrepreneurial stage where common action projects and programmes are initiated. These are held under group ownership and might comprise investing in fruit orchards, afforesting an upper watershed, terracing a hillside, investing in agricultural tools and draught animals for hire to the community, organizing community-run wildlife utilization schemes; establishing workshops and small factories; and building housing for tribal families (Rahman, 1984; D'Souza and Palghadmal, 1990; Shah et al, 1991; Fernandez, 1992; Murphree, 1993). These group activities benefit group members as well as having a wider ecological and social impact.

In Nepal, the Small Farmer Development Project has clearly shown the economic benefits at local level of people working in groups (Rahman, 1984; Uphoff, 1990). Group organizers from a development bank helped form groups of 10–15 people, who agreed to conduct transactions with the bank as a group, to elect a leader, to meet at least once per month and to save Rs5 each month per person to create revolving funds. Loan applications were submitted to the field office of the bank and once the group received funds it was responsible for these as a group. The project has helped to improve agricultural yields; diversity of production has increased; recovery rates for loans are greater than 90 per cent; and social indicators all show general improvements in welfare of the poor.

The groups typically begin by channelling loans to individuals, and then expand their action gradually to take collective and community-oriented action. *'Once the farmers organise themselves into groups for the purpose of acquiring credit, they start gradually realising that there are many things they can do collectively which they had not thought of before.'* (Rahman, 1984)

In one case of intergroup cooperation, seven groups in Tupche borrowed Rs250,000 to instal handlooms in a cottage factory. The factory belongs to all 126 members of the 7 groups, employs 50 previously unemployed local people, and is avoiding the capital intensive urban-based development model by producing a technology corresponding to local skills and needs.

Recreating a Sense of Community

When groups are established for the first time, or resuscitated, then one of the universal benefits of membership expressed by people is the renewed sense of community. This is surprising, as many would assume it is for the economic benefits alone that they have become organized.

In the UK and Australia, members of farmer and community groups commonly state that the important benefits of membership are not so much yield improvement, economic returns and so on, but more the pleasures of problem sharing, friendship and enjoyment of others' company (Wibberley, 1991; Campbell, 1994b,c). This has been a particularly notable effect of the Landcare programme in Australia, where typically independent and 'frontier-spirited' farmers have, in coming together in groups for the first time, achieved significant environmental and social changes (see Chapter 5).

Co-operation and empowerment has proved to be possible in the most unlikely of social settings, and the nature of farming is being transformed by a network of rural community groups committed to the development and dissemination of productive and sustainable farming and land use. The programme has achieved great success, but the factor noticed by commentators, local people and farmers alike is the sense of cohesion brought back into rural communities. New relationships are breaking down mistrust: *'It is the first time in Australian history that I'm aware of that farmers and government are working to the same end. They are usually at each others' throats'* (farmer quoted in Alexander H, 1993). As Andrew Campbell, former national facilitator of Landcare put it: *'the tangible benefits are in a sense misleading, as the most important impacts of landcare are the intangibles – the social cohesion and solidarity, the sharing of stresses, new ideas and intellectual stimuli'*.

A recent study of more than 150 local initiatives in Scotland has illustrated the importance local people give to group action. Groups brought environmental, social and employment benefits in the form of increased conservation of resources, a greater sense of community measured in terms of enthusiasm and commitment, and improved direct rural employment (Bryden and Watson, 1991). But the problems faced by these initiatives suggest they are successful despite, rather than because of, the good intentions of support agencies. In particular, there was a mismatch between the needs of community initiatives and the support offered by agencies, mainly because funding agencies have narrow mandates and are set up to serve different situations. Most important, external agencies routinely undervalued the social benefits of community enterprises. Most support agencies miss this, yet local people put this high on their list of benefits.

LOCAL GROUPS FOR SUSTAINABLE AGRICULTURE

Range of Groups

Six types of local group or institution are directly relevant to the new needs for a more sustainable agriculture (Pretty and Chambers, 1993a, b).

- **Community organizations,** such as for hill resource management in India, agricultural development in Nepal and Pakistan, and soil and water conservation in France.
- **Natural resource management groups,** such as for irrigation tank management in India, for soil and water conservation in Kenya, for irrigation in the Philippines and Sri Lanka, and for soil and water conservation in Australia.
- **Farmer research groups,** such as in Zambia, Botswana, Ecuador and Colombia, and Britain during the agricultural revolution of the eighteenth and nineteenth centuries.
- **Farmer-to-farmer extension groups,** such as for soil regeneration in Honduras and for irrigation management in Nepal.
- **Credit management groups,** such as in southern India, Bangladesh and Nepal.
- **Consumer groups,** such as women's consumer–producer groups in Japan.

These groups are quite different for those arising out of the long tradition of the cooperative movement, in which community wide action has been encouraged through the forming of cooperatives or collectives. Many of these collective approaches to extending technologies to rural people have resulted in inequitable development, with benefits being captured by the relatively well off. Large cooperatives, in which the needs of different members vary enormously and which are too large for widespread participation, have to be managed by small groups, usually comprising the most wealthy, to whom decision making has been delegated. They are thus inevitably less effective in meeting the needs of the poor.

Local Research Groups

The normal mode of agricultural research has been to conduct experiments under controlled conditions on research stations, with the results being passed on to farmers. In this process, farmers have no control over experimentation and technology adaptation. Farmer organizations can, however, help research institutions become more responsive to the diversity of local needs, if scientists are willing to relinquish some of their control over the research process. But this implies new roles for both farmers and scientists, and it takes a deliberate effort to create the conditions for such research-oriented local groups. None the less, there have been successes in both industrialized and Third World countries.

In the USA, the Land Stewardship Project in Minnesota organizes farm families into peer-support and information-sharing groups as part of a Sustainable Farming Association (Kroese and Butler Flora, 1992). These groups encourage farmers to experiment with alternative farming practices on their own farms and at their own pace, and facilitate the exchange of information among nearby farmers about what they have learned. Experiments are done by individual farmers attempting to solve their own particular problems. Other farmers learn from them and extension agents pass on the results.

The data, of course, cannot be aggregated for cross-farm comparisons in the way that results from conventional research are used. But as Ron Kroese and Cornelia Butler Flora put it: '*This is not as problematic for the farmers as it is for academics, as the process of sustainability means adapting technology to the specific conditions of one's own farm. The documentation of the efforts, despite flaws, allows for further dissemination of technology... Participating farmers are most enthusiastic about their current experiments. They question each other in detail, analyse what they have done, or explain why it does or does not work. The lack of competitiveness as people share research results is in contrast to the usual coffee-shop talk of whose yield is highest, whose row is straightest, and who has the fewest weeds*'.

What is equally interesting is the impact farmers say being members of these groups had on themselves. They felt less alone and more a part of a wider effort for agricultural improvement.

In Botswana, farmer research groups have become central to the research strategy of the Ministry of Agriculture, where technologies are tested under both farmer-managed and farmer-implemented conditions (Heinrich et al, 1991; Norman et al, 1989). The key component of the approach is local research-oriented and extension-oriented farmer groups, which have become a powerful means for examining the potential of a range of technologies under farmer management. The process involves researchers presenting a wide range of options gathered from many sources to farmers in villages. Sub-groups of farmers selecting the same options conduct trials, and meet monthly to discuss progress and observations. As harvest approaches, field days are held to share interesting results with farmers outside the groups. The impact of this approach has been to change fundamentally the relationship between the researchers and farmers, increase the linkages with NGOs and to improve crop yields with low input technologies (Box 5.4).

As local people develop the capacity to learn from and to teach each other, so they develop further their own capacity to conduct their own research. There are many recent innovations in farmers' own analyses that point the way to innovative learning and self-spreading (Chambers, 1992b; Guijt and Pretty, 1992; Lightfoot and Noble, 1992). In India,

Box 5.4 The impact of the research-oriented and extension-oriented farmer groups in Botswana

The strong and sustained dialogue between farmers and researchers has:

- given greater flexibility to the research process, as technology options can easily be moved into the testing phase, and researchers respond rapidly to needs and interests of farmers;
- increased the range of topics under joint examination, so increased diversity of options open to farmers;
- led to attitude change in scientists, as they appreciated the benefits to all that could be achieved and enjoyed the personal success;
- developed improved linkages between on-station commodity researchers and FSR teams, as demand for their technologies and feedback from farmers grew;
- increased the total research capacity beyond the available research resources;
- increased linkages with NGOs, as they became involved with the groups;
- led to significant increases in grain (sorghum and millet) yields with low external input technologies – increases over 3 years were 71% for double ploughing, 23% for rowplanting and 56% for small applications of phosphorus (20kg/ha).

Sources: Heinrich et al, 1991; Norman et al, 1989

villagers who have been trained as extension volunteers by the Aga Khan Rural Support Programme are now training the staff of other NGOs in participatory methods. Farmers are both more effective and efficient trainers of other farmers, taking less than 10 per cent of the time that external agents need to train the extension volunteers (Shah, 1994a). In Bolivia, farmers working with World Neighbors have developed innovative ways of conducting their own research on potatoes, as well as trained some 3000 other farmers (Ruddell, 1993; Beingolea et al, 1992). Farmers also work on the radio as broadcasters in Niger and Peru (McCorkle et al, 1988; AED, 1991); and monitor research and conduct surveys (Jiggins and de Zeeuw, 1992).

These approaches all build the capacity of local people to conduct their own investigations and solve their own problems. All have shown that such informal learning is a low cost method of enabling farmer groups to adapt, choose and improve their farming systems. They also provide leadership experience for villagers.

Farmer-to-Farmer Exchanges to Enhance Local Capacity

There is growing experience in farmer-to-farmer extension, visitation and peer training as mechanisms to support agricultural improvement. These can take several forms. Most common are farmer exchange visits, in which farmers are brought to the site of a successful innovation or useful practice, where they can discuss and observe benefits and costs with adopting farmers. Professionals play the role of bringing interested groups together and facilitating the process of information exchange. During the visits, participants are stimulated by the discussions and observations, and many will be provoked into trying the technologies for themselves. For farmers *'seeing is believing'*, and the best educators of farmers are other farmers themselves (Jintrawet et al, 1987).

Such farmer-to-farmer extension has resulted in the spread of *Leucaena* contour hedgerows in the Philippines (Fujisaka, 1989); peanuts after rice and sesame before rice in north-east Thailand (Jintrawet et al, 1987); management innovations for irrigation systems in Nepal (Pradan and Yoder, 1989); post-harvest cassava treatment in Ecuador (CIAT, 1989); agroforestry in Kenya (Huby, 1990); green manures in Honduras (Bunch, 1990); and a range of watershed protection measures in India (Mascarenhas et al, 1991; Shah, 1994a,b).

In irrigation management, it has become increasingly clear that physical improvements do not solve all the problems causing poor performance (Pradan and Yoder, 1989; Yoder, 1991). Farmer-to-farmer extension is now being used to show farmers what they can achieve with good organization and local governance. In Nepal, farmers from a weakly-managed system are taken to visit several well-managed systems, where they have the opportunity to talk to local groups and hear how they manage water. In this way the visitors see that each group has evolved different rules and practices to suit the local conditions, and that these rules and regulations are continually changed (Box 5.5).

Box 5.5 The process and impact of a farmer-to farmer extension exchange in Nepal

The Gadkar Irrigation System in Nuwakot District is 105 ha in size and was constructed under the World Bank-financed Rasuwa-Nuwakot IRDP in 1979. A water users' committee was established but, because the members were district officials and large farmers, the allocation of water was inequitable. The chairman and vice-chairman dominated the committee, and used it to protect their own privileged access to water. Two delegations of 20 farmers and 20 officials were taken to three well-functioning farmer-managed irrigation systems.

As Pradan and Yoder put it: *'Visiting delegates were amazed at the accomplishments of the farmers in constructing and maintaining technically difficult irrigation systems, in their ability to establish fair rules and regulations, and the power of the organization to discipline its members. They saw what they could do as a group, rather than complain about poor management of the system by the agency that had constructed it for them'.*

The impact of the visit was that the previous committee was dissolved at a community meeting, and a new one elected. The incidence of water theft declined; the committee were able to allocate water fairly and impose rules, such as the growing only of maize in the pre-monsoon period rather than thirsty rice; the committee employed water guards; and farmers began to contribute a substantial amount of labour for maintenance.

Source: Pradan and Yoder, 1989

As a result of these exchanges, farmers have, on return to their communities, elected new leaders, collectively made new operating rules, improved canal maintenance, adopted systematic record keeping, held regular meetings at a regular meeting place, changed cropping patterns, and mobilised labour and resources for maintenance (Pradan and Yoder, 1989). All of this has created a greater sense of ownership. Robert Yoder (1991) put it: '*Treating the symptom, that is upgrading the physical system, when the cause of poor performance is governance, may temporarily improve system performance but experience has shown that there is little, long-term gain unless institutions are also strengthened... All the trainees made the connection between effective governance and agricultural productivity.*'

Sometimes expert farmers are hired by farmers' groups. In Ecuador, a cassava farmer from the Colombian north coast was hired to advise Ecuadorian farmers' associations (CIAT, 1989). This farmer-to-farmer approach has been an effective form of extension as, in just 3 years, the number of cassava-drying farmer associations in Ecuador grew to more than 20 with nearly 400 members. The cost of extension and applied research was cut to about one-third of what it had been, mostly by eliciting the cooperation of organized farmers.

But one of the greatest constraints for promoting wider use of farmer-to-farmer exchanges lies in the quality of available facilitators. They must be well acquainted with the farmers; they must know about the different systems and practices present in the various communities; they must be able to facilitate discussions, interjecting where necessary to guide the conversation; and they must be able to stimulate the discussion while not dominating it. They must, therefore, have all the qualities of the new professionalism described in Chapter 6.

Local Credit Management Groups

It has long been assumed that poor people cannot save money. Because they are poor and have little or no collateral, they are too high a risk for banks and so have to turn to traditional money lenders. These inevitably charge extortionate rates of interest and very often people get locked into even greater poverty while trying to pay off debts.

Recent evidence is emerging, however, to show that when local groups are trusted to manage financial resources, they can be more efficient and effective than external bodies, such as banks. They are more likely to be able to make loans to poorer people. They also recover a much greater proportion of loans. In a wide range of countries, local credit groups are directly helping poorer families both to stay out of debt and reap productive returns on small investments on their farms. The Grameen Bank, first established in Bangladesh, is perhaps the best-known example. First established in the 1970s, it has spread to reach 1.6 million members, giving many the opportunity to escape the trap of indebtedness. Its principles are being increasingly widely applied.

In the remote Northern Areas of Pakistan, the Aga Khan Rural Support Programme has helped to establish more than 1700 male or mixed village organizations, and 900 women's organizations, for resource and financial management, catering for some 53,000 households (AKRSP, 1994). Village groups originally organized to help construct a physical improvement, such as an irrigation channel, road or bridge, have also helped local people to save small amounts of money and to create collateral for credit provision. It had long been assumed that the desperate poverty of local people would make such an effort worthless. But with local control and responsibility, groups have been able to save substantial sums.

The success of the AKRSP approach is now being replicated elsewhere in Pakistan. The National Rural Support Programme was established in 1991 to build a countrywide network of grassroots organizations which would enable local communities to plan and undertake their own development (NRSP, 1994). To do this, they help to organize the rural poor into multipurpose community organizations (COs) or sectoral organizations of special interest groups. The key to success has been the mobilization of local resources through a savings and credit programme, combined with technical assistance to improve agricultural production. By mid-1994, NRSP had helped to establish 420 COs with some 12,200 members.

Another notable success has grown up in southern India, where the NGO, Myrada, has shown the value of small groups to credit supply (Fernandez, 1992; Ramaprasad and Ramachandran, 1989). Years of relying on banks and local cooperative societies to supply credit had rarely helped the poor. But when they started to work with small independent groups with members feeling they could trust each other, they noticed that *'not only was the money managed more carefully, there was a far greater commitment and responsibility from the groups towards repaying the amount of money, something that had not unduly bothered them when they were part of the cooperative'* (Ramaprasad and Ramachandran, 1989).

It was realized that members of small groups participated more, had common concerns and needs with others and, once they had developed their own rules and decision making, they expanded their resource base and took up common action programmes. Groups are first organized around a collective need, such as for a drainage system, desilting a tank or even in one case for an elephant trap. They then develop a role in savings and credit management. The strength of the approach is that no two groups end up being alike. There are, however, common principles known to all. All groups evolve their own set of rules, with each deciding its own interest rates to members as well as the types of loans it will permit. All agree that leadership responsibility must be shared, with no office bearing titles. All groups encourage members to save. All hold money in a common fund. All advance loans for consumption as well as production purposes. All can engage in providing or running community services.

What is particularly significant for the programme is that some 95–98 per cent of loans are repaid. This contrasts with just 20–25 per cent for banks making loans under Integrated Rural Development Programmes. In addition, the total advanced far exceeds the total fund size, implying an efficient rolling use of funds. By mid-1992, some Rs108 million (US$3.6m) had been lent out by more than 2000 local groups to their 48,000 members. The total common fund is Rs24 million, implying that each rupee had been lent out and repaid five times. The number of loans made is much greater than what banks can cope with. In four years, groups in Talavadi advanced 26,454 loans, in Bangarpet 5293 and in Holalkere 8376, while one branch of a bank finds it difficult to handle 400 loans in a year. The majority of loans are taken out for consumption purposes, and many of these for very small sums, often less than Rs100 (Table 5.3). Rather than borrowing money from a moneylender to pay for a funeral, marriage or food before the harvest, now rural households are able easily to borrow small amounts of money. Small amounts are vital for small farmers. As Aloysius Fernandez (1992) put it *'while a farmer may be eligible for a loan package that can buy him 20 sheep, what he wants and can manage is only two sheep. A credit group understands such priorities much better than a bank can'.*

Table 5.3 Measures of success of local credit groups programme of Myrada in southern India

Measures of performance	*Talavadi*	*Thally*	*Kamasumudram*
Number of local groups	72	58	110
Number of members	1754	1569	2500
Fund size (Rs million)	1.31 m	0.904 m	0.626 m
Total advanced (Rs million)	nd	1.2 m	2.09
Number of loans advanced	26,454	6515	5293
Proportion of loans for consumption purposes (mostly food)	77%	82%	19%
Proportion of loans for less than Rs 500	98.5%	nd	81%
Proportion of loans for less than Rs 100	38%	nd	12%

nd = no data
Source: Fernandez, 1992; Ramaprasad and Ramachandran, 1989

Perhaps the most significant aspect of Myrada's programme is that its success has led to an important national policy change. The convention in India has been that banks must lend only to individuals. But this has recently changed. The Reserve Bank of India issued a government order and NABARD followed with guidelines instructing all banks to relate to local groups as institutions. This has opened the way for ensuring that many more poor and needy people have access to much needed credit via their local groups.

All of this is in stark contrast to the way credit supply to local groups is normally managed. If only one institution is present in the community, with powers to refuse membership, then the poor and women are liable to be excluded. This has happened in Malawi, where the farmers' clubs established to channel credit to poor farmers have the autonomy to select their own members (Kydd, 1989). No collateral is required and repayment rates are more than 90 per cent. Yet as only one group is formed per community, only 20 per cent of households are in credit groups, most of those excluded having less than one ha of land each (see Table 2.4). The wealthy are those with access to credit.

SCALING UP FROM THE LOCAL LEVEL

Joining Together for Wider Impact

There are a growing number of local successes in community based and participatory planning. But these tend to remain local and so do not spread. Locally based organizations are good at having an integrated view of problems, tend to have a power base with local links and receive ready feedback. But their major difficulties lie both in commanding technical expertise and trying to solve problems arising out of the wider political context, such as product pricing and labour markets (Bebbington, 1991). Local institutions working alone are very unlikely to influence state policies. The problem is that existing platforms for decision making have not been set up for natural resource management, nor do they correspond to ecosystems to be managed (Röling, 1993). A major challenge for sustainable agriculture lies in widening the impact of local groups and ensuring the persistence of successful initiatives.

One way to ensure stability of groups is for them to join with others to work on influencing district, regional or even national bodies. Such intergroup cooperation might involve several groups coming together to federate and pool resources and knowledge. This can open up economies of scale to bring greater economic and ecological benefits. The emergence of groups and federated groups also makes it easier for government and non-governmental organizations to develop direct links with the poor. This can in turn result in greater empowerment of poor households, as they draw on public services.

Scaling up can occur through the establishment of federations or coordinating networks. Smaller organizations can federate to produce larger organizations, which can have a regional lobbying role and can express political concerns to state level. Moving up does not necessarily imply institutional growth, as this can be a threat in itself. But it may help to spread good ideas through a geographical area. At this level, organizations with greater membership carry greater political clout, can begin to influence policy and are able to draw on technical expertise.

The Federation of Free Farmers in the Philippines is a nationwide effort, directly supporting low input and sustainable agriculture (Montemayor, 1992). It has a membership of some 250,000 farmers, organized into village chapters. Local groups are linked to municipal chapters, provincial associations, and then to regional and national offices. A wide range of services are offered to members and, at national level, the federation has been able to influence policies on land tenure and fertilizer supply. The Agriculture Department was convinced, for example, to allow farmer–borrowers to buy organic fertilizer instead of being limited to inorganic products under the government credit programme.

Another advantage for local institutions of these scaled-up networks or federations is that they present a united front to funding agencies and governments. These create the opportunity for more efficient and more effective disbursement of funds by donors, with lower administrative costs. In Burkina Faso, ACORD has helped local groups produce village portfolios that are consolidated into a coherent regional planning document. This has resulted in many more groups having access to external support, as well as the strengthening of existing planning institutions (Box 5.6).

Box 5.6 The advantages of scaling up as part of the approach of the NGO ACORD in Burkina Faso

ACORD has used participatory methods in its programme of support to local groups in Burkina Faso. It aims to assist groups to recognize change as one way to improve their situation, and this is achieved through a continuing cycle of analysis, reflection and action. A survey of past efforts showed that projects failed because villagers did not consider them as their own, but externally imposed; that limited management capacity hindered implementation; and that some village groups had internal problems that were aggravated by the project. At the regional level, there was no overall policy to tackle the particular needs of the area.

Since 1983, this programme has strengthened village groups, encouraged links between them, and facilitated their access to financial and other support from other agencies. The problem for many funders was their inability both to identify suitable groups or projects to support and to follow up on what they funded. ACORD, through its process of participatory animation, built up *village portfolios* that corresponded to the individual needs of the groups into a coherent regional planning document. This allowed donors to invest in the diverse areas of support that had been locally identified. This process, apart from reinforcing local planning capacities, succeeded in channelling an average of between US$1–2.5 million each year to properly identified projects. This guaranteed a better utilization of funds as well as increasing the accessibility of such support to many more groups.

This indicates the importance of strengthening local groups, and at the same time coordinating and directing external support to them.

Source: Roche, 1991

Farmers' Federations in Ecuador

A good example of the success of locally based farmers' organizations comes from the Andean province of Chimborazo in Ecuador (Bebbington, 1991). There is a strong tradition of organization among indigenous farming communities, mainly originating from demands for land, religious rights, affordable transport and better infrastructure. In response to the weakness of government services, federations of groups have now initiated their own research and extension programmes, and, unusually, have attained high levels of control over the research and extension process.

Although this process uses demonstration plots, field days, extension visits, seed multiplication and input distribution systems, farmers in these organizations control, implement and own a large part of these activities. The main activity of federated organizations is to help members conduct trials aimed at raising yields without increasing production costs or risks. These are conducted with the help of agronomists hired by the groups. Through this process, technologies are progressively adapted and, most importantly, information on the changes is made available to members through a variety of extension methods, including training courses, meetings and radio programmes. Farmer extensionists also go on courses given at the national agricultural research institute. They are trained in formal agricultural science, but can now assess this in the context of their own local knowledge. Some become more 'modern' than others, but all end up with an understanding of both formal and informal agricultural science.

Most farmers' groups aim to sustain and enhance rural livelihoods through strong organization, and want to increase local income generation so that they do not have to migrate out in search of work so much. Now that they spend more time at home, they are able to strengthen family and community ties, as well as avoid the deep personal and economic costs of migration. The range of services offered by these federated organizations includes more than just extension. Some also run subsidized seed and input distribution, forestry projects, guinea pig projects, veterinary services, school vegetable projects, artisanal workshops, radio services, community water projects and health education services.

Consumer Groups in Japan

Federated groups of consumers can also be important actors in the quest for a more sustainable agriculture and regenerated rural communities. In Japan, women have formed remarkably successful consumer groups to make direct links with farmers and manufacturers of other goods (Furusawa, 1994, 1992, 1991). These consumer–producer groups come in all sizes, are based on relations of trust, and put a high value on face-to-face contact. There are now some 800–1000 groups in Japan, with a total membership of 11 million people and a turnover of more than US$15 billion each year for all activities.

Some are small ventures in which a few households, say 10 to 30, make a link with a single farmer, who supplies food of a particular quality, usually organic. One medium-sized group is the Young Leaves co-operative, begun by Hiroshi Ohira, who farms in Tokyo. It now has 400 household members and 11 farmers, who supply vegetables, rice, root crops and fruit. Farming is intensively organic. Members buy about 75 per cent of their food through Young Leaves. Prices are decided at an annual meeting of producers and consumers, and there tends to be little year-to-year variation. Sometimes prices are higher than normal market prices, sometimes they are lower.

The largest group is the Seikatsu Club, which has a membership of more than 200,000 households and branches all over Japan. It was set up in 1965 by a housewife living in Tokyo, who wanted to find a way of avoiding the high price of milk. Her idea was to band together with 20 other customers in the neighbourhood and buy milk directly from the distributors. Over the next few years, they also began to purchase food, clothes and cosmetics wholesale. In 1971, club members began to deal directly with farmers and take care of distribution themselves. Soon after, agreements with farmers were reached for rice, meat and fish. Members then began to order soap powder to replace detergents that they felt were polluting rivers and lakes. In 1978 a new headquarters was set up in

Setagaya and the first Seikatsu Club housewife was elected to local government the following year. As groups became frustrated with inaction in local and national institutions, they have increasingly entered the political arena (Box 5.7).

The turnover of the Seikatsu Club is now 40 billion yen (US$320-350 million). The club believes that *'women can begin to create a society that is harmonious with nature by taking action at home'* (Clunies-Ross and Hildyard, 1992). A survey of changes in members' lifestyle after joining one group, the Society for Reflecting on the Throwaway Age, is revealing. Formed in 1973, it now has 1800 members, including 80 farmers, some 72 per cent of whom have noticed a difference in eating habits; 42 per cent take better care of things and do not waste them; 43 per cent have more interest in social affairs; 36 per cent started to recycle used oil; and 20 per cent noticed changes in the way they raised children (Furusawa, 1991).

Box 5.7 The widening influence of members of the Seikatsu Club in Japan

Unlike other consumer groups, the Seikatsu Club has entered politics. It has members and the resources to influence Japanese society in a fundamental way. It has built alternative cultural centres in local areas and carried out national campaigns of potentially far-reaching significance.

There are now 31 members of the Seikatsu Club holding elective positions in local government in the Kanto area. When the campaign against synthetic detergents was in full swing, women members had presented evidence to local government concerning the dangers of such detergents, but they were brushed aside by male officials. This experience made it clear that unless they themselves gained office, their efforts would come to nothing. And so they entered politics as independents, steering free of vested interests and established parties, and emphasizing that politics begins with daily life. With the slogan *Political Reform from the Kitchen*, they have successfully appealed to public concern over issues of safe food, conservation of nature, women's rights, peace and grassroots democracy.

Another campaign aims to transform both people and cities under the banner *From Collective Buying to All of Life*. This has led to the setting up of free schools and workers' collectives so that people can begin to free themselves from the grip of the centralized economy.

Source: Furusawa, 1992

NATIONAL INITIATIVES FOR LOCAL GROUP ACTION

Scaling-up of local efforts may occur in spite of, rather than because of, policies at national level. It is rarer for national policies to be set up explicitly to encourage local action. Two notable examples are from Australia and India.

Landcare Groups in Australia

In Australia, a community-based revolution called Landcare is turning farming and conservation on its head by encouraging groups of farmers to work together with government and rural communities to solve a wide range of rural problems (Campbell, 1992, 1994b,c; Woodhill, 1992). More than 2000 voluntary community groups are currently working to develop more sustainable systems of land use, supported by a national 10-year funding programme. Landcare aims to combine elements of community and environmental education, action research and participatory planning, so as to tackle a range of agricultural production and conservation issues. It is working in a wide range of environments and providing policy makers with the opportunity to react to local needs.

Involvement of farmer groups in soil conservation is not new to Australia. The earliest forms of the current landcare groups were established for soil conservation programmes in Western Australia and Victoria in the early 1980s. Their activities broadened in the mid-1980s to focus on soil, water, flora and fauna, rather than just soil conservation, and taking a more bottom-up and group-oriented approach. These programmes grew much faster than expected with a minimum of resources, and were credited with enhancing the extent and the quality of land-user involvement in land conservation activities.

The level of attention to landcare increased dramatically in mid-1988, when an historic partnership was forged between the National Farmers Federation (NFF) and the Australian Conservation Foundation (ACF). The NFF and ACF jointly developed a National Land Management programme, which proposed a ten-year programme of funding support for landcare groups. Andrew Campbell, former national facilitator for Landcare, described the impact of this on policy and the public: *'The joint thrust of two powerful lobby groups, unlikely bedfellows from opposite ends of the political spectrum, presented a fascinating image to the media. The potent political ingredients of timing, a discrete package with broad voter appeal, against a background of exponential growth in community awareness of environmental issues, ensured that landcare became 'flavour of the month'* (Campbell, 1994).

The prime minister announced in mid-1989 that the 1990s would be the Decade of Landcare and outlined a A$340 million funding programme based to a large degree on the NFF–ACF document. By October 1989, the total number of landcare groups in Australia was about 350, a number which doubled by July 1990. Despite tough economic conditions in rural communities, the explosive growth of the landcare movement has continued, now with over 2000 groups, comprising more than one-third of the farming community.

Landcare groups usually form when farmers at a local level perceive a problem (such as salinity, erosion, weeds, rabbits or tree decline) requiring cooperative efforts and decide to form a group to take practical action. The concerns of landcare groups typically evolve from a focus on the immediate problem which catalysed the formation of the group, to a broader range of environmental, social and economic issues as groups mature. The term 'landcare' itself has evolved in Australia and is now used both in a narrow sense to refer to voluntary local land conservation groups, and more broadly to refer to an emergent philosophy of participatory approaches improving land management planning, policy making, research, extension and education. Landcare groups are now complemented by a spectrum of initiatives including participatory education programmes (Land Literacy), and group projects targeting specific issues such as conservation cropping (SoilCare), and farm profitability and business management (Farm Advance and Farm Management 500).

Some of the most far-reaching and powerful of the landcare initiatives are the land literacy programmes, as these involve children, who have a critical role to play in the future if some of the existing institutional, political and economic barriers to sustainable agriculture are to be shifted (Campbell, 1994b,c). Land literacy refers to activities designed to help people read and appreciate the signs of health (and ill-health) in a landscape, and to understand the condition of and trends in the environment around them. Many of the most important land degradation problems are complex, insidious and not visually obvious. For land degradation problems, it is wise to assume that prevention is cheaper and more effective than cure. But it is difficult to get people excited about prevention, if they cannot see or appreciate the problem. There is a wide range of land literacy programmes complementing the activities of Landcare groups in Australia (Box 5.8). These are described in more detail in Campbell (1994b).

Box 5.8 Land literacy in Australia: the example of Saltwatch

Saltwatch began in Victoria in 1987 as a participatory community education initiative conceived by the Victorian Salinity Bureau. It is now taking place in five States. By 1992, more than 900 schools and 50 Landcare groups were involved in gathering and analysing tens of thousands of water samples from creeks, rivers, reservoirs, irrigation channels and bores. Each school or community analyses its data and sends it to a central agency for processing, receiving in return a computer-generated overlay map of water quality in the district. This is often displayed in the school, store, hall or the pub. Data are stored on school computers as well as in government agencies, and groups are encouraged to look at trends over time within their district. The composite maps are used for interpretation, discussion and planning further action such as excursions, rehabilitation projects and interpretative displays. Schools and community groups have access to education kits, manuals and curriculum materials, and training programmes for teachers in land literacy have been developed.

Source: Campbell, 1994b

It is still too early to measure many of the physical impacts of landcare. But there have been major individual and institutional changes. Many people involved in landcare are learning more about their own land, about the land in their district and about issues they may have rarely considered in the past. Group leaders in particular have gained great satisfaction from seeing other people get involved, from influencing others through their interaction in the group and occasionally from group projects. Extension staff have also changed, becoming more than providers of information. They are evolving into facilitators of learning and are being trained to work with groups, helping them become self-reliant.

Some groups have already created a climate of opinion more favourable to the adoption of resource-conserving practices and some have achieved notable successes in land management improvements particularly suited to group action, such as controlling rabbits and weeds. Landcare, by involving committed people closest to the land, has the potential to be the first step in evolving new land use systems, and new relationships between people and land, which build upon human resources instead of discounting them or seeing them as part of the problem.

But the programme is not without problems (Campbell, 1994b,c). The learning and satisfaction at local level is often tempered by growing frustration: about the level of knowledge and resources available seriously to tackle problems; about the few people who really understand what needs to be done and the amount of poor land management still occurring; and about the bureaucracy, paperwork and politics of landcare, particularly for project funding. Many professionals have little training in people skills or participatory methods and they find it hard to be accountable to local people. They tend to constrain and hold back progress. A key constraint, therefore, remains the existing institutional cultures, which are yet to be oriented towards genuine community involvement and self-reliance.

Joint Forest Management in India

As the result of a series of laws and policies evolved over the past century which have nationalized community and private forest lands, and gradually eroded the rights and concessions of surrounding forest communities, the state governments' forest departments now own 95 per cent of India's forest lands. These agencies have an historic mandate to maximize revenue and protect the forests from expanding local populations. Yet they have been largely unsuccessful: less than half of India's forest lands remain under closed canopy forests and the remaining forest lands are in various stages of degradation (SPWD, 1992; Singh, 1990).

During the late 1970s, and 1980s, enlightened officials in several states began to realize they could never hope to protect forests without the help and involvement of local communities. They helped to establish local forest protection committees (FPCs) or hill resource management societies, which were given the responsibility of protecting degraded land and granted rights to the use of a range of timber and non-timber forest produce. Success in the form of biological regeneration and increased income flows was so spectacular (Dhar et al, 1991; Pandit, 1991; Campbell, J 1992), that the national government issued an order on 1 June 1990, requesting all states to undertake participatory forest management. This also encouraged the involvement of NGOs as intermediaries and facilitators. Many states have now passed their own orders and regulations, *'outlining rules for reversing decades of confrontation between forest departments and local communities, and pointing the way to a new form of joint forest management undertaken in partnership with local communities'* (Campbell, J 1992).

By 1992, the area managed by nearly 10,000 formal and informal forest protection committees was some 800,000 ha, including 100,000 in Madhya

Pradesh, 150,000 in Orissa, 200,000 in Bihar and 300,000 in West Bengal (SPWD, 1992). One survey of 12 FPCs in Midnapore District of West Bengal revealed that of 214 wild plants species observed in regenerating sal forests, 155 were now used by local people. The mean income to tribal and caste households was some Rs 2500, contributing between 16–22 per cent of total farming income (SPWD, 1992). Old attitudes are changing, as foresters appreciate the remarkable regeneration of degraded lands following community protection and the growing satisfaction of working with, rather than against, local people.

The lack of tenure is a potential future problem, as community groups may feel less secure in their commitment without clear time horizons. Forest agencies may also be inclined to take back forests once they are regenerated and now productive. In some parts of Bihar and Orissa, however, the local committees have grown in significance such that new political movements have developed out of forest protection and utilization. As V K Bahuguna recently (1992) put it: '*The only solution to the present day crisis of depletion of forest resources, and the circumstantial alienation of people, is to opt for people's forests by involving local people in forest protection and development*'.

An example of the types of local rules developed by the forest protection committees is shown in Box 5.9. These rules are a sign of strong local institutions with rights and access to resources. They are the foundation for sustainable development. In some cases, fines have been imposed not only on villagers but also on forest guards and, in others, communities have taken action on social issues, punishing anti-social drinking and abuse. In Madhya Pradesh, the benefits have included improvements in fuelwood,

Box 5.9 Examples of rules formulated by forest protection committees in Madhya Pradesh in India

Grazing

'It was resolved by the committees that all those areas where the trees are marked with red paints along the boundary are closed for grazing and hence all of us unanimously resolve not to take our cattle for grazing in these areas, nor allow the villagers of other villages to do so. We shall keep our cattle at home and all cases of violation would be reported to the forest officer'.

Protection of trees

'It was unanimously resolved that we shall not girdle any tree nor allow others to do so. We shall have some strict watch over illegal cutting of trees'.

Goats

'It is resolved that all those villagers who are having goats with them must sell them within a period of 3 days, otherwise action will be taken'.

Firewood

'No villager would carry the fuelwood head load for sale outside the village. The defaulters would be charged Rs51 per head load'.

Source: Bahuguna, 1992

grass and crop yields; reduced poaching of elephants and other animals; changed relations between forest officials and local people; and the creation of democratic local organizations (Bahuguna, 1992).

SUMMARY

Although there are many potentially productive and sustainable technologies available to farmers, a transition to a more sustainable agriculture will not occur without the full participation and collective action of rural people. The development of a more sustainable agriculture depends not just on the motivations of individual farmers, but on the action by groups or communities as a whole. This makes the task facing agriculture exceptionally challenging.

The problem with national and international institutions is that they have tended to substitute for local action, so smothering any existing initiatives or institutions. As local groups and institutions have been ignored, so many have disappeared entirely. This has led both to increased degradation and to decreased capacity in local people to cope with environmental and economic change.

It is increasingly well established, however, that when people who are already well organized, or who are encouraged to form new groups, they are more likely to continue activities after project completion. There are six types of local group relevant to the needs for a more sustainable agriculture. These are community organizations; natural resource management groups; farmer research groups; farmer-to-farmer extension groups; credit management groups; and consumer groups.

The process of establishing and/or supporting self-reliant groups is not easy. It should not be rushed into nor forced on local people. It needs external catalysts or facilitators, and should focus on building the capacity of people to develop new ways of learning and new forms of leadership. These groups have led to direct economic benefits for many rural people, as well as improvements in natural resources. What many comment upon, however, are the more intangible benefits of rediscovered social cohesion and solidarity. As confidence grows with success, groups evolve new roles and responsibilities, often joining with other groups to achieve a wider impact.

As yet, the scaling-up of local efforts has occurred largely in spite of, rather than because of, policies at national level. However, there are emerging examples of national policies designed to encourage these approaches, in which local groups do not substitute for local government services, but are seen as partners. If these local institutions are not to be suffocated or coopted, as they have been in the past, then external institutions must begin to play a role quite different from the norms of the past. They will have to focus much more on facilitating change in others. This means they will have to become enabling institutions.

6

EXTERNAL INSTITUTIONS AND PARTNERSHIPS WITH FARMERS

'Tea rooms were constructed of materials that could be found easily and near at hand. Rare and expensive materials were avoided. A log or branch from a nearby grove of trees, a stone by the roadside, were collected and incorporated into the final design. The original spirit of tea-room architecture is the same. It is an architecture built by gathering things close at hand... As a result, the tea room seems not to have been designed but built through a process of natural accretion.'

Kisho Kurokawa, 1991.

THE CONVENTIONAL INSTITUTIONAL CONTEXT

The previous two chapters have given details of the resource conserving technologies and the local groups and institutions necessary for agriculture to become sustainable. The third essential element is the way external institutions are organized and the way they work with other institutions and farmers. In the process of agricultural modernization, external institutions have tended to ignore and so suffocate local knowledge and initiative.

The complexities involved in achieving a diverse and productive sustainable agriculture mean that organizations will have to adopt new ways of working. This implies greater multi-disciplinarity, more structured participation with farming communities in research, extension and development activities, the evolution of learning processes in organizations and the development of a whole new agricultural professionalism itself.

Why Learning is so Difficult

For many reasons, existing agricultural institutions, whether universities, research organizations or extension agencies, find it difficult to learn from farmers and rural people. This is because of their internal structures, the way they develop their staff and the ways these staff interact with people outside their institutions (Korten, 1980; Chambers, 1992a; Roche, 1992; Pretty and Chambers, 1993a,b).

The first problem is that organizations are characterized by restrictive bureaucracy and centralized hierarchical authority. Their staff spend more time looking inwards and upwards towards seniors than outwards towards clients. They follow long-established norms of behaviour, filtering and passing information up to seniors. If they have ideas about how changes can be made to improve performance, it is difficult for them to get these heard. This is chiefly because staff are afraid to make mistakes, as they expect to be punished rather than rewarded for invention. Mark Easterby-Smith (1992) describes a typical organization: *'a combination of power culture and highly centralized controls, with rigidly designed systems and procedures, produces behaviour amongst managers that makes learning almost impossible. In particular, the tendency to make scapegoats out of those who made mistakes leads to a general aversion to taking risks, and managers, afraid of being punished as harbingers of bad news, tend to concentrate on providing only good news to their superiors'*. In such contexts, new initiatives are bound to fail.

The second problem is that the majority of agricultural professionals are specialists. They see only a narrow view of the world, yet are encouraged to continue to work in this way by internal reward systems and incentives. The performance of researchers, for example, is commonly measured by the number of scientific papers they have published in prestigious journals. Without a good publishing record, they will be unable to get promotion (McRae et al, 1989). Whether the research has had a positive impact on farmers' livelihoods is mostly irrelevant. As Patrick Madden and Thomas Dobbs (1990) put it: *'Disciplinary work generally receives greater recognition and acceptance than does multidisciplinary work in peer-oriented professional journals, in university tenure and in promotional processes'*.

But specialist professionals tend to have higher status than those working more closely with farmers and rural people (Chambers, 1985). Specialists in agriculture, such as genetic engineers and biotechnologists, focus on controlled environments with organisms or small parts of them, such as cells or genes. Changes to crops and animals are made without regard to the real-world context of these crops/animals. By contrast, those with the lowest status and pay in agriculture are the community development and extension workers, who work with rural people using a wide range of social and technical skills. They work in the complex and uncontrollable real world.

This is also true of many other professions. In medicine, it is the transplant and micro surgeons, dealing with their patients as machines, who have the highest status. They do not need to know anything about the social context of their patients. By contrast, community health workers come into contact with sick people in their environment, with all its complexity and uncertainty. They are probably more concerned with preventative medicine. Yet they have low status and are certainly paid much less than surgeons.

The third major problem restricting or preventing learning about the complexities of a changing world is that organizations commonly get misleading feedback from their peripheries. This is partly because of the methods and approaches they use to gather information and measure performance. Both questionnaire surveys and brief development tourism visits are deeply flawed because of the selective nature of information coming from them. Senior staff are left with falsely favourable impressions of the impact of their work and so they themselves have few reasons for initiating or encouraging change.

Self-Deception and Questionnaire Surveys

A major reason why agricultural professionals and institutions have had only a partial view of rural realities relates to the widespread reliance on questionnaire surveys. The formal survey with a preset questionnaire has long been the standard choice for those needing information on rural resources and people. The questionnaire is commonly designed by senior professionals and given to enumerators who interview a sample of people selected from a larger population. As each informant is asked the same set of questions, it is assumed that the interviewer does not influence the process. Many informants are selected to account for all variation and the resulting data are statistically analysed. Surveys are used at practically all levels, from the large-scale census to small-scale village level research; by governments and NGOs; and for planning, research and extension.

But questionnaire surveys do not always produce useful and relevant information. This is because of the structure of the questionnaire forms themselves, the perceived need to interview large numbers of people, and the nature of the interaction between the outsider and local people.

The questionnaire designer has to determine well in advance what questions will be included on the form. But those who design these instruments, themselves outsiders, do not know in advance what issues are important for local people. So they tend to add more and more questions, to ensure all relevant issues are covered. This leads, in some cases, to forms of absurd length, with several hundred questions taking hours to administer. Such questionnaires, therefore, eliminate the possibility of capturing the unique and spontaneous insights which might arise in the course of a conversation or interview.

Rarely is attention paid to the nature of the interviewing process. Questioning and answering are ways of speaking that depend on culturally shared and often tacit assumptions about how to express and understand beliefs, experiences, emotions, and intentions. Yet in the structured survey, many of the contextual grounds for understanding are systematically removed or ignored. As a result, surveys have lost the capacity to understand what respondents mean by what they say. In the drive for standardization, the multiple perspectives on problems and issues that relate to local context are lost, and the whole process of learning is impoverished.

Another problem is the tendency for people to want to please interviewers by giving them what they want. The stranger tends to be looked on as a guest and the duties of the host are often regarded as sacrosanct. Not understanding the real purpose of the survey, the respondents try to please their guest by giving what is assumed to be the required answer. Very often, the ill-trained enumerator makes this all too easy by prompting with suggested answers. The enumerator also has a quota to fulfil, but the respondent does not know that. As Gerry Gill (1993) has put it: *'The stranger then produces a little board, and clipped to it, a wad of paper covered in what to the respondent are unintelligible hieroglyphics. He then proceeds to ask questions and write down answers – more hieroglyphics. The respondent has no idea of what is being written down, whether his or her words have been understood or interpreted correctly. The enumerator, being simply a data-gatherer, has no way of knowing – and no responsibility to know – whether the answers being given are correct or whether they make sense to the broader framework of the survey. The interview complete, the enumerator departs and is probably never seen again'*.

Over the past 30 years, the structured questionnaire has developed into *'an industry'* in which practitioners have become *'slaves to the methodology'* (Ashby in Rhoades, 1990). Despite many critiques of this mode of data gathering (Chambers, 1983, 1992c; Fowler and Mangione, 1990; Rhoades, 1990; Gill, 1993), official surveys, such as sample censuses of agriculture or household expenditure surveys, are as popular as ever.

Self-Deception and Rural Development Tourism

At the other end of the spectrum to the over-structured approach of the questionnaire are the brief field visits made by development professionals, which they use as their basis for understanding complex rural life. Such 'rural development tourism', though, is beset by invisible biases that ensure that professionals not only see a small and selected portion of rural life, but also that they believe that this is an accurate picture. Robert Chambers (1983) characterized these biases of rural development tourism into four main types. These are:

- spatial biases, in which it is the better off communities, and people living near to roads and services that are visited, and those who are remote and poorer missed;
- time biases, in which visits are made during the seasons when roads are open and people are better off, rather than say during the wet season when people are starving and desperate; and in which visits are made during office working hours, when rural people are busy in the fields, rather than in the early mornings or evenings;
- people biases, in which development tourists speak only to rural leaders and articulate people, who tend to represent the elite, dominant and wealthy groups, and so are not exposed to the perceptions of women, the poorest, the weakest and so on;
- project biases, in which a showcase village or technology is repeatedly selected to show to outsiders, who assume this is typical of all efforts.

What all this implies is that institutions come to believe that this selective information represents a complete picture. They misunderstand the poor and non-elite, and so are surprised when technologies they develop are rejected. Contrary to learning about local conditions, development professionals have tended to impose their own criteria and constructs. As local criteria are almost always more diverse, the result is the reduction of diversity to simplicity. This may result in a fair representation of some people's views, but certainly will not be fair for a whole community of diverse needs and values.

A classic example of this comes from the way official definitions of poverty differ from local people's perceptions. Standard definitions of poverty arrive at a poverty line based on external concepts of welfare. This is often described as the minimum amount of goods and services necessary to live a decent life. But the common focus on money income for measuring poverty has major flaws because local perceptions are ignored in the process (Chambers, 1993; Mukherjee, 1992; *RRA Notes*, 1992; Glewwe and van der Gaag, 1990).

Most people themselves do not characterize well-being so strictly and simply. N S Jodha's work over a 20-year period (1988) with people of 2 villages in Rajasthan showed that they had 38 local criteria for economic status. For those that had become poorer by official measures since the early 1960s, it was found that they had actually become better off in all but one of their own criteria. These improvements included fewer households working as attached labourers, fewer residing in the landlord's yard, fewer marketing produce only through landlords, fewer with members having to migrate out seasonally to search for work, more making cash purchases during the festival season, more eating green vegetables, more where maternity feeding to mothers provided up to a month or more, and more with sturdier housing. Similar examples of such diversity of local indicators of well-being have been described in many other countries (*RRA Notes*, 1992; Grandin, 1987).

Another example relates to the way local people judge modern crop varieties. They do not always see them in the same way as researchers and extension workers. Their criteria for evaluating and making choices are frequently so different that sometimes the best products of research services are rejected, while others judged inappropriate are chosen by farmers as favourable (see Chapters 2 and 3).

In Colombia, a high yielding variety of bush beans (*Phaseolus vulgaris*) was rejected by farmers because the variable colour made marketing difficult; another variety rejected by researchers for its small bean size was acceptable because, as one farmer put it *'is good for consumption purposes because it swells to a good size when cooked – it yields in the pot'* (Ashby et al, 1987). In the Philippines, sweet potato varieties bred for high yield and sweet taste were rejected by upland farmers who preferred rapidly vining varieties that prevented weed growth and rain-induced soil erosion. They also selected tolerance to weevil damage during the underground storage phase as an important characteristic, as this meant the potatoes could be harvested only as required (Acaba et al, 1987). In Andhra Pradesh, India, women farmers working closely with agricultural scientists from ICRISAT appreciated the most productive and pest-resistant characteristics of the researchers' most favoured variety of pigeonpea, yet declined to grow it because of its bitter taste (Pimbert, 1991).

The lack of understanding of local perceptions and needs is not necessarily restricted to modern, high-external input agriculture. If the resource conserving technologies and social organizations described in Chapters 4 and 5 are forced on rural people, then they too will go the way of 'modern' agricultural technologies. The emerging danger is that agricultural professionals, in promoting new technologies that are low cost, sustainable and productive, will forget the diverse conditions and needs of rural people. If this occurs, then the widespread adoption of resource-conserving technologies will remain as remote as ever.

'PARTICIPATION' IN DEVELOPMENT

Multiple Interpretations of Participation

There is a long history of community participation in agricultural development, and a wide range of development agencies, both national and international, have attempted to involve people in some aspect of planning and implementation. Two schools of thought and practice have evolved. One views community participation as a means to increase efficiency, the central notion being that if people are involved, then they are more likely to agree with and support the new development or service. The other sees community participation as a right, in which the main aim is to initiate mobilization for collective action, empowerment and institution building.

In recent years, there have been an increasing number of analyses of development projects showing that 'participation' is one of the critical components of success in irrigation, livestock, water and agriculture projects (Montgomery, 1983; USAID, 1987; Baker et al, 1988; Reij, 1988; Finsterbusch and van Wicklen, 1989; Bagadion and Korten, 1991; Cernea, 1991; Guijt, 1991; Kottak, 1991; Pretty and Sandbrook, 1991; Uphoff, 1992a; Narayan, 1993; World Bank, 1994b).

As a result, the terms 'people's participation' and 'popular participation' are now part of the normal language of many development agencies, including NGOs, government departments and banks (Adnan et al, 1992; Bhatnagar and Williams, 1992). It is such a fashion that almost everyone says that participation is part of their work. This has created many paradoxes. The term 'participation' has been used to justify the extension of control of the state, and to build local capacity and self-reliance; it has been used to justify external decisions, and to devolve power and decision making away from external agencies; it has been used for data collection and for interactive analysis. But *'more often than not, people are asked or dragged into participating in operations of no interest to them, in the very name of participation'* (Rahnema, 1992).

One of the objectives of agricultural support institutions must, therefore, be greater involvement with and empowerment of diverse people and groups, as sustainable agriculture is threatened without it. The dilemma for authorities is they both need and fear people's participation. They need people's agreements and support, but they fear that this wider involvement is less controllable, less precise and so likely to slow down planning processes. But if this fear permits only stage-managed forms of participation, distrust and greater alienation are the most likely outcomes. This makes it all the more crucial that judgements can be made on the type of participation in use.

In conventional rural development, participation has often centred on encouraging local people to sell their labour in return for food, cash or materials. Yet these material incentives distort perceptions, create dependencies and give the misleading impression that local people are supportive of externally driven initiatives. The confusion is complete when technical work is known to be completely inappropriate. Norman Hudson (1991) describes the reaction of a visitor to a project where food for work was used as an incentive for people to participate: *'But you are planting the wrong species at the wrong time in the wrong place, and the survival rate will be almost zero.' The project officer said 'I know. It hurts my professional pride too. But there are people starving in this District, and this project brings them food'.*

In another project in Kenya, this time on integrated pest management, farmers were given improved maize, sorghum and cowpea seeds, 50 kg of fertilizer, farm implements, construction materials for new granaries and tractors for ploughing. All were provided free, except for the ploughing, which was at cost, to help trigger the process of adoption. In spite of all this, the project's view was that *'the participating farmers are expected to apply the recommendations after the project ends'* (Kiss and Meerman, 1991).

This paternalism undermines sustainability goals and produces results which do not persist once the project ceases (Reij, 1988; Fujisaka, 1989; Treacy, 1989; Kerr, 1994). Few have commented so clearly and unequivocally as Roland Bunch (1983, 1991) on the destructive process of giving things away to people or doing things for them. He suggests five major problems:

- give-aways blind people to the need for solving their own problems;
- people become accustomed to give-aways and come to expect them;
- give-aways are 'monstrously expensive';
- give-aways hide people's indifference to programme efforts; and
- give-aways destroy the possibility of there ever being a multiplier effect.

Despite this, development programmes continue to justify subsidies and incentives, on the grounds that they are faster, they can win over more people, the people cannot help themselves or that the people are just so poor that justice demands they are given one chance. But as Roland Bunch (1991) put it: *'Obviously, though, programmes must do something for the people. Were they able and willing to solve all their own problems, they would have done so long ago... It should be emphasised that anything we do that people can do for themselves is paternalistic'.*

As little effort is made to build local skills, interests and capacity, local people have no stake in maintaining structures or practices once the flow of incentives stops.

'Participation' in Soil and Water Conservation Projects

Soil and water conservation is one field of agricultural development long characterized by multiple interpretations of participation. For close to a century, rural development policies and programmes have taken the view that farmers are mismanagers of soil and water, and so must be advised, lectured, paid and enforced to adopt conserving practices and technologies (see Chapters 2 and 3). Yet most projects have adopted the rhetoric of participation to describe these activities (Box 6.1). Very impressive physical results have been achieved in the short-term and projects assume, therefore, that maintenance will occur after the project. But, the disappearance of soil and water conservation structures, such as 120,000 ha of earth bunds built in Burkina Faso in the 1960s and 20,000 ha of narrow-based terraces constructed in Kenya in the 1950s, is so common that there would appear to be a sad future for many contemporary efforts based on a similar controlling participation (Reij, 1988; Hudson, 1991; Gichuki, 1991). Three examples from Africa illustrate the confusion.

A major project in Niger was described by the implementing agency in this way: *'People's participation is the power behind the Keita project. From decision-making – to planning – to action: local farmer-livestock owners have been consulted and actively taken part in every step'* (FAO, 1992). Yet some 2.76 million work-days were paid for with World Food Programme rations, which served as *'incentives to participate in land reclamation and training courses offered by the project'*. The project, therefore, believes its own success: *'the techniques for soil and water conservation have been learned readily by local farmers and should continue to be used after the project ends'*. Of course, this may be the case. But history suggests that these structures are unlikely to be sustained.

In Ethiopia some 200,000 km of terracing were constructed during the 1980s with food for work (Mitchell, 1987). But an evaluation indicated that *'the target group is not questioned as to their needs and preferences, nor do they participate in project planning. They implement the project in the sense that they perform the constructing and planting tasks assigned them. Participation is either compulsory via peasant association campaigns or paid through food for work.'* (SIDA, 1984).

This participation was extraordinarily controlling: *'when the conservation work is completed, a technician from the project... comes to inspect. If the work has been carried out in a technically acceptable manner, then full payment is awarded. If not, payment is delayed until the work has been corrected'*. In the same document, the project indicated that it *'expected that peasants will, in future, bear the costs of whatever maintenance is carried out'*, yet also that *'the use of food for work... has diminished farmers' commitment to the maintenance of soil conservation structures'*. In fact, some 40 per cent of terraces were already broken in the first year after construction (SIDA, 1984).

This effort in Ethiopia was described by FAO (1986) as *'one of the largest and most successful soil conservation projects in the world... (with) Peasant Associations able to mobilise organized labour quickly and efficiently'*. A total of 34.3 million person-days of work was devoted to conservation, involving the *'co-operation of some 8000 Peasant Associations'*, according to FAO (1986, in Ostberg and Christiansson, 1993).

Box 6.1 The changing rhetoric over participation and soil conservation

Ostberg and Christiansson (1993) describe how the rhetoric over participation and soil conservation has changed in recent years in FAO documents. The tone struck in a 1986 booklet *Protect and Produce: Soil Conservation for Development* was that precious soil is threatened everywhere. *'One thoughtless action by one human being can remove for ever tens of tonnes of soil from each hectare that he or she farms. In a few days the legacy of thousands of years of patient natural recycling can vanish for good. It is terrifying to consider what is at stake.'*

It is clearly thought that farmers do not know their job: *'when the wrong crop is selected, or the wrong farming technique chose, yield inevitably drops. Erosion follows... The causes of soil erosion are well known. So are the techniques with which to combat it'.*

As Ostberg and Christiansson suggest *'the FAO publication offers advice on how to inform and train farmers... The technical fixes are there. With the right incentives farmers can be persuaded to switch to new farming practices and maintain existing conservation work'.*

A later 1990 blueprint for soil conservation contains a chapter entitled 'Encouraging participation'. It suggests that land users themselves are best suited to plan and implement their own solutions. *'But the rest of the chapter is hardly distinguishable from what is said about most development cooperation. Picking verbs from the text it becomes obvious who is considered to be in the know. Some are to motivate, introduce, teach, persuade, alert, make aware'.*

By 1992, it had become time for a revised edition of *Protect and Produce.* The passage quoted from the 1986 edition remained the same, as did the general perspective. Population increase is regarded as the major cause of land degradation.

But a new perspective is emerging, as shown by the following quotes: *'success stories are rare enough to be notable', 'the general approach to soil conservation has been faulty'; 'governments should become facilitators instead of being agencies that implement conservation projects'.* The emphasis, as Ostberg and Christiansson put it, is at last on people finding their own solutions.

Source: Ostberg and Christiansson, 1993

The arrogance of external agents is pervasive. Apparently, *'farmers' participation was shown by their contributions of labour for infrastructure development'*, and the project expected these structures to be maintained because *'training... will help in sustaining activities when the donor pulls out. The privilege of being trained will keep the individuals responsible in the activities he [sic] was trained for'* (reported in Oxfam, 1987).

Another project in Tanzania has completely removed livestock from whole communities, with tens of thousands of animals removed from individual districts (Mndeme, 1992; Christiansson, 1988). Such a policy was only possible *'after mustering the cooperation of the ruling party and government machinery at village, district, regional and national levels. Inevitably some of the actions necessary to reverse soil degradation processes are a bitter pill to swallow*'. Despite this, the project believes that: '*the favourable results of destocking have sparked an interest in taking similar measures, particularly in the region's other districts*' (Mndeme, 1992).

Most soil and water conservation projects have paid and continue to pay local people in cash or food for their 'participation' (Kerr, 1994). But this is self-defeating. According to Chris Reij (1988): '*practice shows that where people are paid for soil and water conservation, the end of the project almost invariably leads to a stop in the construction of conservation works*'.

Types of Participation

Although there are many ways that development organizations interpret and use the term participation, these resolve into seven clear types. These range from passive participation, where people are involved merely by being told what is to happen, to self-mobilization, where people take initiatives independent of external institutions (Table 6.1). It is clear from this typology that the term 'participation' should not be accepted without appropriate qualification. The problem with participation as used in types 1 to 4 is that the '*superficial and fragmented achievements have no lasting impact on people's lives*' (Rahnema, 1992). The term participation can be employed, knowing it will not lead to action. If the objective of development is to achieve sustainable development, then nothing less than functional participation will suffice. All the evidence points towards long-term economic and environmental success coming about when people's ideas and knowledge are valued, and power is given to them to make decisions independently of external agencies.

But the dominant applications of participation are almost always at best instrumental. A recent study of 230 rural development institutions employing some 30,000 staff in 41 countries of Africa found that people participate at different stages of the project cycle and in very different ways (Guijt, 1991). External agencies rarely permitted local groups to work alone, some even acting without any local involvement. External agencies usually controlled all the funding, though some did permit joint decisions. Participation was more likely to mean simply having discussions or providing information to external agencies. Rarely were components of functional or interactive participation present.

Another study of 121 rural water supply projects in 49 countries of Africa, Asia and Latin America found that participation was the most significant factor contributing to project effectiveness, maintenance of water systems and economic benefits (Narayan, 1993). Most of the 121 projects, however, referred to community participation or made it a specific project component, but only 21 per cent scored high on interactive participation. Clearly, intentions did not translate into practice. It was when people were involved in decision making during all stages of the project, from design to maintenance, that the best results occurred. If they were just involved in information sharing and consultations, then results were much poorer.

According to the analysis, it was quite clear that moving down the

Table 6.1 A typology of participation: how people participate in development programmes and projects

Typology	*Characteristics of each type*
1. Passive participation	People participate by being told what is going to happen or has already happened. It is a unilateral announcement by an administration or project management without any listening to people's responses. The information being shared belongs only to external professionals.
2. Participation in information giving	People participate by answering questions posed by extractive researchers using questionnaire surveys or similar approaches. People do not have the opportunity to influence proceedings, as the findings are neither shared nor checked for accuracy.
3. Participation by consultation	People participate by being consulted and external agents listen to views. These external agents define both problems and solutions, and may modify these in the light of people's responses. Such a consultative process does not concede any share in decision making and professionals are under no obligation to take on board people's views.
4. Participation for material incentives	People participate by providing resources, for example labour, in return for food, cash or other material incentives. Much on-farm research falls in this category, as farmers provide the fields but are not involved in experimentation or the process of learning. It is very common to see this called participation, yet people have no stake in prolonging activities when the incentives end.
5. Functional participation	People participate by forming groups to meet predetermined objectives related to the project, which can involve the development or promotion of externally initiated social organization. Such involvement does not tend to be at early stages of project cycles or planning, but rather after major decisions have been made. These institutions tend to be dependent on external initiators and facilitators, but may become self-dependent.
6. Interactive participation	People participate in joint analysis, which leads to action plans and the formation of new local institutions or the strengthening of existing ones. It tends to involve interdisciplinary methodologies that seek multiple perspectives, and make use of systematic and structured learning processes. These groups take control over local decisions and so people have a stake in maintaining structures or practices.
7. Self-mobilization	People participate by taking initiatives independent of external institutions to change systems. They develop contacts with external institutions for resources and technical advice they need, but retain control over how resources are used. Such self-initiated mobilization and collective action may or may not challenge existing inequitable distributions of wealth and power.

Source: Pretty, 1994, adapted from Adnan et al, 1992

typology moved a project from a medium to highly effective category. Deepa Narayan (1993) summarized the study in this way: *'The good news is that beneficiary participation in decision making is critical in determining project effectiveness, maintenance of water systems, environmental effects, community empowerment and strength of local organizations. The bad news is that so far relatively few externally supported projects have achieved meaningful beneficiary participation. Even fewer have empowered women'.*

Great care must, therefore, be taken over both using and interpreting the term participation. It should always be qualified by reference to the type of participation, as most types will threaten rather than support the goals of sustainable agriculture. What is important is to ensure that those using the term participation both clarify their specific application and define better ways of shifting from the more common passive, consultative and incentive-driven participation towards the interactive end of the spectrum.

Alternative Systems of Learning and Action

There has been in recent years a rapid expansion of new participatory methods and approaches in the context of agricultural development. These have drawn on many long-established traditions that have put participation, action research and adult education at the forefront of attempts to emancipate disempowered people. To the wider body of development programmes, projects and initiatives, these approaches represent a significant departure from standard practice. Some of the changes under way are remarkable. In a growing number of government and non-government institutions, extractive research is being superseded by investigation and analysis by local people themselves. Methods are being used not just for local people to inform outsiders, but also for people's own analysis of their own conditions (Chambers, 1992b,c; Pretty and Chambers, 1993a,b).

The interactive involvement of many people in differing institutional contexts has promoted innovation and ownership, and there are many variations in the way that systems of interaction have been put together. There are many different terms, some more widely used than others (Box 6.2). Participatory Rural Appraisal (PRA), for example is now practised in at least 130 countries, but Samuhik Brahman is associated just with research institutions in Nepal. But this diversity and complexity is a strength. Despite the different ways in which these approaches are used, there are important common principles uniting most of them. These are as follows.

- **A defined methodology and systemic learning process.** The focus is on cumulative learning by all the participants and, given the nature of these approaches as systems of inquiry and interaction, their use has to be participative.
- **Multiple perspectives.** A central objective is to seek diversity, rather than characterize complexity in terms of average values. The assumption is that different individuals and groups make different

Box 6.2 A selection of terms and names for alternative systems of learning and action

Agroecosystems Analysis (AEA), Beneficiary Assessment, Development Education Leadership Teams (DELTA), Diagnóstico Rurale Participativo (DRP), Farmer Particpatory Research, Groupe de Recherche et d'Appui pour l'Auto-Promotion Paysanne (GRAAP), Méthode Accélérée de Recherche Participative (MARP), Participatory Analysis and Learning Methods (PALM), Participatory Action Research (PAR), Participatory Research Methodology (PRM), Participatory Rural Appraisal (PRA), Participatory Urban Appraisal (PUA), Planning for Real, Process Documentation, Rapid Appraisal (RA), Rapid Assessment of Agricultural Knowledge Systems (RAAKS), Rapid Assessment Procedures (RAP), Rapid Assessment Techniques (RAT), Rapid Catchment Analysis (RCA), Rapid Ethnographic Assessment (REA), Rapid Food Security Assessment (RFSA), Rapid Multi-perspective Appraisal (RMA), Rapid Organizational Assessment (ROA), Rapid Rural Appraisal (RRA), Samuhik Brahman (Joint trek), Soft Systems Methodology (SSM), Theatre for Development, Training for Transformation, and Visualization in Participatory Programmes (VIPP).

evaluations of situations, which lead to different actions. All views of activity or purpose are heavy with interpretation, bias and prejudice, and this implies that there are multiple possible descriptions of any real-world activity.

- **Group learning process.** All involve the recognition that the complexity of the world will only be revealed through group inquiry and interaction. This implies three possible mixes of investigators, namely those from different disciplines, from different sectors, and from outsiders (professionals) and insiders (local people).
- **Context specific.** The approaches are flexible enough to be adapted to suit each new set of conditions and actors, and so there are multiple variants.
- **Facilitating experts and stakeholders.** The methodology is concerned with the transformation of existing activities to try to bring about changes which people in the situation regard as improvements. The role of the 'expert' is best thought of as helping people in their situation to carry out their own study and so achieve something. These facilitating experts may be stakeholders themselves.
- **Leading to sustained action.** The learning process leads to debate about change, and debate changes the perceptions of the actors and their readiness to contemplate action. Action is agreed, and implementable changes will therefore represent an accommodation between the different conflicting views. The debate and/or analysis both defines changes which would bring about improvement and seeks to motivate people to take action to implement the defined changes. This action includes local institution building or strengthening, so increasing the capacity of people to initiate action on their own.

These alternative systems of learning and action imply a process of learning leading to action. A more sustainable agriculture, with all its uncertainties and complexities, cannot be envisaged without all actors being involved in continuing processes of learning.

Participatory Methods

In recent years, the creative ingenuity of practitioners worldwide has hugely increased the range of participatory methods in use (see *RRA Notes*, 1988–95; IDS/IIED, 1994; Pretty et al, 1995; Chambers, 1992b,c; Mascarenhas et al, 1991; KKU, 1987; Conway, 1987). Many have been drawn from a wide range of non-agricultural contexts and were adapted to new needs. Others are innovations arising out of situations where practitioners have applied the methods in a new setting, the context and people themselves giving rise to the novelty. The methods are structured into four classes, namely those for group and team dynamics, for sampling, for interviewing and dialogue, and for visualization and diagramming (Table 6.2). It is the collection of these methods into unique approaches, or assemblages of methods, that constitute systems of inquiry or interaction.

Participation calls for collective analysis. Even a sole researcher must work closely with local people (often called 'beneficiaries', 'subjects',

Table 6.2 Participatory methods for alternative systems of learning and action

Group and team dynamics methods	*Sampling methods*	*Interviewing and dialogue*	*Visualization and diagramming methods*
Team contracts	Transect walks	Semi-structured interviewing	Mapping and modelling
Team reviews and discussions	Wealth ranking and well-being ranking	Direct observation	Social maps and wealth rankings
Interview guides and checklists	Social maps	Focus groups	Transects
Rapid report writing	Interview maps	Key informants	Mobility maps
Energizers		Ethno-histories and biographies	Seasonal calendars
Work sharing (taking part in local activities)		Oral histories	Daily routines and activity profiles
Villager and shared presentations		Local stories, portraits and case studies	Historical profiles
Process notes and personal diaries			Trend analyses and time lines
			Matrix scoring
			Preference or pairwise ranking
			Venn diagrams
			Network diagrams
			Systems diagrams
			Flow diagrams
			Pie diagrams

'respondents' or 'informants'). Ideally, though, teams of investigators work together in interdisciplinary and intersectoral teams. By working as a group, the investigators can approach a situation from different perspectives, carefully monitor one another's work and carry out a variety of tasks simultaneously. Groups can be powerful when they function well, as performance and output is likely to be greater than the sum of its individual members. Many assume that simply putting together a group of people in the same place is enough to make an effective team. This is not the case. Shared perceptions, essential for group or community action, have to be negotiated and tested in a complex social process. Yet, the complexity of multidisciplinary team work is generally poorly understood. A range of workshop and field methods are used to facilitate this process of group formation.

In order to ensure that multiple perspectives are both investigated and represented, practitioners must be clear about who is participating in the data-gathering, analysis and construction of these perspectives. Communities are rarely homogenous and there is always the danger of assuming that those participating are representative. Those missing, though, are usually the poorest and most disadvantaged. Sampling is an essential part of these participatory approaches and a range of field methods is available.

Sensitive interviewing and dialogue is a third element of these systems of participatory learning. For the reconstructions of reality to be revealed, the conventional dichotomy between the interviewer and respondent should not be permitted to develop. Interviewing is, therefore, structured around a series of methods that promote a sensitive and mutually beneficial dialogue. This should appear more like a structured conversation than an interview.

The fourth element is the emphasis on diagramming and visual construction. In formal surveys, information is taken by interviewers, who transform what people say into their own language. By contrast, diagramming by local people gives them a share in the creation and analysis of knowledge, providing a focus for dialogue which can be sequentially modified and extended. Local categories, criteria and symbols are used during diagramming, which includes mapping and modelling, comparative analyses of local perceptions of seasonal and historical trends ranking and scoring to understand decision making, and diagrammatic representations of household and livelihood systems. Rather than answering questions which are directed by the values of the researcher, local people are encouraged to explore creatively their own versions of their worlds. Visualizations, therefore, help to balance dialogue, and increase the depth and intensity of discussion.

Local people using these methods have shown a greater capacity to observe, diagram and analyze than most professionals have expected. Yet the view that 'they may have worked in country x, but they will not work here' is extraordinarily common. It is almost always wrong, with the problem being the conventional attitudes of the professionals exposed to the methods.

The Trustworthiness of Findings

It is common for users who have presented findings arising from the use of participatory methods to be asked a question along the lines of *'but how does it compare with the real data?'* (see Gill, 1991). It is commonly asserted that participatory methods constitute inquiry that is undisciplined and sloppy. They are said to involve only subjective observations and so respond just to selected members of communities. Terms like informal and qualitative are used to imply poorer quality or second-rate work. Rigour and accuracy are assumed, therefore, to be in contradiction with participatory methods.

This means that it is the investigators relying on participatory methods who are called upon to prove the utility of their approach, not the conventional investigator. Conventional research uses four criteria in order to persuade their audiences that the findings of an inquiry can be trusted (see Lincoln and Guba, 1985; Guba and Lincoln, 1989). How can we be confident about the 'truth' of the findings (internal validity)? Can we apply these findings to other contexts or with other groups of people (external validity)? Would the findings be repeated if the inquiry were replicated with the same (or similar) subjects in the same or similar context (reliability)? How can we be certain that the findings have been determined by the subjects and context of the inquiry, rather than the biases, motivations and perspectives of the investigators (objectivity)? These four criteria, though, are dependent for their meaning on the core assumptions of the conventional research paradigm (Lincoln and Guba, 1985; Kirk and Miller, 1986; Cook and Campbell, 1979).

Trustworthiness criteria were first developed by Guba (1981) to judge whether or not any given inquiry was methodologically sound. Four alternative, but parallel, criteria were developed: credibility, transferability, dependability and conformability. But these *'had their foundation in concerns indigenous to be conventional, or positivist, paradigm'* (Lincoln, 1990). To distinguish between elements of inquiry that were not derived from the conventional paradigm, further 'authenticity' criteria have been suggested to help in judging the impact of the process of inquiry on the people involved (Lincoln, 1990). Have people been changed by the process? Have they a heightened sense of their own constructed realities? Do they have an increased awareness and appreciation of the constructions of other stakeholders? To what extent did the investigation prompt action?

Drawing on these and other suggestions for 'goodness' criteria (Marshall, 1990; Smith, 1990), a set of 12 criteria for establishing trustworthiness have been identified (Pretty, 1994) (Box 6.3). These criteria can be used to judge information, just as statistical analyses provide the grounds for judgement in positivist or conventional science. An application of an alternative system of inquiry without, for example, triangulation of sources, methods and investigators and participant checking of the constructed outputs, should be judged as untrustworthy.

However, it should be noted that it will never be possible to be certain about the trustworthiness criteria. Certainty is only possible if we accept

Box 6.3 A framework for judging trustworthiness

1. *Prolonged and/or intense engagement between the various actors*
   For building trust and rapport, learning the particulars of the context and to keep the investigator(s) open to multiple influences.
2. *Persistent and parallel observation*
   For understanding both a phenomenon and its context.
3. *Triangulation by multiple sources, methods and investigators*
   For cross-checking information and increasing the range of peoples' realities encountered, including multiple copies of sources of information; comparing the results from a range of methods; and having teams with a diversity of personal, professional and disciplinary backgrounds.
4. *Analysis and expression of difference*
   For ensuring that a wide range of diferent actors are involved in the analysis, and that their perspectives are accurately represented.
5. *Negative case analysis*
   For sequential revision of hypotheses as insight grows, so as to revise until one hypothesis accounts for all known cases without exception.
6. *Peer or colleague checking*
   Periodical reviews with peers not directly involved in the inquiry process.
7. *Participant checking*
   For testing the data, interpretations and conclusions with people with whom the original information was constructed and analysed. Without participant checks, investigators can make no claims that they are representing participants' views.
8. *Reports with working hypotheses, contextual descriptions and visualizations*
   These are 'thick' descriptions of complex reality, with working hypotheses, visualizations and quotations capturing peoples' personal perspectives and experiences.
9. *Parallel investigations and team communications*
   If sub-groups of the same team proceed with investigations in parallel using the same system of inquiry, and come up with the same or similar findings, then we can depend on these findings.
10. *Reflexive journals*
    These are diaries individuals keep on a daily basis to record a variety of information about themselves.
11. *Inquiry audit*
    The team should be able to provide sufficient information for a disinterested person to examine the processes and product in such a way as to confirm that the findings are not figments of their imaginations.
12. *Impact on stakeholders' capacity to know and act*
    For demonstrating that the investigation has had an impact, including participants having a heightened sense of their own realities, and an increased appreciation of those of other people; the report could also prompt action on the part of readers who have not been directly involved.

the positivist paradigm (see Chapter 1). The criteria themselves are value-bound and so we cannot say that 'x has a trustworthiness score of y points', but we can say that 'x is trustworthy because certain things happened during and after the investigation'. The trustworthiness criteria should be used to identify what has been part of the process of gathering information and whether key elements have been omitted. Knowing this should make it possible for any observer, be they reader of a report or policy maker using the information to make a decision, also to make a judgement on whether they trust the findings. In this context, it becomes possible to state that the 'data no longer speak for themselves'.

FARMER PARTICIPATORY RESEARCH AND EXTENSION

Farmer Experimentation

Research organizations have a poor record when it comes to participation with farmers. As has been shown in Chapter 2, the central feature of agricultural modernization has been to impose simple technologies on complex environments. If we are to be serious about the development of a sustainable agriculture, it is critical that local knowledge and skills in experimentation are brought to bear on the processes of research.

The problem with agricultural science is that it has poorly understood the nature of 'indigenous' and rural people's knowledge (Scoones and Thompson, 1994). For many, what rural people know is assumed to be 'primitive' and 'unscientific', and so formal research and extension must 'transform' what they know so as to 'develop' them. An alternative view is that local knowledge is a valuable and under-utilized resource, which can be studied, collected and incorporated into development activities. Neither of these, though, is satisfactory. The former is characteristic of the modernizing tendencies in agriculture that emphasize the 'transfer of technology' (Chapter 2), and the latter of the more populist debates about indigenous technical knowledge and 'farmer first' approaches, which seek to ensure that local knowledge is at least given credit and value (Chambers et al, 1989; Röling and Engel, 1989; Warren, 1991; Reijntjes et al, 1992).

More recently, there has been a wider recognition that neither local knowledge nor western science can be considered as unitary bodies of stock of knowledge. Instead they are just different epistemological constructions within particular social, economic and ecological settings (Chapter 1). Knowledges are socially constructed, and so constantly changing and evolving within society. Interactions and changes thus depend on the dynamic interplays between actors and institutions, and the power relationships between them (Long and Long, 1992; Röling, 1988, Scoones and Thompson, 1994). Within this context, understanding processes of agricultural innovation and experimentation has been an important focus.

Farmers have always experimented to produce locally adapted technologies, practices, crops and livestock (Chambers et al, 1989;

Brouwers, 1993; Scoones and Thompson, 1994). They are continuous adaptors of technology and their systems are rarely static from year to year. Paul Richards (1989, 1992) has likened this process of adaptation to a performance, in which the actors change the nature of the performance according to the specific conditions they experience.

The problem is that researchers commonly do not understand or even accept that farmers are experimenters. They assume that farmers are conservative and bound by tradition. Static and unchanging practices can, therefore, upon investigation at a particular time, be characterized, analyzed and so 'developed'. But such an analysis can give nothing better than a snapshot of a complex and changing reality. It is important, therefore, to begin to see technologies in a different light, not as fixed prescriptions but as indicators of what can be achieved. What agriculture needs is a willingness among professionals to learn from farmers. As Robert Rhoades (1987) put it: *'the farming profession requires experimenters, risk takers, innovators; intensifiers and diversifiers, colonisers or pioneers; addicts for new information; and practitioners of great common sense'*.

Another important aspect of change is illustrated by the fact that when farmers are faced with a new technology or practice, they rarely reject all their existing practices. Rather, they take the new and experiment with it. Perhaps a new variety is grown first in the kitchen garden or in a single row along a field boundary. They watch and observe. If the variety proves itself, the farmer increases production. All the while, they maintain their own bank of germplasm and existing practices. Such an approach to experimentation is inevitably to be more adaptive and holistic than normal agricultural science.

David Millar (1993) has described the many different ways that farmers in northern Ghana conduct experiments and how these are determined by, and in turn influence, their way of seeing the world. He describes curiosity experiments, in which farmers see something interesting elsewhere, such as a combination of cassava cropping or the use of camphor to control pests in stored sorghum, and so set up various tests to compare the new with the old. They conduct problem solving experiments, to deal with problems of Striga weeds or post-harvest losses of yams. One farmer, Nafa, said: *'I encountered the problem and I have adopted crop rotations to find out which rotation best fights gill (Striga). I with my three brothers found out that a continuous cultivation of millet on the field for three or more successive years would kill gill. With other farmers, we are trying to see how long it would take gill to come back if other crops are grown after millet'*.

They also conduct adaptive research, modifying the crop technologies passed to them by government research, and peer pressure research dictated by religious and cultural values.

The Case of Farmer Experimentation in Eighteenth-Century Britain

Farmers were the driving force of the agricultural revolution that occurred in rural Britain during the seventeenth to nineteenth centuries (Pretty, 1991). During a period in which there was no government ministry of agriculture, no national agricultural research or extension institutions, no radio or television, no pesticides or inorganic fertilizers, and poor rural transport infrastructure, aggregate cereal and livestock production increased to unprecedented levels. In the 150 years after 1700, wheat production grew four fold, and barley and oats three fold; the numbers of cattle supplied to markets tripled and of sheep doubled (Beckett, 1990; Mingay, 1989; Holderness, 1989; Chartres, 1985). This remarkable achievement was brought about in two ways: the extension and experimentation by farmers of new technologies that intensified on-farm resource use; and the conversion of common pastures and woodlands to private farming.

New crops offered diversified opportunities to farmers by allowing intensified use of land. Increased fodder supply meant more livestock and so increased supply of manures improved soil fertility. Selective breeding of livestock produced more efficient converters of feed to meat, so permitting slaughter at an earlier age and higher stocking rates. New labour-saving machinery released farmers from the labour bottlenecks at cereal and hay harvests; and new tools and techniques improved the efficiency of seed sowing. Underfield drainage increased cropping options on marginal land; and irrigation of watermeadows increased the supply of fodder, particularly during the late winter shortage. Complementarities with urban and industrial growth, the British population having tripled between 1700–1850, also meant increased soil fertility as agriculture assimilated industrial and human wastes.

Until the last two decades, orthodoxy has held that the British agricultural revolution began about 1760 and ended in the early 1800s (Ernle, 1912). Credit for progress was given to a few, now famous, innovators: Tull for his corn drill; Townshend for turnips; Coke for the Norfolk Four Course rotation; Bakewell for livestock breeding; and Young for promoting all of these. The conventional view is that, once exposed to these innovations, the majority of farmers adopted them and the revolution occurred. However, claims for innovation rapidly driving production growth have not survived scrutiny.

What is now clear is that Tull, Townshend, Coke, Bakewell and Young were simply good popularizers rather than innovators. All 'their' innovations were being practised by some farmers 50 to 100 years before they were born. The lasting fascination for 'inventors' has diverted attention away from the process of experimentation, technology diffusion and local adaptation. Yet in the British agricultural revolution farmers were centrally involved in all three processes. Farmers made diffusion active rather than passive through farmer-to-farmer extension mechanisms; and there is considerable evidence that technologies, once adopted, were the focus of experimentation so as to make the appropriate adaptations to suit local conditions.

Farmers conducted field trials to test the efficacy of various manure and nutrient treatments on soils; they tested corn drills against other methods of seed sowing; they introduced new crops into rotations on some fields, while leaving others unchanged; they tested irrigated against dryland meadows; and they tested new methods of pest control. As Caird (1852) put it *'the detail is everywhere varied by the judicious agriculturalist to suit the necessities and advantages of the particular locality'*. Farmers were concerned with integrating the results of experiments into their farm economies and so analyzed results to discover which were the most profitable options. To many, experiments were seen as a necessary part of farming.

Arthur Young said that *'experiment is the rational foundation of all useful knowledge: let everything be tried'* (Young, 1767). He published *Experimental Agriculture* in 1770, comprising some 900 pages of detailed results of 5 years of experiments on 120 ha of various soils. He had begun confidently expecting conclusive answers, but concluded the task in a different mood: *'I entered upon the following experiments with an ardent hope of reducing every doubtful point to certainty; and I finished them with the chagrin of but poorly answering my own expectations. Where I imagined 2 or 3 trials would have proved decisive, 40 have been concluded in vain.'* (Young, 1770)

Robert Bakewell's approach to experiment was open-minded: *'I would recommend to you and others who have done me the credit of adopting my opinions to pursue it with unremitting zeal as far as shall be consistent with prudence and common sense, always open to conviction when anything better is advanced.'* (Bakewell, 1787 in Pawson, 1957)

And George Culley, in a letter in 1801, wrote, *'I often say that we have a deal to learn yet. And every wise humble man will learn every year and every day'* (Macdonald, 1977).

And yet these considered comments of farmers seem to have been very largely forgotten since the end of the agricultural revolution. They conflict with the predominant view of the agricultural experiment, namely that it is the domain of scientists and takes place solely on the research station or in the university. Many take the view that 'scientific' agriculture began with the establishment of the Royal Agricultural Society of England in 1838 and Rothamsted Experimental Station in England in 1843. Despite the immense benefits to agriculture they have brought, they have also served to hide the experimental practices of farmers. The result is now a deeply held belief that the first scientific experiments occurred only after the 1840s. A recent text on the history of agricultural science in Britain begins at 1840 (Rossiter, 1975); and two earlier books by E John Russell, a former director at Rothamsted, suggest that the 'first experiments' began in earnest at Rothamsted, before which any experiments were conducted by academics working alone. In neither of his seminal books is the role of farmers once mentioned (Russell, 1946, 1966).

Farmer Adaptations to Scientists' Designs

The problem with modern agricultural science is that technologies are finalized before farmers get to see them. Clearly, if the technologies are appropriate and fit a particular farmers' conditions or needs, then they stand a good chance of being adopted. But if they do not fit and farmers are unable to make changes, then they have only the one choice. They have to adapt to the technology or reject it entirely. And such rejection is common. The history of development interventions is littered with examples of bright new technologies rapidly tarnished by lack of widespread adoption or maintenance. Ask any farmer or development professional and they will tell you of tractors inoperable for the sake of a key spare part, of terraces broken or degraded, of irrigation systems in disrepair and so on.

The alternative to these scenarios is to seek and encourage the involvement of farmers in adapting technologies to their conditions. This constitutes a radical reversal of the normal modes of research and technology generation, as it requires interactive participation between professionals and farmers. The term participatory technology development (PTD) has been applied to the process and methodology by which various partners cooperate in technology development (Jiggins and De Zeeuw, 1992; Reijntjes et al, 1992; Haverkort et al, 1991). It is a process in which the knowledge and research capacities of farmers are joined with those of scientific institutions; while at the same time strengthening local capacities to experiment and innovate. Farmers are encouraged to generate and evaluate indigenous technologies, and to choose and adapt external ones on the basis of their own knowledge and value systems.

But, of course, researchers and farmers participate in different ways, depending on the degree of control each actor has over the research process (Table 6.3). The most common form of 'participatory research' is researcher designed and implemented, even though it might be conducted on farmers' fields. Many on-farm trials and demonstration plots represent nothing better than passive participation. Less commonly, farmers may implement trials designed by researchers. But greater roles for farmers are even rarer. Sam Fujisaka (1991a) describes researcher-designed experiments on new cropping patterns in the Philippines. Even though farmers 'participate' in implementing the trials, there was widespread uncertainty about what researchers were actually trying to achieve. Farmers misunderstood experiments, and rejected the new technologies. The reason, as he explains, was that *'co-operation between farmers and researchers implies two groups continually listening carefully to one another. Claveria farmers are avid listeners to... researchers. The challenge is for all on-farm researchers to complete the circle'*.

Where the technology is not indivisible, farmers are more likely to try it, adapting it through experimentation to their conditions. A now classic case is the diffused light stores for potatoes developed with farmers by Robert Rhoades and colleagues at CIP in Peru. In the mid-1970s, a CIP seed specialist, Jim Bryan, observed farmers in Kenya, Peru and Nepal storing potatoes in diffused light. He assembled a collection of slides on traditional storage practices and convinced colleagues at CIP to investigate these practices. After considerable on-station and on-farm research, the technologies developed were introduced into some 25 countries (Rhoades and Booth, 1982).

But Rhoades and Booth were surprised to find that *'adoption had not proceeded as we expected and certainly not as the sociological adoption literature indicated'* (Rhoades, 1987). Out of some 4000 cases checked, some 98 per cent of farmers had changed the basic technology to adapt it to their own farming conditions, household architecture and budgets. In particular,

Table 6.3 Types of participatory research

Designed by	*Implemented by*	*Comments*
Researcher	Researcher	The most common form of research: on farm trials and demonstration plots
Researcher	Farmer	The most common form of 'participatory' research
Farmer	Researcher	Very rare
Farmer	Farmer	The mode of farmers' own research and experimentation; very rare in programmes; some village or community organizations doing themselves

Source: adapted from Biggs, 1989

they found that farmers did not drop their old storage practice (seeds kept in darkness), but simply incorporated the use of diffused light storage alongside existing practices. They frequently adopted only the elements of the package that interested them, actively playing a role in the design and alteration of stores to their conditions. Partial adoption was, therefore, not a failure or 'incomplete', but effective if it worked for the farmer.

Even if changes are not permitted, farmers will try anyway. The Dumoga Irrigation Project in Indonesia illustrates what happens in conventional irrigation projects, in which technical information on soils, landform and natural waterways is used to construct the design. Local knowledge on prior use of waterways, farmers' own structures and boundary patterns is rarely incorporated. Douglas Vermillion (1989) describes what happened during implementation: *'The farmers interviewed frequently reported approaching construction labourers or supervisors in the field to suggest changes and were usually told that the design has been established by the government and could not be changed. Often farmers relocated the construction markers when the crews had left. Others waited until construction was finished and the contractors had moved on, before altering the structures'*.

Farmers made many kinds of alterations. They relocated channels, diverted or ponded streams, abolished project channels, redirected channels into streams, made new flumes, destroyed project flumes and made use of existing structures. One of the farmers' main objectives was to minimize the number of channels and maximize the reuse of water. Farmers frequently redirected water into natural streams, which were checked to make ponds, so that the water could be reused downstream. But still the project did not permit this, as it defined *'all natural streams as drainage ways. Every six months it routinely destroyed farmer-built brush weirs along small streams and natural depressions within the command area with the intent of 'normalizing the drainageways' to prevent obstruction of drainage'* (Vermillion, 1989).

When farmers are able to modify technologies, adapting them to their local conditions, then they are often able to make significant improvements. It was the active participation of farmers in irrigation design and implementation that made the National Irrigation Administration in the Philippines so successful in its work (see Case 18, Chapter 7).

Even when technologies are successful, farmers may still want to change them. In the Philippines, farmers adopting rice–fish farming systems developed by researchers have been able to increase rice yields by 4 per cent per crop, raise vegetables on the banks of raised dikes and so raise their annual incomes from US$142 to $578 per hectare (de la Cruz et al, 1992). Despite this, half of the cooperating farmers were not satisfied with the growth rate and size of the tilapia fish at harvest: *'they thought they would be able to improve the size of the fish by themselves, after the project was over'*. They subsequently made improvements and modifications to the type of fish, to the sizes and locations of ponds and trenches in the fields, and to the techniques for pest management that avoided fish mortality.

Impact of Participation on Research Systems

It is not just farmers who benefit when research is participatory. Researchers benefit too. They learn more about technologies, as farmers are able to test them in a wide variety of conditions. They have the satisfaction of knowing that technologies they produce really are what farmers want. They also develop better lines of communication. Once researchers appreciate that there are multiple sources of innovation, then they greatly increase the opportunity of helping to improve farmers' livelihoods (Biggs, 1989; Bebbington, 1991). In this sense, change can come from joint learning that challenges perceptions, thoughts, and actions of both the researchers and farmer participants.

Research conducted at the Pakhribas Agricultural Centre in the hills of eastern Nepal showed that lentils can be successfully grown after rice or relayed under rice in the irrigated lowland (*khet*) in mid-altitude areas (Chand and Gurung, 1991). Packages of lentil seeds with instructions were distributed among farmers in seven districts. From his experiments, one farmer discovered not only that lentils did poorly on paddy land but was able to provide feedback to researchers on new environments for lentil cultivation (Box 6.4).

A little bit of structured learning can result in wholesale changes in the way research institutions focus their research. Scientists of the Tamil Nadu Agricultural University recently discovered that farmers prefer red rice varieties over white (Manoharan et al, 1993; TNAU/IIED, 1993). Years of research have resulted in the release of about 100 varieties from research stations, of which only 2 were red. Questionnaire surveys had regularly 'confirmed' that farmers preferred white rice. Yet when scientists began using participatory methods, especially matrix scoring, to understand local preferences for rice, they discovered that white rice was recognized to be higher yielding, but disliked for taste, fineness and lack of nutritive value. Villagers said that *'their physique had come down due to consumption of white rice'* (Manoharan et al, 1993).

Farmers therefore still grew red rice, even though it is poor yielding.

Box 6.4 Farmers experimenting in the hills of Nepal and providing new information for researchers

A farmer who received 500 g of lentil seed in a package from Pakhribas Agricultural Centre research station planted one-third of the seed on *khet* land and on paddy bunds in July. He thought it would grow well on the paddy bund, as do other legumes such as soybean and black gram. Unfortunately, the farmer discovered that lentils cannot be grown during the summer; his crop was heavily infested with summer weeds and his plants did not grow well in heavy summer rain. The farmer then intercropped one-third of the seed with potato during January in a high-altitude maize system. The crop did not grow well again, this time because of the cold and, at a later stage, damage by pre-monsoon rain.

The farmer continued his experimentation and planted the remaining seed during the first week of September, after harvesting potato in a potato–maize cropping pattern. The growth of the lentil was good and the crop utilized residual moisture for its development. The farmer thus was able to harvest lentil successfully during February. In this way, the farmer not only discovered the proper planting time for lentil, he also provided feedback to the researchers that lentil can be grown successfully at high altitudes where a potato–maize system is practised and land is kept fallow during the winter season.

Farmers growing lentil also learned to mix lentil biomass with kitchen waste to feed to cattle and buffaloes, the milk yields of which increased by 20%.

Source: Chand and Gurung, 1991

They said they needed a single red variety with bold red grain, high grain and straw yields, with resistance to pests and diseases. Research efforts on red rice improvement have now intensified to meet these needs. The director of research, Dr S Chelliah recently described the success in this way: *'This is one example of the success of the participatory rural appraisal (PRA) approach and I am confident that this would lead to the identification of many new research priorities which will be location-specific and field-oriented reflecting farmers' needs and preferences'*.

Such improved communication between scientists and farmers led directly to improved returns from sheep and goats in Brazil (Baker et al, 1988). With the use of Regular Research Field Hearings (RRFHs), an approach that emphasized close and regular contact with farmers' groups, mutual trust and understanding grew. With this better dialogue, farmers learned more about the uses and necessary adaptations of the technology, which comprised drenching, vaccinating against diseases, umbilical cord cutting and treatment with iodine, castrating males not needed for reproduction and regular visits by a veterinarian. Scientists also learned more about the farmers' conditions. Where these RRFHs were used, the daily weight gain of the animals was 24 per cent greater than in groups who received the technical package without the dialogue. Farmers were also more satisfied with the performance of the animals and were willing to pay some 36 per cent more for the animals.

In recent years, there have been many similar mechanisms developed in national agricultural research and extension systems that have systematically increased connectivity and collective learning (Box 6.5). However, these changes have rarely been spread through large institutions. In their analysis of national agricultural research systems, Merrill-Sands and Collion (1992) concluded that *'it is fair to assert that although farming systems and farmer participatory methods have in many cases led to more client-responsive research scientists, they have in few cases resulted in more client-responsive research organizations.'*

From Directive to Participatory Extension

A similar approach to involving local people is needed in extension systems, where the challenge is just as great as in research institutions. Extension has long been grounded in the 'diffusion' model of agricultural development, in which technologies are passed from research scientists via extensionists to farmers (Rogers, 1962). Farmers who choose not to adopt are often labelled by extensionists as 'laggards' with attitudinal barriers (Russell et al, 1989; Chambers and Ghildayal, 1985). In the late 1980s, there were some 540,000 extension personnel worldwide (Swanson et al, 1990). But most of these work in systems that ignore local groups and institutions. The tendency has been to deal with individual farmers or households, who are selected on the basis of likelihood of adopting new technologies. They are, in turn, expected to encourage further adoption in their community through a demonstration effect.

This approach is exemplified by a type of extension that came to be known as the training and visit (T and V) system. It was first implemented in Turkey in 1967 and later widely adopted by governments on the recommendation of the World Bank (Benor et al, 1983; Roberts, 1989). It was designed to be a management system for energizing extension staff, turning desk-bound, poorly motivated field staff into effective extension agents. Extension agents receive regular training to enhance their technical skills, which they then hope will pass on to all farmers through regular communication with the smaller number of selected contact farmers. Between 1977–92, the World Bank disbursed US$3000 million through 512 projects for extension systems along the lines of the T and V model (World Bank, 1994). Although a substantial sum, this represents just 5 per cent of the World Bank's lending to the agricultural sector during this period.

But as the contact farmers are usually selected on the basis of literacy, wealth, readiness to change and 'progressiveness', this often sets them apart from the rest of the community. The secondary transfer of the technical messages, from contact farmers to community, has been much less successful than predicted and adoption rates are commonly very low among non-contact farmers. Without a doubt, T and V is now widely considered as ineffective (Box 6.6) (Axinn, 1988; Howell, 1988; Russell et al, 1989; Moris, 1990; Röling, 1991; Antholt, 1992, 1994; Hussain et al, 1994).

What is not clear is whether all these problems were due to the T and V system itself or to the way it was institutionalized (Antholt, 1992, 1994). T and V was usually associated with large increases in staff, yet extension

Box 6.5 Selection of successful innovations in national agricultural research and extension systems

A selection of innovations include:

- working groups, research teams and joint interdisciplinary treks based in Lumle and Pakhribas Agricultural Centres, Nepal;
- farmer field schools in Indonesia, Philippines and Honduras;
- catchment approach to participatory planning and implementation of soil and water conservation, Ministry of Agriculture, Kenya;
- Adaptive Research Planning Teams and village research groups, Ministry of Agriculture, Zambia;
- farmer groups for technology research and extension in Ministry of Agriculture, Botswana;
- innovator workshops in Bangladesh and India, in which farmers come to workshops attended by researchers to talk about their innovations;
- linking with farmer and community groups for Landcare, Australia;
- policy analysis network of universities in Nepal, coordinated by Winrock International;
- National *Azolla* Action Programme, Philippines;
- participatory planning and research design, Pakistan Agricultural Research Council;
- teams of female bean experts working with plant breeders in Rwanda;
- participatory research teams, Tamil Nadu Agricultural University, India;
- farmers working in groups and feeding information directly to local radio programmes in Peru;
- group farming in Kerala, India;
- farmer groups for technology adaptation and extension, Narendra Deva University of Agriculture and Technology.

Sources: Chand and Gurung, 1991; Mathema and Galt, 1989; Kiara et al, 1990; MALDM, Kenya, passim; Pretty et al, 1994; Sikana, 1993; Drinkwater, 1992; Heinrich et al, 1991; Abedin and Haque, 1989; Campbell, 1994b; San Valentin, 1991; Guijt and Pretty, 1992; TNAU/IIED, 1993; Shereif, 1991; Maurya, 1989; Sperling et al, 1993; AED, 1991

departments have rarely had the resources to keep them in the field working with farmers. In Tamil Nadu, for example, the number of village extension officers increased from 1730 to 4000 with the adoption of T and V, but the resources available for demonstrations in 1991 amounted to about only US$1 per year per extension worker.

More importantly, though, T and V has tended deeply to institutionalize extension's top-down hierarchy, so preventing extension systems from being learning organizations. Bureaucrats liked the system, because it could be used to hold staff accountable. But higher level staff and research scientists have severely inhibited the upward flow of information, despite early intentions to do so. As Charles Antholt (1991) put it: *'Time-bound, centrally-determined, highly-concentrated work programmes can sometimes, but not always make sense under homogenous conditions. But the realities of most agricultural systems... are rather different. Given the seasonality of workloads, the heterogeneity of agroecological systems, the complex choices facing farm families...*

Box 6.6 The impact of training and visit extension in a range of contexts

- In Somalia, only one non-contact farmer adopted a high-input package for each contact farmer, a ratio much lower than the 10:1 expected; this was despite the fact that maize and sorghum yields were 40–45% greater on contact farmer fields.
- In Kerala, India, non-contact farmers have been found to have very little contact with contact farmers, preferring to consult a wide range of alternative information sources, such as newspapers and the mass media, and fellow farmers.
- In Andhra Pradesh, T and V was found to have had no effect on agricultural productivity.
- In West Bengal, Bihar, Maharashtra, and Tamil Nadu, all of which have had T and V for at least ten years, no causal connection was found between incremental investment in T and V and incremental changes in agricultural production.
- In Nepal, ten years of T and V in the Terai was found to have had no impact on wheat yields.
- In Bangladesh, T and V was not successful in achieving any positive changes in the orientation of extension towards local people, despite this being a major objective when introduced.
- In Indonesia, T and V made no impact on non-rice dryland crops.
- In Pakistan, T and V had no impact in Punjab province, focusing too little on increasing the relevancy of technology for farmers.

Sources: Antholt, 1992, using various World Bank evaluations; Axinn, 1988; Mullen, 1989; Chapman, 1988

extension services must be much more flexible, more timely, and less centralized'.

Important lessons have been learned from the problems associated with T and V, and there is clearly a need to address the systemic issues facing extension (Zijp, 1993; Antholt, 1994). Extension will need to build on traditional communication systems and involve farmers themselves in the process of extension. Incentive systems will have to be developed to reward staff for being in the field and working closely with farmers. There must be a *'well-defined link between the well-being of field officers and the extension system, based on the clients' view of the value of extension's and field workers' performance'* (Antholt, 1992). Achieving such a vision will need a complete overhaul of the notion of extension. It may be that the time has come to abandon the term extension altogether, as it implies passing something from someone who knows to someone who does not. Participation, if it is to become part of extension, must clearly be interactive and empowering. Any pretence to participation will result in little change. Allowing farmers just to come to meetings, or letting a few representatives sit on committees, will be insufficient.

There have been some recent innovations in introducing elements of farmer participation and group approaches into extension, and these have already had a significant impact. Differences in impact between individual and group approaches have been well documented in both Nepal and Kenya (Sen, 1993; Eckbom, 1992; SWCB, 1994). In Western Nepal, Sen compared the rate of adoption of new technologies when extension worked with individuals or with groups. With groups, there was better communication between farmers and extensionists, and so more adoption. When the individual approach was resumed after the experiment, adoption rates fell rapidly in succeeding years.

In Kenya, the Ministry of Agriculture is increasingly adopting a community-oriented approach to soil and water conservation (see Case 12, Chapter 7). This is steadily replacing the former individual approach of the T and V system. One particularly important study compared the impact of the catchment approach with the individual T and V in two neighbouring communities in Trans Nzioa. For a wide range of indicators, farmers' livelihoods were more improved where the community approach was implemented (Table 6.4). Such impacts have been confirmed by other ministry self-evaluation and monitoring studies (SWCB, 1994; Pretty et al, 1994; MALDM, passim). Where extension staff interact closely with communities in developing joint action plans and local people freely elect members to a local catchment committee, then the impact on agricultural growth is immediate and sustained. Strong local groups mobilize the interest of the wider community and sustain action well beyond the period of direct contact with external agents (see Chapter 5).

TOWARDS LEARNING ORGANIZATIONS

A systematic challenge for agricultural research, extension and planning institutions, whether government or non-government, is to institutionalize approaches and structures that encourage learning.

Narayan's study of the importance of participation in water projects indicated that whether people participated or not was influenced as much by factors in external institutions as local needs or interests (Narayan, 1993). Those that succeed were characterized by a prior orientation or value towards local people, responsiveness during implementation and giving up to or sharing decision making control with local communities. In particular, if the agencies actively used local knowledge, made participation a goal, and then monitored and rewarded it, they were more likely to succeed.

There is much we can learn from the private and corporate sector (see, for example, Thompson and Trisoglio, 1994; Easterby-Smith, 1992; Peters, 1987; Peters and Waterman, 1982; Argyris and Schön, 1978). It is increasingly recognized that organizations that succeed in a changing and increasingly complex world are also those that have the ability to learn from their experiences, and adapt quickly. The central difference between the private and public sectors is that if a private company fails to learn from its clients (those who pay for the products or services) then it will close down. Institutions unable to learn do not recognize incrementally changing circumstances until a crisis suddenly manifests itself. But, at the moment, if an agricultural research organization fails to provide

Table 6.4 Comparison between the impact of the catchment approach and the individual farmer approach in neighbouring catchments at Geta Farm, Cherangani Division, Trans Nzoia, Kenya

	Catchment approach	*Individual farmer approach*
Annual value of crop production, 1990 (KSh)	14,260	13,470
Variable costs, 1990 (KSh)	3910	3450
Gross margin, 1990 (KSh)	8100	7860
Gross margin per person-day (KSh)	13	8.3
Annual increase in productivity	12%	8%
Net present value (4% discount rate) in KSh	114 million	54 million
Benefit/cost ratio	1.75	1.27
Increase in land value: 1986–91	62%	29%
Increase in land value: Annual	10.1%	5.3%
Change in leasehold prices, 1986–91	+ 8%	–10%
Increase in average livestock holding	45%	15%
Milk consumption ratio	1.6	1

Source: Eckbom, 1992

technologies that farmers desire, it will not close down.

There are three areas in which agricultural institutions can improve their ability to learn. They can promote experimentation; promote connectivity and group work based on roles rather than disciplines; and develop monitoring and self-evaluation systems to improve learning and awareness.

Promote Experimentation and Diversity

An experimental approach often involves the taking of risks, and so tends to fly in the face of the instincts of senior staff and the procedures of most organizations. Most try to select people who will fit in with their particular cultures. But too much homogeneity makes it more difficult for new ideas to emerge. It is well established that creativity and innovativeness tend to be generated where diversity of experience is encouraged, particularly at lower levels (Nonaka, 1988; Easterby-Smith, 1992; Williams and Antholt, 1992).

A good sign is that agricultural research and development has become increasingly diverse in recent years, with a growing number of disciplines engaged. Robert Rhoades (1989) characterizes four overlapping periods of steadily shifting emphases. These stages are as follows.

- **Production stage** (roughly 1950–75), in which the pioneering disciplines were breeding and genetics, and farmers were seen as recipients of technology.
- **Economic stage** (roughly 1975–85), in which Farming Systems Research was pioneered by economists and agronomists, and farmers were seen as sources of information for technology design.

- **Ecological stage** (roughly 1985–95), in which anthropology, agroecology and geography are pioneers, and farmers contribute their indigenous knowledge, and are seen both as victims and causes of unsustainable development.
- **Institutional stage** (roughly 1995 onwards), in which the pioneering disciplines will be management specialists, psychologists, organizational sociologists, political scientists, training specialists and educators; in which farmers will be full collaborators in research and extension; and in which alliances will be developed between different institutions.

Even though each wave of enthusiasm for a new approach has grown out of antecedents, there has been a tendency for those who pioneer and embrace each new direction to play down the accomplishments of earlier approaches, and so *'the 'old' always argue that the 'new' is not so new ('we were doing it all along') while the 'new' fiercely defends what it perceives to be the wave of the future'* (Rhoades, 1989). Precisely how the new disciplines and skills would work together is often unclear, but as Rhoades put it; *'this should not be a cause for alarm, given that early in every new stage no one was able to appreciate the vast bodies of methods and theories available in disciplines still marginal to the agricultural research and development establishment'*.

But increasing diversity may not alone be sufficient, as changes in attitudes will be essential for any long-term change. The problem for senior officials with creativeness and innovation is that they must recognize that many new initiatives are bound to fail. It is therefore crucial that people do not get punished for taking risks that might fail. Easterby-Smith quotes the chief executive of a large corporation, who said that staff have *'the freedom to do things and fail. We accept that... But at the same time we expect any mistakes to have originality. We can afford any mistake – once'*.

In the long term, innovative behaviour will only persist if it is rewarded. The challenge is for institutions to have rewards, whether financial or promotional, so that those who do not take risks are not rewarded. Similarly, those who make mistakes should not be disadvantaged. Such incentives are needed in research and extension institutions if scientists are to be encouraged to adopt alternative methods of working with farmers.

Promote Connectivity and Group Work

Learning organizations need collective analysis and good connections between their various parts. By working in groups, staff can approach a situation from different perspectives, carefully monitor one another's work and carry out a variety of tasks simultaneously. Groups can be powerful and productive when they function well, and their outputs are likely to be greater than the sum of their individual members.

Although groups generally produce fewer ideas than individuals working separately, they usually generate better ideas as each is discussed and thought through more deeply. Groups are more likely to identify errors of judgement before action is taken. Discussion stimulates more careful thinking and leads to consideration of a wider range of ideas. Rather surprisingly, groups take riskier decisions than the individuals comprising them would have done if acting independently. Individuals are more adventurous as they can take courage from their fellow group members.

But several people brought together to work on a single research or development activity do not necessarily make a productive team. Charles Handy (1985) has suggested that before a group of people can function well as a team, they must pass through a series of stages: 1. forming; 2. storming; 3. norming; and 4. performing. First, various individuals come together, sometimes as strangers, sometimes as colleagues, to create a new group for some stated purpose. In this early *forming* stage, they are still a collection of individuals, each with their own agenda and expertise, and little or no shared experience. As these individuals become more familiar with one another, the group will enter a *storming* phase. Now personal values and principles are challenged, roles and responsibilities are taken on and/or rejected, and the group's objectives and mode of operating will start to be defined more clearly. If there is too much conflict and discord within the group, it will collapse. But if some common ground can be found, the group will gain greater cohesion and a sense of purpose.

As the group members begin to understand their roles in relation to one another and establish a shared vision or goal, they will develop a clearly discernable identity and group-specific norms of behaviour. At the *norming* stage, the group has settled down. People know each other better, they have accepted the rules and probably developed little subgroups and friendship pairs. Once these norms have been established, the group will be ready for action and will enter into the *performing* phase. It is in this phase that they will work most effectively as a *team*. This team has a life of its own; its power to support learning will be quite considerable. The confidence level of the team members will have reached the point where they are willing to take significant risks and try out new ideas on their own. It is also the stage when learning will be greatest.

Recent research by Meredith Belbin has led to a greater understanding of the mix of roles needed to make groups work (Box 6.7). It is assumed that nobody is perfect, as each of us has at least one allowable weakness. But a team can be almost perfect, as it frees individuals to concentrate on their strengths. The best teams have a wide mix of roles and functions represented, while teams consisting wholly of one type, however brilliant the individuals, can be disastrously ineffective. For example, a team with several *shapers* could stay locked in conflict, whereas too many *fixers* will produce a team good at gathering information and making contacts but poor at reflecting and implementing.

Two factors make it hard for institutions to produce effective teams based on a full mix of roles and functions. The first is the tendency to select individuals that match the image of the organization. This results in too many people of the same type. For a team leader or manager, it may be necessary to select people they may not like as individuals, but who will be key team members. The second is that all too often individuals are selected according to qualifications, their apparent eligibility for the position, rather than their suitability in terms of their potential team contribution.

Box 6.7 The nine roles required for a 'perfect' team

The Coordinator: The team's natural chairperson; confident, talks easily, listens well; promotes decision making; able to elicit contributions from all team members; need not be a brilliant intellect.
Allowable weakness: somewhat manipulative.

The Spark: The team's vital spark and chief source of ideas; creative, unorthodox, imaginative.
Allowable weakness: lacks practicality, a bit of a handful, up in the clouds.

The Implementer: The team's workhorse; turns ideas into practical actions and gets on with them logically and loyally; disciplined, reliable, conservative.
Allowable weakness: can only adapt if told why, lacks imagination.

The Fixer: The team's extrovert; amiable, good at making and using contacts; an explorer of opportunities.
Allowable weakness: undisciplined, short attention span.

The Shaper: Usually the self-elected leader; dynamic, positive, outgoing, argumentative, a pressurizer; seeks ways round obstacles.
Allowable weakness: not always likeable, tendency to bully, provokes opposition.

The Monitor Evaluator: The team's rock, strategic, sober, analytical, introvert; capable of deep analysis of huge quantities of data; rarely wrong.
Allowable weakness: an unexciting plod, lacks imagination.

The Teamworker: A counsellor and conciliator; social, perceptive, accommodating, aware of undercurrents and others' problems; promotes harmony; most valuable at times of crisis.
Allowable weakness: indecisive.

The Completer Finisher: The team's worrier and stickler for detail, deadlines and schedules; has relentless follow-through; chief catcher of errors and omissions.
Allowable weakness: reluctant to let go, worries about small things.

The Specialist: The team's chief source of rare knowledge and skill; a single-minded loner; self-starting, dedicated and makes the occasional dazzling breakthrough.
Allowable weakness: contributes on a narrow front.

Source: adapted from Meredith Belbin, 1992

The challenge for organizations involved in developing a more sustainable agriculture is to incorporate these principles of group work in the mechanics of participation and institutional collaboration. To do this, there needs to be good connectivity between different components of institutions, so that learning can be shared and distributed. Such connectivity and the development of institutional memory is enhanced through mechanisms for monitoring and self-evaluations.

Monitoring and Self-Evaluations to Improve Learning

Most institutions have mechanisms for identifying departures from normal operating procedures. This is what Argyris and Schön (1978) call single loop learning. But most institutions are very resistant to double-loop learning, as this involves the questioning of, and possible changes in, the wider values and procedures under which they operate. For organizations to become learning organizations, they must ensure that people become aware of their own processes of learning from both mistakes and successes.

Institutions can, therefore, improve learning by encouraging systems that develop a better awareness of information. What they need is to be assured of the quality of external information. The best way to do this is to be in close touch with external environments, such as farms, villages and communities, that generate this information. Professionals must spend time out of the office and in the field. They should also be aware of the fact that we all tend to focus on good rather than bad news.

Good external information must be supported by good internal handling of information. What happens to the information? Who has access to it? Who accepts or rejects it? Many institutions get locked into an unproductive cycle of information manipulation that has been characterized by Chris Argyris in the following way: senior managers are fed with the positive that they want; this is used to confirm the views they already hold; there is therefore little open testing of ideas; and so people become cynical and defensive, and become less willing to provide accurate information (in Easterby-Smith, 1992).

Breaking this cycle can only be done with a genuine commitment to participative decision making, combined with participatory analysis of performance. Most development institutions rely on conventional evaluations to monitor performance. As the subjects of the evaluation are not involved in the process, they feel defensive and hide any errors. Instead of honestly reflecting on the past, they tend to be forced on to the defensive. They then feel little or no ownership in the final recommendations and corrective measures. And so there is little or no learning.

Alternatives to this common scenario are the increasingly important fields of self-evaluation and participatory monitoring. These approaches are designed to improve learning, and have been effective in the Philippines, India, Sri Lanka and Thailand (Lauraya et al, 1991; SDC, 1992; Uphoff, 1992b; Sommer, 1993). Norman Uphoff (1992b) describes what happens when such mechanisms were put in place in the context of irrigation management: *'Government personnel started working more conscientiously and effectively once they came to know the real conditions at the village level through a systematic monitoring and evaluation system. Moreover, there was a dramatic change in local people's collective and individual behaviour once they knew with some precision, and in a comparative way over time and across jurisdictions, how well they were meeting basic needs'.*

But such feedback through regular self-monitoring has been easier in NGOs than in national agricultural systems. Recent studies of some 30 on-farm research and FSR projects in a wide range of countries have found that learning was generally weak (Merrill-Sands et al, 1991; Byrne, 1989; Tripp, 1991). Despite *'a decade of rhetoric about feedback'*, few agricultural institutions have managed to become learning organizations (Merrill-Sands et al, 1991).

One successful application has been in a project in southern India involving partnerships between a range of institutions (Box 6.8; and see Case 10, Chapter 7). A self-evaluation process was designed as an inter-institutional learning process, so that each group of institutions and individuals could reflect on both their own domains of responsibility and the organization of the project, and then develop options for individual and collective change. External evaluations of complex development programmes commonly create stress and uneasiness among those being evaluated. Instead of following this common route, the process led to new commitments and policies to help future practice and consolidate existing partnerships.

Box 6.8 The self-evaluation process in the Participative Integrated Development of Watersheds (PIDOW) project, Karnataka, India

For close to a decade, four sets of partners have worked together in the PIDOW project: the NGO Myrada, the Drylands Development Board of the Karnataka government, the village level sanghams and the donor, Swiss Development Cooperation. The project works in semi-arid Gulbarga District on the rehabilitation of micro-watersheds, focussing on community development approaches to soil and water conservation, afforestation, dryland agriculture and credit support. The main issues for analysis during the self-evaluation were financial viability, community participation, technical appropriateness, institutional collaboration, equity, sustainability and replication. The self-evaluation involved field-based analyses using participatory methods, plus a range of workshops with villagers and policy makers.

The impacts of the self-evaluation process were as follows:

- the process released many critical and self-critical feelings among all the partners;
- it gave an opportunity to reconfirm past achievements;
- it came up with the proposed emergence of Apex-Sanghams for farmers to air policy concerns;
- it was a forum for exploring conflicts over technical issues. Boulder bunds, for example, had been imposed rigidly by government staff, but farmers said their trapezoidal design was better. They said it could be raised if needed, it required less land and material, and was just as stable if well constructed. This traditional design was recognized by the government at the policy workshop: *'the indigenous practices when found useful may be incorporated into micro-watershed planning in conjunction with recommended government practices'*;
- it led to commitment from government to revise the legal framework regarding management and access to resources on the upper catchment public lands in favour of local communities.

Source: Sommer, 1993

A New Institutional Context

The implications of all this are fundamental for most agricultural and rural organizations. The new learning organizations or clumsy institutions (Shapiro, 1988) should be more decentralized, with an open multi-disciplinarity, flexible teams and heterogeneous outputs responding to demand-pull from farmers. They should be able to operate in turbulence (Roche, 1992). Personal promotion and institutional survival should depend more on external achievement. These new institutions will have to have realistic and rapid feedback flows, so as to make adaptive responses to change. This learning environment should focus on problem solving, and so be interactive and field based. Time will have to be built in for professionals to reflect on what has happened and how they might do things differently next time. This means that the multiple realities and complexities will be better understood through multiple linkages and alliances, with regular dialogue and participation between all actors (Table 6.5).

TOWARDS A NEW PROFESSIONALISM FOR AGRICULTURE

Learning and Teaching

The central concept of sustainable agriculture is that it must enshrine new ways of learning about the world. But learning should not be confused with teaching. Teaching implies the transfer of knowledge from someone who knows to someone who does not know. Teaching is the normal mode of educational curricula and is also central to many organizational structures (Ison, 1990; Argyris, 1991; Russell and Ison, 1991; Bawden, 1992, 1994; Pretty and Chambers, 1993a). Universities and other professional institutions reinforce the teaching paradigm by giving the impression that they are custodians of knowledge which can be dispensed or given (usually by lecture) to a recipient (a student). Where teaching does not include a focus on self-development and enhancing the ability to learn, then '*teaching threatens sustainable agriculture*' (Ison, 1990).

But teaching itself can impede learning. Professionals who are to work for a more sustainable agriculture must be able to let go of certain ideas, and adopt new ones as situations and they themselves change: '*No one learns who claims to know already in advance*' (Rahnema, 1992). But the existing policy culture '*gives credibility to opinion only when it is defined in scientific language, which may not be adequate to describe human and social experience, and this has alienated people. This is not usually the fault of scientists themselves; it is a function of the form of science, including social science, that has been allowed to dominate.*' (Wynne and Mayer, 1993)

Because of the widespread failure of the formal educational sector to provide the necessary learning environments for the development of new professionals, it has been other institutions which have led the way. These have chiefly been NGOs from both the North and the South. Enlightened individuals in government organizations, NARSs and CG institutes, and

Table 6.5 Comparison between old and new institutional settings

	From the old institutional setting	*To the new institutional setting*
Mode of decision making	Centralized and standardized	Decentralized, flexible and participatory
Mode of planning and delivery of technologies or services	Single design, fixed packages, supply-push	Evolving design, wide choice, demand-pull
Response to external change	Collect more data before acting	Act immediately and monitor consequences
Mode of field learning	Field learning by 'rural development tourism' and questionnaire surveys; error concealed or ignored	Learning by dialogue and systems of participatory learning; errors not punished
Mode of internal learning	Single-loop learning at best; misleading feedback from peripheries gives falsely favourable impressions of impact	Double-loop learning with time for reflection on experience; use of participatory monitoring and self-evaluations
Importance of creativity	Suppressed if a threat to existing structures and procedures	Experimentation encouraged and original mistakes not punished
Connectivity, linkages and alliances	Institutions work in isolation; individuals in institutions work alone	Institutions linked formally and informally to each other; individuals linked in task forces and informal groups

farmers have also played their part. The investment is not in knowledge, in the formal sense, but in attitudes, behavioural changes and facilitation skills. Training is centred on learning by doing and bringing scientists, extensionists and farmers together to negotiate and learn from each other on personal level. This is quite different to the way normal universities and colleges work.

What learning organizations should seek is to ensure the generation of timely, relevant, and agreed information and knowledge that will support the quest towards a sustainable agriculture. This will occur when we can find ways of developing both new institutional arrangements and alliances to encourage wider involvement, and a new professionalism with greater emphasis on the process of learning (and unlearning) itself.

From the Old to the New Professionalism

A move from a teaching to a learning style has profound implications for agricultural development institutions. The focus is less on *what* we learn, and more on *how* we learn and *with whom*. This implies new roles for development professionals, leading to a whole new professionalism with new concepts, values, methods and behaviour (Table 6.6). Typically, normal professionals are single disciplinary, work largely in agencies remote from people, are insensitive to diversity of context, and are concerned with themselves generating and transferring technologies. Their beliefs about people's conditions and priorities often differ from people's own views. The new professionals, by contrast, are either multidisciplinary or work in close connection with other disciplines, are not intimidated by the complexities of close dialogue with rural and urban people, and are continually aware of the context of interaction and development.

The problem with characterizing an old and a new professionalism in this way is the risk of depicting complex relationships as simple polarizations, in this case the bad and the good, whereas true sensibility lies in the way opposites are synthesised. A distinction is needed here between the strengths of normal science as bodies of knowledge, principles and methods, and the weaknesses of the beliefs, behaviour and attitudes which often go with it. It is mainly the beliefs and values which present problems, and which the new professionalism seeks to change (Pretty and Chambers, 1993a,b).

However, it is also important to note that the old is really the modern and before the modern there were practices more akin to the new. Extensionists in the USA in the last century and early part of this were clear about the way they worked with farmers. One definition of extension education was given as *'working with people, not for them: of helping people become self-reliant, not dependent on others; of making people become the central actors in the drama, not the stage hands or spectators'* (A H Maunder in Campbell, 1994b).

It is clearly time to let go of some of the old paradigm of positivism for science and embrace the new alternatives. This will not be easy. Many existing agricultural professionals will resist. But as the science writer, Arthur Clarke (1973), put it *'When a distinguished but elderly scientist states that something is possible, he is almost certainly right. When he states that something is impossible, he is very probably wrong'*. It is only when some of these new professional norms and practices are in place that widespread change in the livelihoods of farmers and their natural environments is likely to be achieved.

When challenged about the work of the Latin American Consortium for Agroecology and Development (CLADES), Miguel Altieri (1992) recently said: *'I don't believe in objective true, universal, neutral science. We use agroecology within a development paradigm that must have a certain direction in terms of social change. Our approach has been marginalised from academia and we have always been called the radicals or the dreamers. It is interesting that academia is now knocking on our doors. I don't argue with anybody any more about whether the set rotation works for seven years; I'll take you there and you can argue with the farmers because the proof is theirs'*.

The prospects of a sustainable agriculture that is built on this new professionalism may be too much for many existing scientists. Thomas Kuhn (1962) indicated that new paradigms inevitably mean some

Table 6.6 Changing professionalism from the old to the new

	From the old professionalism	*To the new professionalism*
Assumptions about reality	Assumption of singular, tangible reality	Assumption of multiple realities that are socially constructed
Scientific method	Scientific method is reductionist and positivist; complex world split into independent variables and cause-effect relationships; researchers' perceptions are central	Scientific method holistic and post-positivist; local categories and perceptions are central; subject-object and method-data distinctions are blurred
Strategy and context of inquiry	Investigators know what they want; pre-specified research plan or design. Information is extracted from respondents or derived from controlled experiments; context is independent and controlled	Investigators do not know where research will lead; it is an open-ended learning process. Understanding and focus emerges through inter-action; context of inquiry is fundamental
Who sets priorities?	Professionals set priorities set priorities together	Local people and professionals
Relationship between all actors in the process	Professionals control and motivate clients from a distance; they tend not to trust people (farmers, rural people etc) who are simply the object of inquiry	Professionals enable and empower in close dialogue; they attempt to build trust through joint analyses and negotiation; understanding arises through this engage-ment, resulting in inevitable interactions between the investigator and the 'objects' of research
Mode of working	Single disciplinary – working alone	Multi-disciplinary – working in groups
Technology or services	Rejected technology or service assumed to be fault of local people or local conditions	Rejected technology or service is a failed technology
Career development	Careers are inwards and upwards – as practitioners get better, they become promoted and take on more administration	Careers include outward and downward movement; professionals stay in touch with action at all levels

Source: Pretty and Chambers, 1993a,b

destruction of the old and so it should not be surprising that many who oppose sustainable agriculture will fear the loss of career prospects. Rod MacRae and colleagues (1989) put it this way when considering the professional barriers to the development of a sustainable agriculture: '*For those who are well recognised in their field there is fear of irrelevancy. Since many scientists have produced research over the years that is irrelevant to sustainable approaches at best, and destructive at worst, they would have to reject the value of an entire lifetime of work; an especially difficult task because many scientists, particularly during their younger years, have allowed their self-esteem to become tied up entirely in their research work... Naturally, they are resistant to approaches, such as sustainable agriculture, that challenge the orthodoxy that has helped them achieve their present position*'.

The use of participatory methods must be combined with action to create both appropriate institutional contexts for them to flourish and appropriate learning environments for individuals to develop their own problem-solving capacities. Where all three combine, namely new systems of participatory learning and action for partnerships, dialogue and analysis; new learning environments for professionals and local people to develop capacities; and new institutional settings, including improved connections both within and between institutions; then widespread and persistent change is more likely (Pretty and Chambers, 1993a; Roche, 1992). It is only with all three in place that a sustainable and productive agriculture can be developed.

SUMMARY

For many reasons, existing agricultural institutions, whether universities, research organizations or extension agencies, find it difficult to learn from farmers and rural people. This is because they are characterized by restrictive bureaucracy and centralized hierarchical authority; their professionals are specialists and so see only a narrow view of the world; and they have few systemic processes for getting feedback on performance. The widespread reliance on questionnaire surveys, supplemented by short rural visits, gives a distorted picture of rural reality. The tendency is for rural complexity to be simplified.

There is increasing recognition that 'participation' between agricultural professionals and rural people is essential for sustained agricultural change. But the term participation is interpreted in many different ways, most of which are characterized by no giving up of control to local people. They may be passive participants, listened to or even consulted, but rarely do they fully interact with the opportunity to take control. Interactive participation can be ensured through the use of alternative systems of inquiry. These are defined methodologies and systemic learning processes; they seek multiple perspectives and diversity; they use group processes of learning and they are context specific. The experts' role is best thought of as a facilitator of local people's analysis and the whole process should lead to local institution building or strengthening, so increasing the capacity of people to take action on their own. Many methods can be used and their rigour is ensured through the use of criteria for trustworthiness.

Research organizations have a poor record when it comes to participation with farmers. If we are to be serious about the development of a sustainable agriculture, it is critical that local knowledge and skills in experimentation are brought to bear on the processes of research. The problem with agricultural science is that it has poorly understood the nature of indigenous knowledge and farmers' capacity to experiment. When given the opportunity, farmers have been innovative at adapting technologies to their own conditions, often having a significant impact on research and extension institutions in the process.

The wider challenge is for agricultural organizations to become learning organizations. To do this, they will have to promote experimentation; promote connectivity and group work based on roles rather than disciplines; and develop monitoring and self-evaluation systems to improve learning and awareness. The central concept of sustainable agriculture is that it must enshrine new ways of learning about the world But learning should not be confused with teaching. A move from a teaching to a learning style has profound implications for agricultural development institutions. The focus is less on *what* we learn, and more on *how* we learn and *with whom*. This implies new roles for development professionals, leading to a whole new professionalism with new concepts, values, methods and behaviour.

It is clearly time to let go of some of the old paradigm of positivism for science and embrace the new alternatives. This will not be easy, as many professionals will resist. But it is only when some of these new professional norms and practices are in place that widespread change in the livelihoods of farmers and their natural environments is likely to be achieved.

7

THE TRANSITION TO SUSTAINABLE AGRICULTURE: LINKING PROCESS TO IMPACT

'These improvements would impart financial benefits, besides they would add to the beauty of the prospects from the house'
Mr Brolter, farm adviser and valuer, 1790, Britain

SOURCES OF EVIDENCE

Throughout this book, it has been consistently argued that the development of a more sustainable agriculture requires attention both to resource-conserving technologies and practices, and to local groups and institutions, with external organizations and professionals working in partnerships with local people. When the process of agricultural development focuses on these three components, then sustainable agriculture can bring both environmental and economic benefits for farming households, communities and nations. The elements of the process have been defined in Chapters 4, 5 and 6. This chapter focuses on the existing evidence for impacts in farming communities.

Although the amount of evidence is not yet extensive, it does show that sustainable agriculture can lead to significant improvements in yields in complex and diverse agriculture, maintaining of yields in Green Revolution areas and maintaining of profitability coupled with a fall in yields in industrialized agriculture.

There are several types of evidence that can be drawn on to indicate the impacts of resource-conserving technologies and practices. Not all, though, are equally trustworthy when it comes to suggesting the possible or even likely impacts of a sustainable agriculture. The majority of evidence is from research stations and on-farm trials. Some also comes from the analysis of the achievements of individual farmers. Much less is available to illustrate the impacts on whole groups or communities of farmers. None yet exists for the effects on whole regions or even countries.

As has been well demonstrated by many commentators worldwide, the highly controlled ecological, social and economic conditions on research stations make them quite unlike farmers' own conditions. Such research commonly produces technologies and practices that poorly fit farmers' needs and so, unless farmers are willing to change their environments, will not be adopted. Even most on-farm trials result in the development of technologies not immediately relevant for nearby farmers. This is not to dismiss both these types of research. They are essential to the development of a sustainable agriculture. They illustrate what is technically possible. But, unless they become much more interactively participative and responsive to farmers' needs, they do not show what can be achieved in farmers' fields.

In the field of soil and water conservation, for example, there are huge numbers of scientific papers reporting the positive impacts of conservation technologies on crop yields or soil erosion. Enormous effort goes into analysis of tiny runoff plots, from which results are generalized for farms, regions and even countries. Such research tells us about what can be achieved on research stations and on small runoff plots, but tells us little about what can be achieved in real communities.

At the next level up, there is considerable evidence from individual farms. This is obviously stronger. If one farming household is using resource-conserving technologies, and is surviving in the face of existing market and economic pressures, they are showing what is possible for other farmers. Many of these farmers are innovators, making path-breaking advances and absorbing the costs of experimentation. However, as is the case for contact farmers used by some extension systems, they may be seen by other farmers as a special case, perhaps because they are better off or because they are seen as experimenters. Again the technologies may not spread.

Most of the evidence from Europe and the USA is at this level, with relatively small numbers of individual farmers having changed their practices in recent years to produce economically and environmentally viable enterprises. These individual farmers are like most farmers in the Third World in only one respect – they are individual decision making units. In terms of area, a single farm in Europe or North America can be equal to a whole community in the Third World. They may, therefore, have control over sufficient soil, water, predator and vegetation resources to achieve the coordinated action necessary for sustaining resource conservation.

But even this evidence is not as convincing as that emerging from whole communities in countries of Africa, Asia and Latin America. If several tens or hundreds of farming households have changed their farming practices, are working together in groups and are linking with enabling external institutions, then the impacts observed must be convincing. Where this does happen, there is clear evidence to suggest the improvements will also be maintained and that regeneration of local economies can occur as a result of agricultural regeneration. However, most of the successful improvements at this level have been achieved despite the existing policy environment.

For these reasons, projects or initiatives that rely on heavy subsidies, incentives or other forms of coercion do not constitute reasonable evidence (Kerr, 1994). This means that some past 'successes', as presented in the literature, have now to be seen in a different light. The Majjia Valley Windbreak Project in Niger appeared in the late 1980s as a major success in the greening of degraded semi-arid environments (Harrison, 1987). By the end of 1988, 463 km of windbreaks had been planted, 4000 ha protected and crop yields improved by some 15–20 per cent. But later analyses and reflections (Kerkhof, 1990; Leach and Mearns, 1988) indicated that the farmers were not consulted about where the windbreaks of neem should be planted, and farmers lost considerable crop area to the trees. Only 2 per cent of the farmers thought the windbreaks belonged to them, rather than the government forestry department. A project manager said *'if we were to leave Majjia Valley now, I wouldn't be surprised if there were no windbreaks left in 5 years' time'* (in Kerkhof, 1990). The project clearly demonstrates that major tree planting can be successful at a technical level. But the challenges to adapt the programme approach to bring about local participation are still great.

At the regional or national level, as would be expected, there is no evidence yet of significant change. This level implies changes in both the three circles of Figure 1.2 and in the policy environment. Supportive policies for regenerating rural and agricultural areas have been established by some governments, but these are still rare (see Chapters 8 and 9).

THE IMPACTS OF SUSTAINABLE AGRICULTURE

This section summarizes the impacts that have occurred at farmer or community level in the three main types of agriculture. It does not include any research station data, though there is some evidence in Europe and North America drawn from research farms.

The evidence demonstrates that a more sustainable agriculture leads to:

- stabilized or lower yields in industrialized countries, coupled with substantial environmental improvements;
- stabilized or slightly higher yields in Green Revolution lands, with environmental benefits;
- substantially increased agricultural yields in complex and diverse lands based mostly on available or local resources.

Sustainable Agriculture in Industrialized Systems

A great deal of evidence to support the economic success of regenerative agriculture comes from the USA, where relatively well-developed policies, combined with considerable research effort and state financial support, are being translated into new productive practices by farmers. The publication *Alternative Agriculture*, the result of a nationwide study by the National Research Council, was enormously influential (NRC, 1989). This demonstrated, using 11 case studies of individual enterprises, together with wider evidence, that the economic performance of regenerative agriculture could regularly match or better neighbouring conventional farms. This evidence is supported by many other studies (see, for example, Lockeretz et al, 1984; Liebhart et al, 1989; NRC, 1989; Faeth et al, 1990;

Hanson et al, 1990; Reganold et al, 1990; Dobbs et al, 1991; Faeth, 1993; and regular journals, such as the *American Journal of Alternative Agriculture*, the *Journal of Sustainable Agriculture* and local newsletters).

A summary of a range of enterprises is shown in Table 7.1. These show that, in most cases, yields are above the local county averages. Inputs are substantially lower and have been eliminated entirely on some farms. There are other benefits not represented in this table. Of a sustainable

Table 7.1 Impacts of sustainable agriculture in USA

Location	*Resource-conserving technologies*	*Crop yields (% in comparison with local averages)*	*Input use: pesticides (%)*	*Input use: fertilizers (% or kg/ha)*
Ohio (1 farm)[1]	mixed organic farm with rotations, legumes, manures	Mz: 132% Soy: 140% Wt: 105%	zero	zero
Iowa (1 farm)[1]	mixed livestock, ridge tillage, rotations	Mz: 111%	some herbicides only	30–40 kg /ha
Virginia (1 farm)[1]	mixed farm, rotations, weeds harvested, multiple cropping	Mz: 128%	herbicides: 28%	55 kg/ha
Pennsylvania (1 farm)[1]	mixed farm, crop rotations, multiple cropping, chicken manures, legumes	Mz: 127% Soy: 121%	24%	29%
Iowa[1]	mixed farm, crop rotations, ridge tillage, municipal sludge, manures, green manures, soil testing, cover crops	Mz: 104–121% Soy: 112–138% Hay: 117–147%	antibiotics 0 herbicides 0 (except spot spraying)	33 kg N/ha
Iowa, Illinois, Nebraska, Missouri, Minnesota (20 farms)[2]	crop rotations, legumes, reduced tillage	Mz: 92% Soy: 95% Wt: 57%	zero	zero
Washington State[3]	rotations and conservation tillage	Mz: 99% Wt: 92%	limited use	zero
California (1 farm)[4]	organic almonds with intercropped vetch	89%	variable costs: 58%	zero
Nebraska[5]	mixed farm, rotations, import of manures	~100%	variable costs: 40%	zero
Nebraska[5]	mixed farm, soil and water conservation, rotations, manures	> 100% in dry < 100% in wet years	herbicides 0	zero

Mz=maize Soy=soybeans Wt=wheat

Sources: 1 NRC, 1989; 2 Lockeretz et al, 1984; 3 Reganold et al, 1990; 4 Shirley, 1991; 5 Parr et al, 1990

farm in Washington State, John Reganold and colleagues (1990) say: *'the soil contained significantly more organic matter, nitrogen and biologically available potassium than on the conventional farm. It had a better capacity for storing nutrients, a higher water content, a larger micro-organism population... The soil also had better structure and tilth and 16 more centimetres of topsoil'.*

Several studies have compared the profitability of alternative with conventional agriculture. Some suggest that alternative agriculture is much less profitable (Crosson and Ostrov, 1988; Dobbs et al, 1991); others that it is unprofitable during just a short transitionary period (Liebhardt et al, 1989); while many illustrate that the cutting of costs can add substantially to net returns (Dobbs et al, 1991; Hanson et al, 1990; Koenigstein et al, 1990; Reganold et al, 1990; Smolik and Dobbs, 1991; Halmers et al, 1986). Evidence cannot be unequivocal, however, as research results are often strongly influenced by assumptions about prices, yields, technological options and future practice (Batie and Taylor, 1989). None the less, the individual farmer evidence is showing that, with the substitution of knowledge and management skills, farmers can cut inputs while remaining economically viable.

Agricultural policies, though, still discriminate against alternative agriculture, as subsidies and support mechanisms favour monocultures and high input strategies (Faeth, 1993) (see Chapter 8). But even without major policy changes, the gap between conventional agriculture and its sustainable alternatives has narrowed in recent years. Says the director of Iowa State University's Leopold Center for Sustainable Agriculture, Dennis Keeney (in Holmes, 1993), *'what we used to call conventional agriculture in Iowa is pretty much sneered at by the conventional farmer today. Very few of them plough anymore. Conservation tillage and cutting back on chemical use are becoming bragging points in coffee shops'.*

Sustainable agriculture in the USA is increasingly well established as a viable alternative to both organic farming that relies on premiums and industrialized farming that relies heavily on external inputs. Such shifts have not yet occurred in Britain, where there is considerably less experience of sustainable agriculture on farms. It is only recently that the debate has moved away from a polarisation between high input and organic agriculture (Pretty and Howes, 1993). Comparative studies of the two, though, have generally shown that organic farming is unable to compete with conventional farming unless it receives extra returns in the form of premiums paid by consumers (Lampkin, 1992; Vine and Bateman, 1983; Rhône-Poulenc, 1992–4; IFOAM, *passim*).

Until recently, few had considered the potential for a regenerative or sustainable agriculture that is situated somewhere between organic and very high input agriculture. In contrast to organic systems, such an agriculture, if profitable, would be immediately available to all farmers, as it does not rely on consumer premiums.

There is growing economic evidence to indicate that farmers can reduce external input use significantly without losing out on gross margins. Following the adoption of resource-conserving technologies, yields and variable costs fall, and so gross margins can be matched or bettered. Generally, the loss in yield per hectare is some 5–10 per cent for crops and 10–20 per cent for livestock (see Tables 7.2 and 7.3). Livestock perform less

Table 7.2 Economic indicators for performance of crop components and complete farms of sustainable agriculture in UK and continental Europe as proportion (%) compared with conventional

Description	*Crop yields*	*Pesticide use*	*Fertilizer use*	*Gross margins*
Integrated crop rotations of cereals and legumes, west England[1]	89–92%	10–81%	69%	103–116%
Environmentally benign sugar beet production systems, east England[2]	78%	nd	nd	102%
Alternative rotation and low input systems, west England[3]	SBn 88%	0–50%	nd	96%
	Oa 105%	50%	50%	112%
Standard rotation low input approach, southern England[3]	99.6%	insecticide 0%	nd	101%
		fungicide 50%	nd	101%
Reduced fungicide use on spring barley, Scotland[4]	102%	12–25%	nd	126–152%
Supervised low input pesticides on wheat and rape, east England[5]	Wt 88%	15–54%	nd	105%
	OSR 105%	0–96%	nd	139%
Integrated farm, Netherlands[6]	94%	26–50%	26%	120%
Biodynamic farms, Baden Würtemburg, Germany[7]	75–91%	zero	zero	80–101%
Switzerland 8 biodynamic farms[8]	95–100%	zero	zero	100%
Individual organic and 'semi organic' farms, data for wheat, UK[9]	53–114%	zero	zero	57–138%
57 organic farms, 1986–8, Switzerland[10]	73%	P 6%	22%	83%
200 whole mixed farms, 1986–91, Germany[11]	Wt 66%	P 6%	3%	105%
	Rye 67%			
	Pot 61%			
Lautenbach integrated and conventional farm, Germany[12]	99–105%	zero	75%	104–109%

Note: where a value is 100%, this means that the yield, input use, or gross margins are the same for both the sustainable agriculture and conventional agriculture. A value of 116% in the gross margins column means that the sustainable agriculture outperforms the conventional by 16%. A value of 89% in the yield column means that the sustainable agriculture yields 11% less than the conventional comparison.

nd = no data Wt=wheat Pot=potatoes SBn=spring beans Oa=oats OSR=oil seed rape

Sources: 1 Jordan et al, 1993; Jordan, 1993; 2 Jaggard, 1993; 3 Ogilvy, 1993; 4 Wale, 1993; 5 Greig-Smith et al, 1993; 6 Vereijken, 1990; 7 MELU, 1977; 8 Karch-Türler, 1983, in Lampkin, 1992; 9 Vine and Bateman, 1983; 10 Muehlebach and Naef in Lampkin, 1992; 11 BMELF, in Lampkin, 1992; 12 El Titi and Landes, 1990; Zeddies et al, 1986

Table 7.3 Economic indicators for performance of livestock component of sustainable agriculture in UK and continental Europe as proportion (%) compared with conventional

Description	*Yields (kg milk or meat per animal)*	*Variable costs*	*Gross margins per animal*	*Gross margins per hectare*
Beef cattle in organic grass/white clover system with 24% lower stocking rate, compared with a conventional, intensively fertilized at 270 kg N/ha/yr system, Scotland[1]	100% slaughter weight	107% (per head)	100% 131%*	76%–100% 101%*
Sheep on white clover pastures with no N and 22% lower stocking rate, compared with 150–180 kg N/ha/yr ryegrass systems, Scotland[2]	122%	nd	nd	95–100%
Organic milk herds, with 20% lower stocking rate, UK[3]	92%	116%	111%* 98%	89%* 79%
Alternative dairy, Germany[4]	83%	72%	86%	75%
Organic dairy, 26 farms, Switzerland[5]	88%	93%	98%*	90%*
Biodynamic dairy, Germany[6]	64%	64%	113%*	69%*
•Dairy cattle in low input (no inorganic N) grass/clover compared with 350 kg N/ha/yr systems, Scotland[7]	99.8% milk sales, litres per cow	nd	90%	90%

* = gross margins with organic premiums
nd= no data
Sources: 1 Younie, 1992, FW 1993a; 2 Vipond et al, 1992; 3 Redman, 1992; 4 Gravert et al, 1992; 5 Steinmann, 1983; 6 Jochimson, 1982, in Lampkin, 1992; 7 Bax and Fisher, 1993; Bax, 1992

well, mainly because of the substantially lower stocking rates necessary for clover-based pastures. Grassland in Britain has very large amounts of nitrogen fertilizer added and it is almost impossible to match returns when switching to clover pastures. But there is good evidence to suggest that animals are better off. In Germany, cows in 'alternative' herds are more fertile and live longer (Gravert et al, 1992).

Most farmers, researchers and policy makers have assumed that although resource-conserving practices might be environmentally beneficial, the reduction in gross output would inevitably mean reduced profitability. It is increasingly clear that this is not the case. However, this evidence is not yet as good as that from the USA.

The principal differences between Europe and the USA appears to be that crop and livestock yields on sustainable farms cannot match the currently high levels under conventional practice. However, this yield 'penalty' is only a problem if it is viewed from the perspective of needing to maximize food output. This is, of course, not now the case. As European Union and individual state policies are now targeted to reduce output through the use of the set-aside mechanism, this yield penalty could now be seen as an alternative means of achieving this goal.

Sustainable agriculture in industrialized systems appears economically and environmentally viable. It also brings indirect benefits of value both to farmers, ecosystems, and the public and society as a whole. These include:

- better quality of products, such as better tasting meats and fruits;
- the amenity value associated with more diverse and wildlife-rich landscapes;
- the maintenance of environmental quality, such as uncontaminated aquifers and surface water;
- the sustaining of resources for future generations;
- increased wild bird populations and numbers of territories;
- reduced soil erosion;
- increased numbers of beneficial insects;
- lower livestock stocking rates;
- improvements in animal welfare through cutting of routine use of unnecessary drugs, eg antibiotics to promote growth.

Sustainable Agriculture in Green Revolution Lands

The Green Revolution lands are the current 'bread basket' for many Third World countries, supporting some 2.3–2.6 billion people. Although these systems have good soils and reliable water, and are close to roads, markets and inputs, the hope that they will be the source of future agricultural growth is being undermined by emerging evidence of stagnating yields coupled with increasing environmental and health costs (see Chapters 1, 2 and 3). If, as seems likely, the micro-environments for crops are being disrupted by the high input regimes, then a more realistic hope would be to stabilize yields in these areas while reducing environmental impacts.

To date, there has been relatively little effort to introduce sustainable agriculture into Green Revolution lands. The best evidence comes from the remarkable integrated pest management (IPM) for rice programmes in South and South East Asia, and is supplemented by various localized and individual efforts elsewhere. Generally, this evidence from 12 countries shows that input use can be cut substantially if farmers substitute back knowledge, labour and management skills. Yields can be maintained or even slightly improved (Table 7.4).

Most of the organic examples survive with consumer premiums, usually paid by consumers in industrialized countries. These are,

Table 7.4 The impact of sustainable agriculture in Green Revolution lands

	Resource-conserving technologies	*Crops*	*Input use (%)*	*Yields (%)*	*Spread*
Bangladesh[1]	fish-culture and IPM	rice	0	124%	58 farmers
Brazil[2]	composts, manures	sugar cane, organic	0	76%	9 farms, 2200 ha
Burkina Faso[3]	IPM	rice	50%	103%	1037 farmers
China[4]	waste recycling, composts, rice – fish culture	rice	30% for N	110%	1200 ecofarms
China[5]	IPM farmer field schools	rice	46–64%	110%	14 counties
Dominican Republic, Azuca Valley[6]	manures, neem sprays, composts	bananas, organic	66%	124%	60-70 farmers on 100 ha
India, Andhra Pradesh[7]	IPM	ground-nuts	0	100%	1 community
India, Tamil Nadu, Karnataka and Pondicherry[8]	agro-forestry, green manures, multiple cropping, legumes, manures	rice sorghum ragi	25% for N	123% 151% 84%	7 ecological farms paired with 7 conventional
India, Tamil Nadu[9]	composting, green manures, trenching, agroforestry, resistant vars	tea, organic	0	111%	310 ha estate
Indonesia, nationwide[10]	IPM farmer field schools	rice	37–48%	107%	110,000 farmers graduated
Mexico, Chiapas[11]	IPM, green manures, agroforestry, manures from cattle	coffee, organic	0	66–72%	1 estate of 320 ha
Mexico, Baja California Sur[12]	rearing predatory wasps, green and livestock manures, fish meal and bat guano, botanical sprays	tomatoes, organic	nd	83%	160 farmers
Philippines[13]	IPM farmer field schools	rice	62%	110%	nd
Philippines[14]	*Azolla* network	rice	50% N	100%	nd
Philippines[15]	Irrigation improvement	rice	100%	116–119%	nationwide

Table 7.4 *Continued*

	Resource-conserving technologies	*Crops*	*Input use (%)*	*Yields (%)*	*Spread*
Sri Lanka[16]	IPM farmer field schools	rice	23%	135%	nd
Togo[17]	IPM (compulsory package)	cotton	50%	90–108%	160,000 farmers on 80,000 ha
Turkey[18]	Green manures, *Bt, Trichogramma* sprays, wasps	cotton, organic	zero	50–80%	2 farms

Sources: 1 Kamp et al, 1993; 2, 6, 8, 9, 11, 12, 18 UNDP, 1992; 3, 17 Kiss and Meerman, 1991; 4 Laird, 1992; 5, 10, 13, 14, 16 Kenmore, 1991; 7 ICRISAT, 1993; 8 van der Werf and de Jager, 1992; 10 GoI, 1992; Winarto, 1993; 15 Bagadion and Korten, 1991

therefore, likely to be niche markets and so not available to all farmers. Several cases record wider community benefits, particularly where extra labour was required to offset the fall in input use. For more discussion of details of the technologies and practices, see Chapter 4. For details of specific cases, see Cases 11, 18 and 20 below.

Sustainable Agriculture in Complex and Diverse Agricultural Systems

It is in the complex and diverse agricultural systems, where external input use is low or non-existent, that there is considerable evidence for the impact of sustainable agriculture (Table 7.5).

It is now not uncommon to find two to three-fold increases in agricultural yields following community-wide adoption of resource-conserving technologies and practices. Increased crop yields from 500–800 kg/ha to 1000–2500 kg/ha, and sometimes higher, have been achieved without the use of fertilizers and pesticides in programmes focusing on soil and water conservation, land rehabilitation, nutrient conservation, raised field agriculture, green manuring and IPM.

For details of impacts not contained in this table, such as changes in wage labour rates, increased local employment opportunities, changed groundwater levels, livestock yields etc, see the 20 following case studies.

CASES FROM 20 COMMUNITIES

A Cautionary Word About 'Success'

The next section contains 20 cases from communities in Brazil, Burkina Faso, Honduras, India, Indonesia, Kenya, Lesotho, Mali, Mexico, Peru, Philippines and Sri Lanka.

Presenting case studies in this way always carries an element of risk, as it can appear to suggest that these 'successes' are complete. This does not imply they will always be successes. It is possible that some of these will fail. This may be because external conditions will change and so undermine local efforts, such as national economic policies, international

Table 7.5 Impact of resource-conserving technologies and practices in complex and diverse agricultural systems

Country and location	*Resource-conserving technologies*	*Yields achieved (t/ha)*	*Increase in yields (%)*	*Scale*
Brazil, Santa Catarina[1]	green manures and cover crops	Mz: 3–5	nd	38,000 families
BotswanaSu[2]	double plough, row planting	So/Cowp: 0.33–0.49	130–169%	nd
Burkina Faso: Bam and Passoré[3]	rock bunds, contour planting, composting	So: 1.65–2.0	189–230%	farmers in 197 communities
Burkina Faso: Yatenga[4]	planting pits, rock bunds, composting	So/Mi: 1.2 So/Mi: 0.97	** 140%	8000 ha (by 1991) 4542 farmers trained
Burkina Faso, Zabre[5]	composts, manures	So/Mi: 1.0–1.2 Veg: 30 to 70	285–300% 240%	10,000 members of group
Burkina Faso, Mossi Plateau[6]	rock bunds, grass strips, manures, agroforestry	So/Mi: 1.4 Gn: 0.88	275% 180%	500 families
China, Jiangxi Province[7]	soil conservation and watershed management	nd	152%	3200 families
Honduras, Cantarranas[8]	green manures, soil conservation	Mz: 2.55	300%	600-1000 families
Honduras, Guinope[9]	green manures, soil conservation	Mz: 1.2	300%	1200 families
India, Gujarat[10]	soil and water conservation, biogas	nd	R: 253% S: 117% Ppea: 222% Co: 153%	55 households in one community (programme working in 100 communities)
India, Tamil Nadu[11]	contour bunds, percolation tanks, gully checks, agroforestry	R: nd Gram: 0.4	++ **	1 community of 100 households (programme in 45 villages)
India, Maharashtra[12]	soil and water conservation	So (dry): 0.7 So (irrig): 2.2	350% 176%	one community of 168 ha
India, Haryana[13]	soil and water conservation, social fencing	Grass: nd	400–600%	50 communities
India, Rajasthan[14]	grass strips, field and contour bunding	So: 0.33-0.46 Mi: 0.72-0.93	210–292% 120–154%	nd

Table 7.5 *Continued*

Country and location	*Resource-conserving technologies*	*Yields achieved (t/ha)*	*Increase in yields (%)*	*Scale*
Kenya, Trans Nzoia, Bungoma, West Pokot, Nyeri, Kirinyaga[15]	soil and water conservation, manures, legumes	Mz: 1.1–2.5	150–300%	6 communities (programme nationwide)
Kenya, Trans Nzoia[16]	biointensive agriculture, composting, botanical sprays	Veg: nd	200%	3000 families in 195 communities
Mali, Menaka Oasis[17]	earth bunds, dams	So: 1.35	**	400 ha in one community, then 8 communities
Mali, Koutiala[18]	stone bunds, grass strips, trees	Co: 2.0 So/Mi: 1.0	154% 125%	20 villages
Mexico: Oaxaca[19]	composts, contour planting, terracing	Cf: 0.9	257%	3000 families
Paraguay, Chaco Boreal[20]	green manures, *B t*, IPM	Gn: 0.8	100%	33000 ha
Paraguay, Parana[21]	*B t*, manures, green manures, Baculovirus	Soya: 1.8	90%	36 farms, 4000 ha
Peru, Colca Valley[22]	rehabilitation of ancient terraces	Mz: 2.98 Ba: 1.95 Po: 17.2	165% 143% 141%	one community
Peru, Huatta District[23]	raised fields, green manures, manures	Po: 10.0	333%	30 communities
Rwanda, Nyabisindu[24]	agroforestry with legume trees, contour bunds, intercropping, composting	Mz: 1.4 Be: 0.8 SPo: 3.35	117% 107% 140%	30–90% families using soil and water conservation and composts
Senegal[25]	composts, contour bunds	Gn: 0.53 Mi: 0.6	141% 300%	11 villages

nd = no data ** % not possible to calculate, as programme involved rehabilitation of lands with former yields of zero. ++ same yield, but new second crop harvested

Co = cotton Cf = coffee Cowp = cowpeas Gn = groundnut
Mi = millet Mz = maize Ppea = pigeonpea Po = potato
R = rice So = sorghum Veg = vegetables

Sources: 1 Bunch, 1993; 2 Heinrich et al, 1992; 3 GTZ, 1992; 4 Gubbels, 1994; Reij, 1991; 5, 19, 20, 21 UNDP, 1992; 7 ASOCON, 1991; 8, 9 Bunch, 1990, 1987; 10 Shah et al, 1991; Shah, 1994a; 11 Devavaram, 1994; 12 Lobo and Kochendorfer-Lucius, 1992 13 Mishra and Sarin, 1988; 14 Krishna, 1994; 15 Pretty et al, 1994; SWCB, 1994; 16 Kisian'gani, pers. comm; 17 Rands, 1992; 18 Critchley,1991; Wardman and Salas, 1991; 22 Erickson and Candler, 1989; 23 Treacy, 1989; 24 Kerkhof, 1990; 25 Diop, 1992;

commodity price changes or wars. All of these would disrupt some key components of success. It is much more important to focus on the procedures and processes that brought about the current impacts. It is from these that lessons can be learned.

It is also important to note that no case is 'perfect'. There is always room for adaptations in the processes that could lead to further improvements.

Shortage of space is another problem. Justice cannot be done to the innovative efforts of local people and external professionals in these programmes. Years of experience, of hard work, of celebrated success, of reversals and of remarkable achievements simply cannot be reflected in these short summaries. Each alone would be a book to rivet, enchant and even entrance the reader. References are, of course, provided to some of the published materials.

Threats to Sustained or Spreading Improvements

We still do not know whether any of these will be sustainable. In ten years' time, some may have failed, some may have spread and expanded. No doubt some will be internationally renowned. None have lasted long enough for us to be certain. There are many challenges relating to each case.

There are also systemic problems that each will face. As benefits accrue to farmers and communities, what will happen to those who are excluded or do not benefit so much? Relative deprivation could lead to new conflicts.

Another issue relates to the growing success of community-led initiatives. As they become more strong and more capable of drawing on external resources, and even of influencing higher level decision making, what will happen when this brings them into conflict with existing power structures? Will there be accommodations or will these successes in the hands of local people be dashed?

A critical issue relates to national policies. Most of these cases are still relatively localized. What is lacking in almost every case is supportive policies. It is the currently disabling policies that most threaten the successes. Most policy frameworks still actively encourage farming that is dependent on external inputs and technologies, and these will have to change if successes are not to be undermined (see Chapters 8 and 9).

Finally, there is the problem of the norms of practice of agricultural professionals. Participatory processes are clearly a central element of all these cases – it is these that make individual successes happen at community level. There are two problems. One is that the 'empire will strike back' against these alternative approaches and philosophies. The positivist model will try to reassert its dominance. The other is that only the rhetoric of participation will be adopted, with no wider systemic change in institutions and professionals. If this happens, it can be confidently stated that this will not lead to improvements in people's livelihoods and environmental quality.

CASE 1: EPAGRI IN SANTA CATARINA, BRAZIL

The state government extension and research service, EPAGRI (Empresa de Pesquisa Agropecuária e Difusâo de Technologia de Santa Catarina), works with farmers in the southern Brazilian state of Santa Catarina, from the flat coastal areas in the east to the rolling highlands and mountains of the centre and west. It is involved in working at a micro-watershed level with local farmers to develop low input and productive systems of agriculture. Each member of staff works in about one to three micro-watersheds of about 150 families intensively for a period of two years, playing an important social as well as technical role.

The technological focus is on soil and water conservation at the micro-watershed level using contour grass barriers, contour ploughing and green manures. Farmers use some inorganic fertilizers and herbicides, but there has been particular success with green manures and cover crops. Some 60 species have been tested with farmers, including both leguminous plants such as velvetbean, jackbean, lablab, cowpeas, many vetches and crotalarias, and non-legumes such as oats and turnips. For farmers, these involve no cash costs, except for the purchase of seed. These are intercropped or planted during fallow periods, and are used in cropping systems with maize, onions, cassava, wheat, grapes, tomatoes, soybeans, tobacco and orchards. Farmers use animal-drawn tools to knock over and cut up the green manure/cover crop, leaving it on the surface. With another farmer-designed, animal-drawn instrument, they then clear a narrow furrow in the resulting mulch into which the next crop is planted. As a result, that many farmers no longer plough.

Many farmers have used subsidies to construct housing for pigs and chickens. Manures are now concentrated, fermented in pits and then applied to the fields. This has reduced pollution of waterways, as well as cut the dependency on inorganic fertilizers.

The major on-farm impacts have been on crop yields, soil quality and moisture retention, and labour demand. Maize yields have risen since 1987 from 3 to 5 t/ha and soy beans from 2.8 to 4.7 t/ha. Soils are darker in colour, spongy to the step, moist and full of earthworms. The reduced need for most weeding and ploughing has meant great labour savings for small farmers. From this work, it has become clear that maintaining soil cover is more important in preventing erosion than terraces or conservation barriers. It is also considerably cheaper for farmers to sustain.

EPAGRI has reached some 38,000 farmers in 60 micro-watersheds since 1991. They have helped more than 11,000 farmers develop farm plans, supplied 4300t of green manure seed and supported the construction of 1540 piggeries. Perhaps the most important impact, however, has been on the local municipalities. EPAGRI has worked to involve them fully in the process of participatory technology development and extension, and now many municipalities employ their own agronomists to help in the process.

Source: Bunch, 1993; Irene Guijt, personal communication

CASE 2: THE PROJECT AGRO-FORESTIER, YATENGA PROVINCE, BURKINA FASO

The Project Agro-Forestier (PAF) was begun by Oxfam in 1979 to address environmental degradation in Yatenga Province in Burkina Faso. In the early stages, it focused on promoting micro-catchments for tree growing. PAF conducted on-farm experiments with groups of farmers who, after observing the way the micro-catchments trapped runoff water, instead planted upland rice in the basins. Sorghum was introduced by accident in the manure. Both crops grew well and so farmers became more interested in food crops than trees. As local people have barely enough drinking water for themselves, mortality among tree seedlings was also massive.

PAF shifted its focus to food crops in accordance with local wishes and began to work on improving indigenous conservation technologies. PAF's major technological contribution has been to develop a cheap water tube level that has enabled farmers precisely to trace the contour for laying out bunds. After two years of joint experimentation, locally adapted versions of contour rock bunds (*diguettes*), combined with crop planting in *zai* holes filled with mulch, proved very effective at harvesting water and improving crop yields. The mulch encourages termite activity, which increases the water penetration when the rains come. The project also promotes the use of improved composting techniques as part of a national composting programme.

Yields have improved by some 40 per cent on treated land. The differential is greatest in low rainfall years. More than 8000 ha have been treated and some 4542 farmers from 406 villagers trained. The rapid rate of adoption and replication of the technologies outside the project indicates farmers' enthusiasm. Farmers cite higher yields and increased food security as the primary benefits. The internal rate of return calculated by the World Bank, using conservative assumptions, is 37 per cent.

Attitudes to tree growing have also changed. In some villages, farmers are planting trees and providing protection against grazing animals during the dry season. Livestock are increasingly stall-fed, with fodder cut and carried to the animals. Although the project began by working with individuals, it became clear that most of these improvements could only be achieved if there was community consensus. This is important for choice of communal sites for contour bunds, protection against grazing, collection of rocks and ensuring that conservation measures are complementary across contiguous farms.

Despite the successes, there are concerns over the uneven distribution of benefits, as rock collection and bund construction makes heavy demands on labour, particularly on women. Rich farmers are also more likely to be able to mobilize and provide labour for the communal groups to build bunds on their land. None the less, PAF has been able to develop a strong relationship with local government agencies and farmers' groups now have a voice in local state-NGO decision making about planning extension work.

Sources: Gubbels, 1993, 1994; Kerkhof, 1990; Reij, 1991; Ouedraogo, 1992

CASE 3: PATECORE IN BAM AND PASSORÉ PROVINCES, BURKINA FASO

The Projet d'Aménagement de Terroirs et Conservation de Ressources (PATECORE) is a government project working on the Mossi Plateau in Burkina Faso to improve village land use and conservation. It is a collaborative effort between a consortium of various ministries (agriculture, environment and tourism, and livestock) and NGOs operating in the field of resource management, and is funded by GTZ. It began in 7 villages in 1988 and has since expanded to work in over 200. Rainfall is some 550 mm/year, but soil erosion can reach 200 t/ha/year. There is much evidence of land degradation in the two provinces.

PATECORE staff do not intervene directly in villages but work through existing local governmental and non-governmental institutions. The project involves local groups in the planning and implementation of soil and water conservation, with the objective being to develop the self-help capacity at local level. The project staff coordinate activities at provincial and district level and train village extensionists (VEs) in technical skills and planning methodology. The VEs facilitate analysis and planning in villages, so as to develop village resource plans. The villagers conduct analyses, plan and implement resource-conserving technologies based on their own needs. One land-use committee is set up in each village and members elected by the community.

The main technologies adopted have been permeable dams, stone bunds (*diguettes en pierres*), contour ploughing, tree planting, the establishment of protected zones for regeneration, composting and increased use of manures. Rocks are collected from outside communities and transported in trucks provided by the project (although ways are being sought on how trucks, too, can be managed locally). The impact on yields is immediate, with sorghum yields increasing from 870 kg/ha to 1650–2000 kg/ha.

The project has a special focus on supporting women in developing income-earning enterprises, and giving them access to time-saving technologies so that they have the opportunity to devote time to conservation activities. Loans are provided for sheep-rearing, beer-brewing equipment, bee-keeping, grain mills and wells with pumps. Other major impacts include rapid replication to neighbouring communities (the project only provides support to villages at their request); decreased flood damage and soil erosion; stabilized yields; increased capacity of villagers to plan and implement changes on their own; and increased understanding between government agencies and NGOs who are able to work together with fewer prejudices and better coordinated activities. An important element of success is the national political will that supports decentralization and *Gestion de Terroirs*.

Sources: GTZ, 1992; Guijt, 1992; Critchley, 1991

CASE 4: WORLD NEIGHBORS IN GUINOPE AND CANTARRANAS, HONDURAS

The Guinope (1981–9) and Cantarranas (1987–91) Integrated Development Programmes were collaborative efforts between World Neighbors, the Ministry of Natural Resources, ACORDE (a Honduran NGO), and the Catholic Relief Services (for Cantarranas only). Both programmes focused on soil conservation in the areas where maize yields were very low (400 kg/ha in Guinope and 800 kg/ha in Cantarranas), and where shifting cultivation, malnutrition and outmigration prevailed. Both illustrate the importance of developing resource-conserving practices in partnership with local people. All the achievements were because farmers themselves were convinced that the changes were in their own best interests.

There were several common factors in the success of these two programmes. All forms of paternalism were avoided, including giving things away, subsidizing farmer activities or inputs, or doing anything for local people. They both started slowly and on a small scale, so that local people could meaningfully participate in planning and implementation. They used a limited technology, mainly green manures, together with some physical soil conservation measures. These technologies were appropriate to the local area, and were finely-tuned through experimentation by and with farmers. Extension and training was done largely by villager farmers who had already experienced success with the technologies on their own farms. This meant that a maximum number of villagers could be reached.

In Guinope, 1500 farmers in 41 villages tripled yields of maize (some have increased by 7–8 fold) after adopting the new technologies. Land fertility has increased with the increased use of chicken manures (700 truckloads are imported per year), green manures, contour grass barriers, rock walls and drainage ditches. Farmers have also diversified crop production: once maize and beans production exceeded family needs, they began to reduce area planted to these crops and to plant others, such as coffee, oranges and vegetables. Sixty local villagers are now agricultural extensionists and 50 villages have requested training as a result of hearing of these impacts. The landless and near-landless have benefited with the increase in labour wages from US$2 to $3 per day in the project area. Outmigration has been replaced by inmigration, with many people moving back from the urban slums of Tegucigalpa to occupy houses and farms they had previously abandoned, so increasing the population of Guinope. The main difficulties were in marketing of new cash crops, as structures did not exist for vegetable storage and transport to urban areas.

In Cantarranas, the adoption of velvetbean (*Mucuna pruriens*), which can fix up to 150 kg N/ha as well as produce 35 tonnes of organic matter per year, has tripled maize yields to 2500 kg/ha. Labour requirements for weeding have been cut by 75 per cent and, where a small amount of herbicides is used, eliminated entirely. The focus on village extensionists was not only more efficient and less costly than using professional extensionists, it also helped to build local capacity and provide crucial leadership experience.

Sources: Bunch, 1987, 1990, Bunch and López, 1994

CASE 5: THE ZAMORANO INTEGRATED PEST MANAGEMENT TRAINING, HONDURAS

Since 1988, scientists at the agricultural college, Escuela Agrícola Paramericana, in Zamorano have been working to build the capacity of small farmers to control pests without pesticides. This is done by holding short courses for farmers to fill in key gaps in their knowledge. Farmers' knowledge is already profound, but there are aspects of pest control they do not know about. They know about, for example, many aspects of the disease maize ear rot, but not about the details of fungal fruiting bodies and spore production. Farmers have many words to describe social wasps, but did not know that solitary parasite wasps exist. They tend not to know of the existence of parasitoids, the wasps and flies that spend their larval stages living inside other insect species. They do know that pesticides are toxic, but equate smell with toxic strength and so have no means of perceiving chronic toxicity.

The key to this learning about pest control without pesticides is the participatory mode of teaching and experimentation. Farmers are taught by scientists and by other farmers using local terminology; they observe fungi under microscopes; they collect insects from the field and watch parasitoids emerge; they observe wasps returning to nests with insect prey; they put caterpillars on maize plants and watch ants carry them off within minutes.

The principal impacts have been on farmers' capacity to experiment. They have developed many new low input technologies based on a synthesis of their traditional knowledge and what was learnt in the college. Scientists from the college have documented a wide range of experiments that farmers have conducted after courses. These include one farmer intercropping amaranth among intercropped vegetables to encourage predators; another observing worms in his stored potatoes and then placing the box on an ant nest – the ants cleaned the pests out and he then transplanted ant nests to his farm. Another farmer described taking parasitic wasp cocoons found on his farm to a neighbour's farm to increase the spread of wasps. And many have transported ants' nests to their fields.

Follow up to the course is coordinated with local NGOs, such as World Neighbors, Catholic Relief Services and ACORDE, and the groups of farmer extensionists, who are visited regularly by college scientists. In this way, many more farmers are reached than those directly trained. These visits also mean that information on farmers' needs is taken back to the Crop Protection Department at Zamorano. Small-scale farmers are thus helping to set scientists' formal research agendas, as well as learning more sustainable farming practices. As Bentley and Melara (1991) put it: *'we depend on farmers to help tell us what to study and to work with us carrying out the experiments in their fields, fine tuning the technologies to their conditions.'*

Sources: Bentley et al, 1993; Bentley, 1994; Bentley and Melara, 1991; Catrin Meir, personal communication

CASE 6: THE AGA KHAN RURAL SUPPORT PROGRAMME, GUJARAT STATE, INDIA

AKRSP is a non-government organization established in 1985 to promote and catalyse community participation in natural resources management. A central part of the work has been to support the emergence of strong village institutions that can design, plan, implement and monitor their own watershed development and run their own extension system.

The first stage of watershed management involves participatory appraisal of local natural and social resources. A sequence of participatory methods are used in joint exercises involving the external team and local people. This collaborative analysis ensures that the analytical capacity, knowledge and innovations of local people form the basis for the watershed programme. In some villages, this takes less than a month; in others, it may take six months. Next, village institutions (VIs) are formed for implementation of the village natural resources management plan. The VI nominates three members as extension volunteers (EVs), who are then provided with basic training in PRA methods, technical skills for soil and water conservation, and project preparation and accounting procedures. The EVs then manage the extension process at village level on behalf of the VI, with the teams dividing up responsibilities between soil and water conservation, dryland farming, credit and other commercial activities. The EVs are compensated by the VIs and most have opted for performance-related payment.

As a result of soil and water conservation with contour cropping, bunds, gully checks and percolation tanks, millet, rice and pigeonpea yields more than doubled, and sorghum and cotton improved by 20–50 per cent. Knowledge about the success of village-based EVs has also spread rapidly, and EVs have been invited by other villages to assist them in conducting participatory appraisals and in developing village institutions of their own. This has reduced the dependence on AKRSP staff for project initiation and training.

The cost of watershed treatment is some Rs1340/ha, compared with the Rs3000 to 7000/ha incurred by nearby government programmes, all of which still give a 100 per cent subsidy to farmers. Many communities have also taken up a number of group operations such as ploughing, plant protection and use of implements and post-harvest equipment, coupled with credit and pooled marketing of produce. This shows that VIs are becoming a conduit for greater economic investment and diversification. This is also reflected in the willingness of banks to advance credit to VIs with a large membership of small farmers. These were earlier considered as too high a risk by bankers.

There have also been changes in migration patterns, with a dramatic rise in the number of individuals and families staying all year in their own communities. This has resulted in higher school enrolment, and improved nutrition and health standards. In some villages, there have been changes in local leadership patterns, with a shift away from traditional leaders (who lead by virtue of lineage, patronage, social hierarchy) towards functional leaders (the EVs and active members of village institutions who lead by virtue of their performance). This could have long-term implications for improving governance and enhancing local democracy.

Sources: Shah, 1994a, b; Kaul Shah, 1993; Shah et al, 1991a, b

CASE 7: THE SOCIETY FOR PEOPLE'S EDUCATION AND ECONOMIC CHANGE, TAMIL NADU, INDIA

SPEECH has been working in Kamarajar District of Tamil Nadu since 1986. This region is known for its acute droughts, erratic monsoons, poor services, and entrenched socioeconomic and cultural division. SPEECH has helped to build and strengthen local groups and institutions in 45 villages. The initial involvement is through the establishment of non-formal education classes. Following discussions, local people choose their own village animators, who then receive training in basic teaching methods, PRA, conflict resolution, songs and stories. They also receive a small honorarium, but essentially remain as farmers or labourers. The animators visit every house, raising the awareness of the common problems faced and the need for effective organisation and participation to resolve them. After a period of discussions, plays and other forms of interaction, local people form a *sangha* or village committee.

Sangha leaders, elected by the members, then attend a 30-day training course spread throughout the year. Key personnel from government organizations and banks are invited to attend and meet the *sangha* leaders. As *sanghas* become more confident, they begin to develop their own capacity, providing for health care, roads, credit and so on. Representatives are elected to a Cluster Level Governing Council, an independent society that provides a platform for local groups to address emerging concerns.

This approach has been successful in the village of Paraikulum. There villagers have rehabilitated 30 ha of the upper watershed, so bringing severely degraded land under the plough for the first time in 20 years. They have constructed contour bunds and structures for harvesting water. They have dug a well, and have developed new arrangements about how labour is shared by both men and women in maintaining these new technologies. More water now percolates into the soil and recharges the wells. Surface water is now better channelled into the tank, which gives villagers a second crop on the irrigated lands. Using only locally available resources, this village of 100 households now produces an extra 100 tonnes of rice every year. At the same time, a group of women runs a nursery for tree seedlings, and keep dairy cows for milk production. Each member saves Rs10 weekly for a revolving fund to help the whole group.

The novelty of the approach is beginning to be recognized by the government. A senior engineer of the agriculture department visited Paraikulum. He was surprised to learn about local maintenance, water management, profit sharing and involvement of women. He agreed that it was important to involve farmers before planning any project for them, and so persuaded the district collector to agree to a training programme in participatory appraisal methods. Now the government has made participatory methods a part of their nearby large watershed project and are paying village motivators from Paraikulum to help them.

Sources: Devavaram, 1994; John Devavaram, personal communication

CASE 8: SOCIAL CENTRE IN MAHARASHTRA STATE, INDIA

Social Centre (SC) has worked in the drought-prone district of Ahmednagar since 1966. The average rainfall is 500 mm but, in some years, 60–70 per cent of this falls in June. Soils are poor and degradation on the common lands is widespread. For the first 16 years, SC provided credit and support to individual farmers, but since 1982 has focused on comprehensive watershed development. This is a participatory approach that brings together a wide range of actors, including local people, voluntary and government agencies, agricultural universities and banks, so that their collective resources can have an impact on village resources.

SC works in about 10 villages with a population of 5500 and a land area of 6000 ha. Pimpalgaon Wagha is one of these villages. It has 879 people and 840 ha of land. Prior to 1988, the hills were bare, the wells ran dry in summer, and most of the fuel and fodder had to be imported. Over several months, SC staff went to the village to listen to local people. Local people were also involved in surveying village resources. A village watershed committee was nominated by consensus, and has improved rainwater harvesting through construction of checkdams and weirs; soil conservation through gully checks, bunds, drains and percolation ponds; water conservation through more efficient irrigation systems; and water supply and sanitation. Many training courses were arranged locally for men and women.

As a result, there have been substantial changes. Before 1988, there were 75 seasonal wells, of which only 40 had an 8 months' supply of irrigation water; now these supply 9–11 months of water, even in dry years. Potable water is available year round. The irrigated area has increased from 60 to 168 ha. Village milk production has increased from 150 to 1400 litres per day; the diversity of crops grown has increased, with many farmers now cultivating pulses along with millet and sorghum. Yields of irrigated sorghum have increased from 1250 to 2200 kg/ha and of rainfed sorghum from 200 to 700 kg/ha. Household grain production has increased by 42 per cent for rainfed farmers and by 100 per cent for those with some irrigation. Local employment has also increased and now the landless have work for nine rather than just three months. Eight families who previously had left the village have returned; during the 1991 drought, there was no migration to canal-irrigated villages.

Many new village institutions have been organized by the people themselves, including a grain bank, a women's group, a youth group, a credit union, a dairy cooperative and an agricultural co-operative society. All are represented on the village watershed committee. Many of these groups have saved substantial sums of money and make loans to their own members. The contributions from the women's group, for example, are Rs5 per month per member and the revolving fund now has some Rs10,000. Private moneylending has now ceased altogether.

Sources: Lobo and Kochendorfer-Lucius, 1992; Crispino Lobo, personal communication

CASE 9: WATERSHED DEVELOPMENT BY THE GOVERNMENT OF RAJASTHAN, INDIA

The Watershed Development and Soil Conservation Department of the Government of Rajasthan (GoR) was set up in 1991 to implement a participatory approach for integrated watershed development. Rajasthan is the most rainfall-dependent state in India. Since the 1940s, groundwater levels have fallen dramatically, forest lands have been overused and community institutions undermined. Since 1957, the GoR had spent Rs705 million on implementing soil conservation works on 586,000 ha. However, these measures were scattered, uncoordinated and executed entirely by government with people only participating as wage labourers. The impacts were poor: *'field observations confirm... near zero maintenance by the beneficiaries'* (Krishna, 1993). The challenge is huge: some 25 million ha of Rajasthan need a watershed development type of treatment. With the high cost of past approaches, combined with the poor maintenance, the GoR has come to appreciate that people's initiatives are essential for success.

Each selected watershed of some 1500 to 2000 ha is treated for a period of 5–7 years, so that government can observe the impacts and change its approach as necessary. The process involves working with local users' committees that are elected by communities. The number in each watershed depends on local wishes – there may be one per micro-watershed of 200–30 ha or one for the entire watershed. The GoR has not imposed any standardization of constitutional matters for these users' committees. The technologies are low cost, and based on indigenous and biological technologies. These include strips of vetiver and other grasses on the contour; contour bunds and contour cropping; field bunds; drainage line treatment; and regeneration of common lands with shrubs and trees. These technologies are developed through a process of participatory planning at village level. Local people and government officials are jointly involved in analysis, technology selection and adaptation, and development of the treatment plan.

Field and contour bunds have more than doubled sorghum and millet yields to 400–875 kg/ha (with no addition of fertilizer); and grass strips have improved yields by 50–200 per cent to 450–925 kg/ha. This means that an investment of Rs800–1000/ha can be paid back in less than two years. Grass production using conservation measures, including agave and vetiver in v-shaped ditches, or v-ditches alone, is improved by 10–20 fold from the very low 25–35 kg dry matter/ha. Some 120,000 ha were treated under watershed development work in both 1992–3 and 1993–4 and the users' committees have fostered a sense of ownership among local people.

The process of joint watershed development also involves close work between GoR and NGOs; the training of four to five people from each village in techniques of watershed development; and effective internal monitoring to ensure that HQ staff understand the diverse needs of local people. There are concerns over the distortions created by subsidies. However, *'the operating principle is that if farmers execute the work and also pay a part of the cost, they will do so because they have positive expectations from the treatments, and not as a way to get labour income from the government... Eventually, the projects will have to be unsubsidised. However, this is bound to take time as the experience of cost-sharing in governments is a new one'*.

Source: Krishna, 1994

CASE 10: THE PIDOW PROJECT, KARNATAKA STATE, INDIA

Participative and Integrated Development of Watersheds (PIDOW) is a collaborative project between the Dryland Development Board of the Government of Karnataka, the NGO Myrada and the Swiss Development Cooperation. It began in 1986, and now covers 29 watersheds in drought-prone Gulbarga District. The project has had to be experimental and flexible, as not only were these partners unaccustomed to working together, but local people expected government to act as they always had in the past. Attitudes and values have had to be challenged in order to ensure the participation of local people as effective stakeholders

The starting point of the process of watershed regeneration is the establishment of relationships of mutual trust. Staff live in or close to the watersheds, and visit people regularly in places and at times convenient to them. Project staff organize traditional street plays, making special efforts to relate to all local groups during the plays. Meetings are then held with each of the groups within the village, the main objective of which is to establish an organized pattern of group discussions. These help project staff to get further insights into village dynamics and interrelationships between the various groups, and whether these are conducive to mutual cooperation for resource management. One or more problems that are common to the groups then become the subject of group analyses through a PRA exercise, the focus of which is to work out a strategy for tackling common problems. After implementation, PRA methods are again used to review and evaluate the experience.

Local people are exposed to credit groups working well elsewhere. Small farmers and landless are deeply dependent on traditional sources for credit, and so the development of local credit management groups can have a profound impact on their livelihoods. It is only now that trust and cooperation have been established that the project begins to work specifically on watershed management. They introduce various simple concepts at group meetings, including the importance of protecting a single drainage system; the notion that local people can take charge of any work in their watershed; and that institutions are needed to support these changes. A wide range of PRA methods are used to develop and construct a treatment plan satisfactory to all. A Watershed Development Association is then formed to oversee the plan.

The process has had a substantial impact in these watersheds. Agricultural yields have increased by 50–100% for sorghum and millet, and 100–200% for rice. Local people have shown their capacity to save money; with project staff they have developed resource-conserving technologies appropriate to their own conditions, including earthen bunds, boulder bunds, land reclamation, diversion drains, gully checks, silt traps and farm ponds. They have formed self-sustaining village institutions, many of which are now federated into higher level apex organizations.

Source: Fernandez, 1994

CASE 11: INTEGRATED PEST MANAGEMENT FOR RICE PROGRAMME IN INDONESIA

In 1986, a presidential decree banned 57 brands of pesticide on rice and established a national IPM programme. A symbol for the programme was created and widely used on placards and T-shirts to increase public awareness. It was recognized that effective pest management required coordinated action at community level and the aim was to make farmers experts in their own fields. They attend farmer field schools (FFSs), which are 'schools without walls'. These are spread over a single rice season, running one morning a week for 12 weeks. In this non-formal learning environment, farmers learn a whole new set of principles, concepts and terms relating to rice, pest and predator management.

Farmers are encouraged to observe their fields carefully, and to use a visual method, called agro-ecosystem analysis, for analysis. Farmers draw the rice plant in the centre of a large piece of newsprint, and include details of tiller number, diseased leaves, water level, rat damage, weed density, and insect pest and predators population density. This drawing is then used as a focus for discussion. After making management decisions for the next week, each presents and defends their summary to the other trainees. This agro-ecosystem analysis allows trainees to integrate their skills and knowledge, and trainers can immediately evaluate trainees' abilities. Dyes are also used in knapsack sprayers to show farmers how much pesticide ends up on them when spraying. Insect life cycles are discovered by rearing insects in what is called the insect zoo, which is used to observe predation and parasitism.

By 1993, 110,000 farmers had completed a full session at a FFS. The evidence already indicates significant changes in behaviour. One survey of 2000 of the field school graduates found that rice yields had increased by 0.5 t/ha on average. At the same time, the number of pesticide applications had fallen from 2.9 to 1.1 per season. The cost of pesticide applications fell by more than half, with the greatest savings for farmers in Sulawesi and Sumatra. About a quarter of all farmers are now applying no pesticides. In some villages, more than half use no pesticides.

Many of the FFSs have continued to be active as farmer IPM groups. These groups meet to discuss farming problems; monitor pest and predator populations in their villages; conduct village-wide campaigns to control rats (something farmers found possible to accomplish before the programme); and extend IPM management skills to neighbouring farmers and villages. Since 1990, some 20 per cent of the farmer training has been self-funded by farmers. The impact is best described in the words of one farmer graduate, who said '*After following the field school I have peace of mind. Because I know now how to investigate, I am not panicked any more into using pesticides as soon as I discover some pest damage symptoms*' (in van der Fliert, 1993).

With a programme of this size, there are obviously some difficulties. These include when extension workers set up curricula for field studies without involving local farmers, which leads to reduced attendance by farmers at such sessions; and the selection of IPM participants, which is currently restricted by government only to official members of existing farmers' groups, who tend to be male owner–operators and cultivators.

Sources: Winarto, 1993, 1994; van der Fliert, 1993; GoI, 1992; Kenmore, 1991; Matteson, 1992; Matteson et al, 1993; Kingsley and Musante, 1994

CASE 12: THE CATCHMENT APPROACH OF THE MINISTRY OF AGRICULTURE, KENYA

Conservation programmes in Kenya have long produced only patchy and unsustainable conservation of soil and water. In the 1980s, the government recognized that the only way to achieve widespread conservation was to mobilize people to embrace resource-conserving practices on their own terms. The Catchment Approach was adopted in 1988 as a way of concentrating resources in a specified catchment (typically 200–500 ha) for a limited period of time (generally one year), during which all farms are conserved. Maintenance would then carried out by the community with the support of local extension agents. All financial subsidies were stopped, and resources allocated instead to training, tools and farmer trips.

The Soil and Water Conservation Branch, supported by SIDA, adapted a range of methods from PRA into the planning phase of the Catchment Approach. Interdisciplinary teams drawn from various government departments, together with staff of local and international NGOs actively working locally, work for about a week in a catchment. There is usually three to four days of intensive fieldwork, in which the teams work with farmers to build up a rich picture of local skills, knowledge and perspectives. On the final day, a public meeting, or *baraza,* is held, during which findings are presented in visual form for those present to comment on and suggest changes. Following these exchanges, a Catchment Committee of farmers is elected as the local institution responsible for coordinating implementation. The team constructs a detailed map of the catchment, and with the committee plans and implements the soil and water conservation measures for each of the farms.

The number of farms fully conserved each year in Kenya with various conservation measures has risen with the implementation of the Catchment Approach from 59,450 in 1988 to 97,650 in 1991–2. In addition, each year some 500,000–800,000 m of cut-off drains and 50,000–100,000 m of artificial waterways are constructed, some 1250–2700 gullies controlled, and 1780–3600 km of riverbanks protected.

It has become clear that where there is mobilization of the community, support to strong local groups, committed local staff and collaboration with other departments in interdisciplinary planning and implementation, there is increased agricultural productivity, diversification into new enterprises, reduction in resource degradation, enhancement of water resources, improvement in the activities of local groups and independent replication to neighbouring communities. These improvements have occurred without payment or subsidy and therefore are more likely to be sustained. Output growth has been highest where there has been interactive participation. Thus, increased yields have been achieved by local people with support from an external institution now concerned with facilitating local efforts rather than directing them.

Sources: Ministry of Agriculture, Livestock Development and Marketing, passim; SWCB, 1994; Pretty et al, 1994; Kiara et al, 1990

CASE 13: THE MANOR HOUSE AGRICULTURAL CENTRE, KITALE, KENYA

Manor House Agricultural Centre was founded in 1984 in response to a three-year drought. The centre's training and research complex includes demonstration gardens and livestock facilities that provide a working model of bio-intensive agricultural systems for trainees, visitors and members of local communities. The centre provides practical training to young people, farmers and staff of government agencies and NGOs, as well as conducting adaptive research of its own. It employs 16 staff, and runs an 18-month certificate-level programme for secondary school graduates, some 25 one-week workshops each year for farmers, and one 3-month course per year for agricultural professionals.

Bio-intensive agriculture is based on the principle that production and sustainability can be enhanced using technologies that use soil, water, plant and animal resources available on most smallhold farms. The key is to improve the use of renewable resources and reduce the use of external inputs. Farmers have been able to raise soil fertility, improve productivity and increase household income, but the technologies producing these benefits do require more labour.

The Centre has trained some 6000 farmers in 185 community groups, of whom 3000 are known to have adopted bio-intensive agriculture. The main impact has been on vegetable production. Many have doubled their yields by adopting double digging and composting, using local methods of pest and disease control (such as planting sunflowers to attract predators, local plants extracts to control maize stalk borer and intercropping to reduce tomato blight). There have been big savings on pesticides, as farmers have cut out their use. Farmers have found phosphorus to be limiting over periods of six years of composting and so bonemeal is being brought in to add to compost. The centre encourages these farmer groups to train neighbouring farmers.

A successful group is the Pondeni Farmers Cooperative. This began when 15 farmers were trained at Manor House. They then deputed a keen local student to go for more training who, on return, found that the support from just 15 farmers was too little to guarantee a livelihood. He then acted as a village extensionist, persuading everyone in three villages to adopt bio-intensive gardening. The cooperative was then formed and this now pays his salary. It is very strong, active and proud of its success, and makes money from two sources: it organizes the sale and marketing of the organic produce (there are no premiums received), and it sells compost, for which there is an increasing demand. The compost is sieved and mixed with bonemeal, packed into 90 kg bags and sold for US$20 per bag.

The centre also works closely with the government at district level, encouraging informal sharing of information. Some local extension agents are being sent to the centre for training. It also passes farmers' innovations in pest and nutrient management to the Kenya Agricultural Research Institute for further testing and development.

Source: Eric Kisian'gani, personal communication, 1994

CASE 14: THE PRODUCTION THROUGH CONSERVATION (PTC II) PROGRAMME, LESOTHO

The PTC II programme is a Ministry of Agriculture initiative supported by SIDA to encourage farmers to achieve improved husbandry of their land, and so achieve better conservation and production. It aims to do this by fostering self-sustaining rural communities, capable of planning the development of their own resources for sustained use. It is currently working in the Districts of Mohale's Hoek, Mafeting and Quthing, and follows up earlier projects that emphasized soil conservation as the primary need. In contrast to past approaches, its guiding principle is that immense potential exists in villages and this should be built upon and strengthened. The extension staff at district level are organized into five or six Area Teams, comprising staff from a range of divisions of the ministry. Each Area Team chooses its own leader, who then represents them on the district agricultural officer's (DAO) District Management Team. These arrangements encourage greater multi-disciplinarity.

The programme promotes institution building from village level upwards, facilitating flows from the village and village development committee to the DAO, the Board of Farmers, the District Development Council and on the various national ministries. The programme specifically sets out to raise interest and create demand among villages, while at the same time increasing the supply of information, goods and services to satisfy these new demands. The initial focus is the Headman Village Workshops, to which all people are invited, and the common interest groups which may emerge during the course of participatory analyses and discussions. The emphasis is on establishing good relations between local people and extension staff, in order to ensure that the hitherto undervalued resources of local knowledge and enthusiasm are tapped.

There are strong feelings among rural people that they do not like plans to be made for them by others and so now that they are fully involved in developing joint plans with the Area Teams, they greatly welcome this aspect of the programme. It had been expected that 'encouragements' would be needed for villagers to implement resource-conserving practices. Instead farm families are now stating what they wish to do, and so pressuring government to give them support and help. Staff appreciate this different role of responding to farmers. Because of the team approaches used by extension, everyone is included in the process, and so shares equally in both praise and blame.

There have already been a wide range of impacts, particularly in the government agencies. Staff are positively motivated by the better relations with farmers and by the better exchanges with their colleagues. Because of the better understanding of local needs, district planning and budgeting can be more targeted on expressed needs, which should lead to more efficiency in future. Other benefits include the elimination of unnecessary journeys and the more efficient use of transport. Although it is too early to see impacts in the field, rural people are enthusiastic about the new approach and are also proving more receptive to the technical suggestions being made.

Source: Shaxson and Sehloho, 1993; Mikael Segerros, personal communication

CASE 15: THE MENAKA OASIS PROJECT, MALI

World Vision (WV) of Mali has worked with pastoral communities in the Menaka district of north-east Mali, where rainfall is now only some 200 mm per year, compared with 300 mm between 1940–65. In the past, the district's abundant grasslands and forests made Menaka the centre of a thriving pastoral economy. But recent droughts, including the catastrophe of 1984 in which half of the region's livestock perished, have led to increased land degradation, loss of forest cover and falling water tables.

Following an approach from the community, and after a series of discussions and community studies, WV began to work in the community of Intadeny on a land regeneration scheme to increase land productivity and recharge the water table. The watershed was characterized by severely eroded rainfall-collector valleys and plains feeding into lower runoff receiving areas. The first phase of the work was the construction of contour earthen dikes on a 40 ha plain near the village. The site was chosen for easy access, moderate slope and for the existing vegetation that could be regenerated. The project began by training several members of the community in water tube level use and the laying out of contours. One month after construction, the rains came and it soon became clear that these initial technologies would have to be adapted by the farmers, as breaches occurred in all dikes. Repairs were made, and the number of spillways doubled, and a protective dike uphill constructed. But as too little water then entered the cropped area, a further two passages were opened up.

As not everyone had animals in the community, the project encouraged the cultivation of sorghum, which had not been grown there before. None the less, it yielded 250 kg/ha in the first year. The next year, a gabion wall was built in the wadi to improve water harvesting. Again, adaptations were made to the technologies, but at the end of the rainy season, the water levels in the wells were 2 m higher than the year before. New attitudes began to take shape, and *'people began to realize they could stop the degradation they previously thought was beyond their control'*. This success began to draw interest from neighbouring communities, and so the project trained eight paraprofessionals from Intadeny to help them develop their own technologies.

New technologies, such as rock bunds and filter dikes, were introduced from Burkina Faso. After three years, several hundred hectares have been protected, with sorghum yields reaching 1.7 t/ha. The wider benefits include the reversal of migration trends, as people have been encouraged back to their village and the rise of the water table by 2m. Now the Intadeny paraprofessionals have formally organized as an NGO, using a locally produced slide show with before and after photographs, folk theatre and flannel graphs as part of the participatory process. They are organized in the construction of shallow wells and induced recharge of shallow aquifers, as well as soil and water conservation.

Sources: Rands, 1989, 1992

CASE 16: THE UNION OF INDIAN COMMUNITIES IN THE ISTMUS REGION, OAXACA STATE, MEXICO

The Union of Indian Communities in the Istmus Region (UCIRI) was organized by farmers from three communities in the state of Oaxaca in 1982. The area is mountainous and coffee is grown mainly on slopes, with maize and vegetables on the flatter lands. The altitude is some 400 to 1250 m; the soils volcanic; and the rainfall some 800–1900 mm between May and September. A common characteristic of the coffee farmers is their dependency on intermediaries, who control credit, buy the coffee and supply the basic necessities. The basic family income is only US$250–450 per year.

The basic aim of the UCIRI was to find ways to commercialize the coffee themselves, obtain better prices and take more control over their own livelihoods. In 1985, the organization decided to move from traditional to organic agriculture. This was partly a political decision to reduce dependency on credit, but it was also hoped that yields would improve too. Contacts with so-called 'fair trade' organizations supported these goals, as they offered received a premium on the organic coffee. Now, some 3000 families in 37 communities are members of the union.

Organic coffee cultivation demands more active management and a higher labour input from farmers. Coffee is grown in the secondary forest and farmers leave leguminous trees that are beneficial to the coffee. It is planted on the contour, and slashed weeds and pruned branches are laid on the contour too. Half-moon shaped terraces are constructed for each coffee tree. Formerly the coffee beans were depulped into waterways, causing significant water pollution. But now organic farmers return the pulp to the fields through composting. Other materials used for composting include animal manures, lime and green plant material. Mulches are used to protect the surface of the soil from erosion.

The organic farmers using this improved system produce 600–1200 kg/ha of coffee beans, an improvement of 30–50 per cent, compared with their earlier practices. These yields, though, are not as great as those on the large coffee estates. Most farmers also now cultivate their maize and beans organically.

The union has been able to build up its own infrastructure for the transport, storage, processing and export of its coffee. The premium received for the organic coffee from the fair trade organizations is used for a range of social and economic purposes, including for improving the educational systems. The union also runs a public transport system into the mountains and a medical insurance system, and owns several shops from which local people can buy basic necessities.

Source: UNDP, 1992

CASE 17: RAISED FIELDS IN LAKE TITICACA BASIN, PERU

The Proyecto Agrícola de los Campos Elevados has been working with Quechua communities in and around the District of Huatta to rehabilitate ancient raised fields. These chinampas or *waru-waru* were used widely in the Lake Titicaca basin by pre-hispanic farmers, but had fallen into disuse. In addition, many thousands of hectares had been destroyed by modern, capital-intensive irrigation projects that then failed to improve agricultural yields.

The basin, located about 3800 m above sea level, is a difficult environment for agriculture because of irregular rainfall, poor and degraded soils, and frequent and severe frosts during the short growing season. Pre-hispanic farmers had developed sophisticated ways to overcome these limitations, by focusing on diverse and intensive cropping on terraces, sunken gardens (*gochas*) and raised fields (*camellones* or *waru-waru*), together with social mechanisms to ensure efficient and collective action to achieve high and secure levels of productivity.

During the 1980s, local farmers' organizations of Huatta, involving some 500 families in 10 communities, began to reconstruct these ancient raised fields. The success of the effort was because of the community participation, and the development of effective teaching and learning materials. Although the technology was new to present-day farmers, it was they who conducted the experiments to adapt the technology to their own conditions. In 1986, the programme was taken on by the Peruvian government and has since expanded to include over 30 altoplano communities.

Raised beds require strong social cohesion for the cooperative work needed on beds and canals. For the construction of the fields, labour was organized at the individual family, multi-family and communal levels. Most of the raised fields were constructed on community-owned land that had formerly lain unused for want of local motivation and presence of appropriate institutions. The labour required for construction was between 200–900 person-days per ha, depending on local physical conditions. The raised fields are surrounded by canals, which trap silt, improve the micro-climate for crops, act as a barrier to pests and grazing animals, and act as a habitat for aquatic animals. This microclimate of beds and canals reduces frost incidence. Soil fertility is maintained by green manuring with aquatic plants, livestock manures and crop-weed residues.

This fertility encourages a highly productive agriculture. Potato yields are 8–14 t/ha without use of pesticides or fertilizers. This compares with an average of 1–4 t/ha using fertilizers for the Department of Puno. Extra crops are grown, with forage crops of oats, wheat and barley now grown in winter. These raised fields are also more resilient: one year hundreds of hectares of mechanically prepared fields of wheat and potatoes were destroyed by flooding, but raised fields adjacent to them were unaffected.

Sources: Erickson and Candler, 1989; Denevan, 1970; Donkin, 1979

CASE 18: THE NATIONAL IRRIGATION ADMINISTRATION OF THE PHILIPPINES

The National Irrigation Administration of the Philippines' government seeks to establish irrigators' associations (IAs) to sustain the operation and maintenance of small-scale irrigation systems that have received construction assistance from the government. These small-scale systems are generally less than 1000 ha, but cover about half of the country's irrigated lands. The remainder are government owned and operated.

During the 1960s and 1970s, the NIA approach was distinctly non-participatory. Engineers planned infrastructure and systems were built with only nominal local consultation. Systems often fell into disrepair, as farmers saw little reason to take on management responsibilities. During the 1980s, though, the NIA adopted a participatory approach to irrigation.

Fundamental changes were made in the NIA to support this new participatory approach. These included the introduction of motivated, mostly female, community organizers; the reorientation of site assessment procedures to reflect locally diverse conditions; the devolution of authority to make the provincial irrigation engineers responsible for overall coordination of irrigation programmes in their respective provinces; and the strengthening of agency accountability to water users.

The approach to institution building was also fundamentally different. In the non-participatory approach, farmers were expected to form IAs only shortly before construction began, when NIA personnel called farmers together to elect their officials. The participatory approach, by contrast, focuses on the association months before construction starts. Full-time organizers reside in the project area and prepare local people to work with the engineers. The organizers also continue to work with the association for at least two crop seasons under the improved system. Farmers are now involved from the very start of the project, including determining the layout of the proposed system, and constructing dams, canals and structures. Once the construction is completed, the NIA turns over full authority for the systems to the IAs.

The NIA developed their participatory approaches experimentally over time. This meant that participatory and non-participatory efforts continued side-by-side, and it has been possible to measure the impact of the participatory element alone. The primary impacts include: rice yields increased by 19 per cent in wet seasons and by 16 per cent in the dry; farmers contributions to costs increased from 54 to 357 pesos/ha; an increase from 27 per cent to 83 per cent in systems in which farmers' suggestions were incorporated in design; a fall by half in the number of NIA-built canals abandoned or rerouted; an increase in association members present at the turnover ceremony; an increase from 50 per cent to 82 per cent in remittance of amortization payments within a year; an increase in time farmers contributed to group maintenance of systems; and an improved capacity of IAs to manage their own affairs.

Sources: Bagadion and Korten, 1991; de los Reyes and Jopillo, 1986; Svendsen, 1993

CASE 19: THE FARM AND RESOURCE MANAGEMENT INSTITUTE, EASTERN VISAYAS, PHILIPPINES

The Farm and Resource Management Institute (FARMI) of the Visayas School of Agriculture works with farmers in 14 upland villages of Matalom on Leyte island. The objectives are to adapt and refine research methodologies for the uplands by working on participatory technology development. The villages are remote and unaccessible by road during the rains. Soil erosion was perceived to be a widespread problem.

FARMI began by learning from farmers about their own approaches to soil conservation. About one third ploughed across the slope, leaving 0.5–1 m unploughed strips every 4–10 m of slope. Some strips were straight, others more or less on the contour. According to the farmers, sometimes the strips were broken by heavy rains and animals. FARMI then took a group of farmers on a study tour to Cebu to see two upland projects of the Mag-uugmad Foundation, a local NGO. There the farmers saw for themselves, and heard from other farmers, about different contouring techniques using various combinations of grasses, trees and shrubs. On return, the farmers arranged for a community meeting, at which they presented what they had seen and learnt. This was then followed by various informal and smaller group meetings, including with the parent–teacher association. Photographs taken during the trip were important in helping to persuade everyone to support the initiation of a local soil and water conservation project using the traditional and reciprocal *alayon* labour exchange system.

The next stage involved *alayon* members testing and experimenting with a range of technologies. These included diking, in which soil is piled on the contour with legumes and grasses planted on both sides; slash and pile, in which cut grasses and shrubs were piled along the lines, and branches used to pin them to the soil; and the establishment of contour hedgerows when the land was fallowed or directly in the crops. Side by side, farmers tested a wide range of indigenous and introduced variations. Interest grew and faded, and grew again, until eventually farmers had chosen the most appropriate technologies for themselves. As a result, yields improved on these conserved fields. There are now 29 farmer *alayon* groups with 300 members in 14 upland villages. By early 1994, more than 100 farmers in neighbouring villages had independently adopted contour technology.

One emerging concern among farmers was their past experience of landlords grabbing improved lands – people were afraid that FARMI would do the same after they had put in all the effort. There have been many ingredients for success: the start with farmers' technology, the cross-farm visits, and the gradual development and adaptation of new technologies, and the personal behaviour and attitudes of the project staff that have been central.

Sources: Balbarino et al, 1992; Balbarino and Alcober, 1994

CASE 20: THE REHABILITATION OF GAL OYA IRRIGATION SCHEME IN SRI LANKA

In 1980, the Sri Lankan government's Irrigation Department, the Agrarian Research Training Institute (ARTI) and the Cornell Rural Development Committee began working with smallholder farmers in the Gal Oya irrigation scheme. At the time, this was the largest and most run-down scheme in Sri Lanka. The approach to rehabilitation was to place young community organizers in the field, who would encourage farmers to form water users' associations, so that they could solve irrigation problems for themselves. The historical context was one of 30 years of conflict and non-cooperation. A senior official in the Irrigation Department said *'if we can make progress in Gal Oya, we can make progress anywhere in Sri Lanka'*.

The organizers were recruited and trained by ARTI to live and work in the communities, with the primary objective of ensuring that all plans were the farmers' own. Water users' groups were formed, but not forced on farmers. The organizers, also called animators, promoters or motivators, worked as catalysts to stimulate and nurture local organization. The usual approach to establishing rural organizations (calling a meeting, passing a constitution and electing officers) was known not to lead to sustainable organizations. Here the approach was to let groups evolve, beginning first with problem identification and collective action, which could lead to formal organization later. The process brought forth more tested and altruistic leadership, who had solid support among their members.

The project has rehabilitated 10,000 ha, with benefits exceeding costs by a ratio of 1.5 to 1. The economic benefits of the project depend primarily on increased water use efficiency, which enabled farmers to increase their cropping intensity and thereby raise production. There were also some increases in yields. These changes in the efficiency and equity of water use have been dramatic and long lasting. The number of complaints received by the Irrigation Department about water distribution fell to nearly zero, as adjustments were made by farmers and field-level staff. Before the project, 80 per cent of channel gates were broken; afterwards this problem practically disappeared. Farmers' organizations have maintained themselves, progressed institutionally and developed their own capacity for dealing with problems.

Farmers' organizations, once established, have used their new capabilities to deal with many other needs, such as crop protection, credit supply, settlement of domestic disputes, land consolidation and reducing drunkenness. Bureaucratic reorientation has been essential for success. This has been promoted among engineers and officials by demonstrations of farmers' knowledge and ability to achieve unexpected improvements. This iterative process has been crucial: *'displays of initiative and intelligence by farmers gained some respect from officials, and this in turn encouraged farmers to show more capability, which again increased the respect accorded them by officials'* (Uphoff, 1994).

Sources: Uphoff, 1994, 1992a

SUMMARY

When the process of agricultural development focuses on resource-conserving technologies, local groups and institutions, and enabling external institutions, then sustainable agriculture can bring both environmental and economic benefits for farmers, communities and nations. Until recently, most of the empirical evidence has been limited to findings from research stations, and on-farm trials and demonstrations. This chapter, however, has focused on the evidence derived from the improvements made by farmers and communities. The most convincing evidence is from countries of Africa, Asia and Latin America, where hundreds, and possibly thousands, of communities have now regenerated their local environments. In the industrialized countries, the evidence is generally at best from individual farms.

What the evidence shows is that a more sustainable agriculture leads to:

- stabilized or lower yields in industrialized countries, coupled with substantial environmental improvements;
- stabilized or slightly higher yields in Green Revolution lands, with environmental benefits;
- substantially increased agricultural yields in complex and diverse lands based mostly on available or local resources.

Twenty case studies from communities in Brazil, Burkina Faso, Honduras, India, Indonesia, Kenya, Lesotho, Mali, Mexico, Peru, Philippines and Sri Lanka show in detail what can be achieved in terms of agricultural growth that does not damage the environment. These cases point the way forward. None, though, is in any way perfect. All will have to change over time, adapting to changing circumstances, new challenges and emerging threats.

All will eventually need supportive national policies if they are to flourish and spread.

8

AGRICULTURAL POLICY FRAMEWORKS AND INSTITUTIONAL PROCESSES

'A farmer should live as though he were going to die tomorrow; but he should farm as though he were going to live forever'
East Anglian proverb, in George Ewart Evans, 1966

THE NEED TO FOCUS ON POLICIES AND INSTITUTIONS

The emphasis throughout this book has been on the practical conditions necessary to achieve a more sustainable and self-reliant agriculture. All successes, including those detailed in Chapter 7, have satisfied to a reasonable extent the three conditions reflected in Figure 1.2 and in Chapters 4, 5 and 6. Each has made use of resource-conserving technologies. Each has encouraged action by groups and communities at local level. Each has had external organizations working in a supportive and participatory way with local people, and in networks and alliances with other organizations.

But most successes are still only small scale. As has been indicated in Chapter 7, probably only a few thousand communities throughout the world have benefited. This is largely because the fourth element, an enabling policy environment, is missing in almost every country. Although there are rare exceptions, most policy frameworks still actively encourage agriculture that is dependent on external inputs, technologies and knowledge. What has been achieved so far, largely at community level, represents what is possible in spite of existing policies.

It will be essential for governments to establish a range of public policies to encourage proper incentives for sustainable and efficient management of agricultural and natural resources. This will mean removing existing disincentives and putting in place new, encouraging measures. This chapter focuses both on what has gone wrong in the past, and on what governments and other institutions are already doing to reverse some of these problems.

The final chapter of this book draws together the policies that are known to work, so as to indicate the many ways that different actors and institutions can proceed should they wish for a more sustainable and productive agriculture.

AGRICULTURAL POLICIES

Types of Support for Agriculture

Governments have long provided public support to their domestic agricultural sectors. Through a wide range of direct and indirect monetary transfers from consumers and taxpayers to farmers and other producers, they have sought to ensure that agriculture provides the food and other products needed by the non-farming population. There are five main categories of agricultural policy measures (OECD, 1993a; FAO, 1993a).

1. Market price support, in which producer and consumer prices are influenced by a range of policies that include levies or tariffs on goods entering the country, so raising the price of imports; guaranteed prices for domestic produce, usually at level above world prices and/or those paid by domestic consumers; quotas on imports; and the subsidization of exports to ensure sales in international markets.
2. Direct payments, in which money is transferred directly from taxpayers to farmers without raising prices to consumers. These include payments to encourage both production and conservation oriented technologies, and can be monetary, or in kind, such as food for work.
3. Input cost reduction, in which measures are taken to lower input costs, so encouraging their greater use. These include subsidies for pesticides, fertilizers, water, electricity and credit.
4. Provision of general services, in which measures are taken to lower the long-term costs of research, extension, education and planning services, so ensuring that farmers have access to new technologies as well as the capacity to adapt them to their own conditions.
5. Other indirect support, in which regions receive rural development support or tax concessions are granted for farmers using particular activities or measures.

These policy measures have been widely used in the last half century to encourage the adoption of modern methods of farming. These have led to substantial increases in food production in many parts of the world (see Chapter 2). The development of agricultural policy, and recent transition towards more conservation oriented support, is described for Britain in Box 8.1.

It should be noted, though, that the type of policies adopted by countries

Box 8.1 The development of agricultural policy in Britain over a 50-year period

The principal aim of agricultural policy since 1945 has been increased productivity. Financial support from the state, and later the European Commission, has been linked to output and markets for produce have been guaranteed. This began in the 1940s when provisions were made under various Agriculture Acts for price subsidies of crop and livestock products, for grants for field drainage and other investment in fixed assets, and for subsidies of fertilizers and lime. Provisions were also made for ploughing grants, per capita payments for beef calves, hedgerow removal and an annual price review procedure to set guaranteed prices.

These grants and subsidies continued into the 1970s, with further provisions to encourage the amalgamation of farms and the early retirement of farmers. It was not until after Britain entered the European Community in 1974 that many of these direct grants and subsidies were discontinued. None the less, the Common Agricultural Policy continued to support agricultural prices, protect markets and provide for export subsidies. Guaranteed prices have generally been well above world market prices. Productivity increased so rapidly that many farmers became proud of achieving cereal yields of 10t/ha, enabling them to enter the so-called '10-tonne club'. By the 1980s, food commodities began to accumulate at an alarming rate in the European Community, producing the first food 'mountains'. In 1993, these surpluses absorbed some 20% of the Common Agricultural Policy budget for storage alone. A further 28% was expended on export subsidies.

In the 1990s, the policy signals for continued increases in production have begun to be reversed with the adoption of set-aside. This mechanism forces farmers to remove from production a set amount of their land (usually 15%) if they wish to continue to receive any other forms of government direct support. Many of these are increasingly able to receive support if they farm in an environmentally sensitive way, and if they provide access to the countryside for consumers.

Sources: Pretty and Howes, 1993; Bowler, 1979; Bowers and Cheshire, 1983

affect the agriculture in other countries. Many countries and country groups of the OECD, for example, have subsidized their agricultural exports to keep down prices and ensure sales. This has tended to keep down international prices. They have also levied tariffs and imposed quotas on imports, so as to protect domestic producers (OECD, 1993a).

It is the intention of the Uruguay Round of the GATT (signed by more than 100 countries and leading to the creation of a new World Trade Organization) and NAFTA (signed by the USA, Canada and Mexico) to 'liberalize' and deregulate agricultural markets, so permitting trade free from these measures of protection. However, there are sharply different views over whether these agreements will hinder or support the development of more sustainable futures.

Levels of Support

A widely used concept for measuring the total amount of assistance to farmers is the producer subsidy equivalent (PSE). This measures all the transfers to producers from consumers that result from agricultural policies.

Two of the ways that the PSE can be expressed are the total PSE, which is the total value of transfers to farmers; and the percentage PSE, which is the total value of transfers as a percentage of the total value of production (valued at domestic prices), and adjusted to include direct payments and to exclude levies (OECD, 1993a). A similar measure is the consumer subsidy equivalent (CSE), which measures the implicit tax imposed on consumers by agricultural policies. In practice, there is a close relationship between PSE and CSE measures. All market price support policies, for example, that separate domestic prices and world prices raise the prices to consumers. Such a transfer from consumers to producers is equivalent to a subsidy to producers and a tax on consumers.

Countries of the OECD provide substantial support to their farming sectors (Table 8.1). The total support is some US$180 billion, which is represents some $16,000 per full-time farmer-equivalent, or $179 per hectare of farmland. There is, though, great variation between countries: from the high level of support in Japan of 71 per cent of the total value of agricultural production to just 3 per cent in New Zealand.

Table 8.1 Agricultural support measured by Producer Subsidy Equivalent (PSE) in selected countries of the OECD, 1992–3

	Total PSE (US$ billion)	*%PSE (support as % of total value of agricultural production)*	*PSE per farmer (US$)*	*PSE per hectare of farmland (US$)*
Australia	1.3	12%	5000	3
Canada	6.0	44%	18,000	115
European Union	85.4	47%	13,000	795
Japan	35.7	71%	21,000	9708
New Zealand	0.1	3%	1000	7
Sweden	2.7	57%	34,000	957
USA	33.8	28%	20,000	97
OECD	179.5	44%	16,000	179

Source: OECD, 1993a

Although there have been calls for reducing these levels of agricultural support, most countries have not so far embarked on substantial reforms (OECD, 1993a). Some countries have begun to reduce support prices and replace them with systems of direct payments. Many of these are linked to a reduction of production incentives so as to lower production.

The contrasts between the countries of the OECD and those of the south are enormous. Agriculture is generally given a low priority in policy terms. Despite concerns about ensuring agricultural growth, many countries have pursued urban-biased policies that have strongly discriminated against agriculture (World Bank, 1993; FAO, 1993a; Kesseba, 1992). By selecting macro-economic policies that ensure high real exchange rates and by protecting industry, many countries have distorted the domestic terms of trade against agriculture. Many have imposed heavy direct taxes on agriculture and held farm gate prices below world prices. One recent study of 18 countries by the World Bank found that, on average, the net effect of these policy interventions has been an enormous monetary transfer away from agriculture (Schiff and Valdes, 1992). Between 1960–84, this averaged 46 per cent of agricultural GDP per year. The effect has been to depress agricultural growth.

Many countries have sought to recompense farmers by also subsidizing farm inputs, such as pesticides, fertilizers, credit and irrigation. These subsidies have clearly increased the use of these inputs and so have contributed to agricultural growth. However, these rarely benefit the small and poorer farmers, and also tend to encourage excessive use and wastage (see Chapters 2 and 3). Subsidized water and electricity, for example, have encouraged the expansion of irrigated agriculture, which has in some places contributed to a decline in groundwater levels and accelerated accumulation of salts in the soil. In addition, not all of these input subsidies go directly to farmers. Many leak out to public and private manufacturers, and deliverers of services and inputs (Singh, 1992).

The financial cost can also be great, with subsidies on pesticides, for example, costing some governments tens to hundreds of millions of dollars each year (Table 8.2). For Colombia, Egypt, Honduras and Senegal, this meant an annual per capita support of between US$2–6 in the 1980s. For China, the per capita support is low, but the total expenditure was nearly $300 million annually (Repetto, 1985).

Table 8.2 Pesticide subsidies in selected countries in the early 1980s (in US$)

Country	*Subsidy as a proportion (%) of the full cost*	*Total cost (million US$)*	*Per capita (US$ per person)*
China	19%	285	0.26
Colombia	44%	69	2.23
Ecuador	41%	14	1.40
Egypt	83%	207	4.12
Ghana	67%	20	1.43
Honduras	29%	12	2.50
Indonesia	82%	128	0.73
Senegal	89%	44	6.29

Note: Indonesia cut pesticide subsidies from 82% to zero during the late 1980s

Source: Repetto, 1985

Expenditure on fertilizers can also be high. The Pakistan government spent some Rs2.45 billion in 1980–1 on fertilizer subsidies, which was some 75 per cent of their total agricultural budget (Leach,1985). And in India, the total expenditure on subsidizing all inputs was some 2.8 per cent of GDP in 1990 (Rajgopalan, 1993). The largest contribution is for fertilizer subsidies, the annual expenditure for which has grown from Rs0.2 billion in 1971 to some Rs36.5 billion ($1.21 billion) in 1990 (Gulati, 1990). In all, some Rs193 billion were spent in almost 20 years. None the less Gulati still concludes that farmers had been 'net-taxed' because of crop pricing and other policy measures.

Recent reforms have been occurring, particularly with the increase in structural adjustment of macro and agricultural sectors promoted by the World Bank. A central condition for loan provision has been the agreement by national governments to adopt a wide range of policy reforms that have sought to increase agricultural output and facilitate the privatization of many agricultural marketing and services functions. The main reforms in these adjustment programmes have been reform of exchange rates and industrial policy to make terms of trade less discriminating against agriculture; removal of state monopolies on input supply, processing and marketing; removal of price distortions for inputs; and making agricultural markets open and competitive. These have also been supported by a variety of measures to reform selected public institutions (World Bank, 1993). None of these programmes, however, have yet sought to facilitate a transition to a more sustainable agriculture and so many policies still strongly discriminate against sustainability.

Policy Discrimination Against Sustainable Agriculture

Most, if not all, of the policy measures used to support agriculture currently act as powerful disincentives against sustainability. In the short term, this means that farmers switching from high input to resource-conserving technologies can rarely do so without incurring some transition costs (see Chapter 4 for transition costs). In the long term, it means that sustainable agriculture will not spread widely beyond the types of localized success described in Chapter 7.

The principal problem is that policies simply do not reflect the long-term social and environmental costs of resource use. The external costs of modern farming, such as soil erosion, health damage or polluted ecosystems, are not incorporated into individual decision making by farmers. In this way resource-degrading farmers bear neither the costs of damage to the environment or economy, nor those incurred in controlling the polluting or damaging activity.

In principle, it is possible to imagine pricing the free input to farming of the clean, unpolluted environment. If charges were levied in some way, then degraders or polluters would have higher costs, would be forced to pass them on to consumers and would be forced to switch to more resource-conserving technologies. This notion is contained within the polluter pays principle, a concept used for many years in the non-farm sector (OECD, 1989). However, beyond the notion of encouraging some internalization of costs, it has not been of practical use for policy formulation in agriculture.

In general, farmers are entirely rational to continue using high input degrading practices under current policies. High prices for particular commodities, such as key cereals, have discouraged mixed farming practices, replacing them with monocultures. In the USA, for example, current commodity programmes inhibit the adoption of these resource-conserving practices by artificially making them less profitable to farmers. In Pennsylvania, the financial returns to continuous maize and alternative rotations are about the same. But the continuous maize attracts about twice as much direct support in the form of deficiency payments. In addition, continuous maize farms use much more nitrogen fertilizer, erode more soil and cause three to six times as much damage to off-site resources (Table 8.3). Putting this together shows that a transition to the resource-conserving rotations would clearly benefit both farmers and the national economy.

Table 8.3 Comparison of the impact of conventional and alternative rotations on off-site damage and yields in Pennsylvania and Nebraska, and the subsidies received by farmers for each rotation

	Continuous maize	*Alternative rotation (maize, barley, beans oats, clover, grass)*
Pennsylvania		
Crop sales	267 $/ha	203–220 $/ha[2]
Production costs	247 $/ha	188–198 $/ha
Returns	20 $/ha	15–22 $/ha[3]
Deficiency payments	145 $/ha	62–89 $/ha
Nitrogen fertilizer use	168 kg/ha/yr	0 kg/ha/yr
On-farm depreciation (–) or appreciation (+)	–62 $/ha/yr	+7 to 20 $/ha/yr
Soil erosion	23 t/ha/yr	8–15 t/ha/yr
Off-site resource costs[1]	230 $/ha	$/ha42–71
Nebraska		
Crop sales	310–470 $/ha	362–391 $/ha
Production costs	169–237 $/ha	156–175 $/ha
Returns	73–300 $/ha	206–218 $/ha
Deficiency payments	124 $/ha	62 $/ha
Nitrogen fertilizer use	95 kg/ha/yr	45 kg/ha/yr
On-farm depreciation (–) or appreciation (+)	–19 $/ha/yr	+ 2 to 10 $/ha/yr
Soil erosion	16 t/ha/yr	5–9 t/ha/yr
Off-site resource costs[1]	30 $/ha	$2–10 $/ha

1 Both soil depreciation and off-farm costs included
2 This rises to $309 after transition period (a period of conversion from high to low input)
3 This rises to $121 after transition period
Source: Faeth et al, 1991

In this context of systemic support for high input agriculture, many countries have sought to 'bolt-on' conservation goals to these policies. These have tended to rely on conditionality, such as 'cross-compliance', whereby farmers receive support only if they adopt certain types of resource-conserving technologies and practices. Some of these pay farmers to remove land completely out of production, such as the set-aside and conservation reserve programmes of the EC and USA, while permitting normal agricultural practices on the rest of their farms. In the USA, various recent farm bills have mandated soil conservation and wetland conservation practices for farmers in certain areas if they wish to receive subsidies.

Farmers have to submit farm reports and plans, and seek official approval before they get support. In the UK, most environmental or conservation-oriented schemes are voluntary. These include the national Environmentally Sensitive Areas and Countryside Stewardship, plus other local schemes, which covered less than 10 per cent of the farming area in 1994. Farmers receive payments for entering into long-term management agreements, with the aim of protecting valued landscapes and habitats. Only the Nitrate Vulnerable Zones are compulsory and farmers must adopt practices that do not lead to nitrate leaching to aquifers.

Such cross-compliance occurs widely in the south too. As has been illustrated in Chapter 2, the process of agricultural modernization has widely involved encouraging farmers to adopt modern practices through the linkage of credit or other benefits. If farmers wish to receive one type of support, they must adopt a particular set of technologies and practices. In many cases, heavy coercion has been used to achieve levels of adoption needed or expected.

The problem with all these cross-compliances is that they do not necessarily buy the support of farmers and rural people. They create long-term resentment and can lead to the complete reversal of practices when policies change or the money runs out. There are also important contradictions in having modernist and production-oriented policies that only seek conservation at the margins. The piecemeal action that focuses on individual aspects of a farm, such as a riverside meadow or chalk grassland, does not encourage integrated farming. As a result, farmers may be encouraged to manage sustainably one particular field, but not the rest of the farm. In this way, the internal linkages and processes essential for sustainable agriculture are not promoted.

For the full benefits of sustainable agriculture to accrue, policies must be more integrated and more directed towards alternative practices (see Chapter 9).

LEGAL ISSUES AND REGULATIONS

Legal Traditions

The first place that such integration for sustainable agriculture can occur is in the provision of appropriate regulatory frameworks. These can be used either to encourage or to discourage particular types of farming practice or technology and can act from local to national levels.

Several traditions of law may exist in one country at the same time. This might include 'formal' law established by the state and customary rights framed at local level. Local laws have long been important in maintaining indigenous management systems throughout the world. These have regulated resource use and ensure a high degree of cooperation among farmers (see Chapter 5). What distinguishes local regulations from national laws is that they tend to reflect local peculiarities and conditions. This means they can be more finely-tuned and more effective (see Table 5.1). Many of these regulations and their supporting institutions, however, have been replaced during modernization by national level institutions (see Chapter 3). This loss of local capacity to manage, regulate and fine resource users has led directly to greater degradation in many societies. This has been particularly acute where there has been the effective conversion of closed-access common resources to open-access ones (Kottak, 1991; Jodha, 1990).

There are two traditions of formal law that are important for agricultural and resource users. These are private law, in which individuals damaged or threatened by eviction or pollution can use to take offenders to court; and public law, which provides various mechanisms for the control and support of public agencies and individual actions (Macrory, 1990; Conway and Pretty, 1991). Private law is concerned with the rights of individuals and how they can obtain redress for damages they have suffered. In theory, it is a powerful deterrent, but in practice its usefulness is limited. The major drawback to private law solutions is the considerable cost of litigation that often arises, particularly when cases involve a large number of victims. In practice, small farmers or pastoralists in most countries cannot seek redress for grievances, however well justified, in courts of law.

More important for them is the role that public law can play in shifting the balance between different types of agriculture and resource use. It can make modern agriculture less favourable through environmental protection and anti-pollution laws, and the setting of standards, and it can make sustainable agriculture more favourable through the granting of appropriate rights and access to local users natural resources. Both regulations and land reform are important components of any policies aimed at supporting a more sustainable agriculture.

Regulations and Standards

Public law plays an important role in establishing whether farmers and local people will be able to adopt resource-conserving technologies. More often than not, such legislation restricts the actions of individual farmers by ensuring that certain things are not done. Direct regulations enforced by penalties are among the most common forms of policy instruments used by governments (Conway and Pretty, 1991). These measures can influence a wide range of aspects of farming practice. These include, for example, restrictions on polluting or harmful practices, prohibition of 'undesirable' practices, licensing agreements, and standards for pesticide production and use. Specific examples include:

- declaring illegal the cultivation of certain types of land, such as on steep slopes, on land by riverbanks, or on government land and in forests;
- banning the use of harmful pesticides for health or pest resurgence reasons;
- seeking to prevent spraying during stated periods before harvest, so as to prevent residue accumulation;
- restricting the use of antibiotics or growth regulators for livestock;
- restricting the cutting of trees – in the name of conservation it is illegal in many countries to fell trees growing on private land;
- establishing upper limits on animal stocking densities;
- certifying crop varieties before multiplication and distribution to farmers;
- establishing standards or limits for the contamination of foodstuffs or drinking water by pesticides or nitrates.

In general, regulations seeking directly to influence farmers' behaviour are both more difficult to implement and less likely to be complied with than those acting on, say, the supply of inputs. Implementation and policing are difficult and costly when compared with other economic activities that are less diffuse, such as industry. National policies are sometimes unfair when followed strictly at local level, as the case of illegal tree planting in a drought-prone village in India shows (Box 8.2). The best level for establishing regulation is at local level by local groups. In this

Box 8.2 Where tree planting is a crime

Gopalpura is a poor, drought-stricken village at the base of the Aravali Hills. For its inhabitants environmental improvement has been a long and unending battle with the law. In 1986, helped by a local voluntary agency, the villagers built some small earthen structures called *johads* on their fields and grazing lands to harvest water and improve percolation to recharge wells, and plug the local nullah. The structures were declared 'illegal' by the State's irrigation department under existing laws, and the agency was served a notice to remove the structures as all drains and nullahs are government property. After a protracted fight, the charges were finally dropped by the administration, especially after it was seen that government structures were washed away while the *johads* were still effective.

In 1987, again assisted by the agency, the villagers decided to plant trees on the ridge at the top of their watershed. To keep animals out, a protection wall was also built and about 25 ha of land was afforested. But when the administration saw that the trees had been planted on revenue land lying within the village boundary, they served another legal notice on the voluntary agency. Even though regeneration was good, the villagers lost their case, and so all the efforts of the villagers to discipline themselves and protect their lands went to waste. As Agarwal and Narain put it: *'the country's land and water resources continue to be governed by 19th century laws even while it is facing 21st century problems of environmental destruction'.*

Source: Agarwal and Narain, 1989

sense, national policies would then provide the enabling framework to permit this to happen, whilst monitoring activities to ensure that higher level resources are not threatened. Local control combined with education is more likely to lead to sustainable management of resources than higher level enforcement and control.

Land Reform

Policies that affect farm size and land tenure are crucial for sustainable agriculture. It is well established that lack of land tenure increases the risk of resource degradation. Landless or tenant farmers are more willing to take the risk of degradation as they know they may not have access to the land in the future. If there is no security of tenure, they are also less likely to invest in resource-conserving practices, such as tree planting, terracing, predator management or soil improvement. Perhaps more than any other factor, appropriate rights to land determines the success of sustainable agriculture (Platteau, 1992; FAO, 1991; Agarwal and Narain, 1989). But even this may not be sufficient. If farmers have title to their own land, but lack usufruct rights to local forest resources and wild foods, their livelihoods may still be undermined (see Chapter 3).

Land reform in Iran, Taiwan, Kenya and Kerala in India has had a substantial impact on agricultural growth and poverty alleviation (World Bank, 1993; Prosterman et al, 1990). In Kenya, programmes of land adjustment and registration for individual title have been under-way in the relatively crowded better quality lands for many years. Those who have obtained individual title to their farms are more likely to invest in soil and water conservation than those farmers in areas not yet adjudicated (MALDM, *passim*; IFAD, 1992). Land title also makes farmers eligible for loans.

Some of these successful land reforms came about through major political upheavals, such as social revolutions or decolonization. Recently, an increasing number of countries, including Brazil, Colombia, Indonesia, Peru and the Philippines have enacted legislation in support of land reform. Many such initiatives have, however, failed to have an impact at local level. Vested interests and political elites have been able to mobilize opposition to reforms that would have provided more equal access to resources. In addition, many national forestry and other resource agencies are actively opposed to such policy changes, as they fear that recognition of tenure will undermine the existing role of foresters and other government officials (Garrity et al, 1993). As a result, land tenure legislation is generally not popular with many policy makers (Kesseba, 1992; World Bank, 1993). Indeed, the current fashion is to concentrate much more on economic instruments for policy reform.

None the less, recent initiatives in Burkina Faso, India and Nepal are demonstrating the importance of national support for local resource users' rights and security of tenure. In both India and Nepal, recent policy changes have established rights for local communities to manage and use forest resources (see Chapter 5). Such 'joint forest management' has spread rapidly and is having a significant impact on local people's capacity to organize themselves for sustainable resource use. Many have formulated their own regulations (see Box 5.9), a sign of strong local institutions.

In Burkina Faso, the Programme National de Gestion des Terroirs Villageois has established the land tenure conditions necessary for widespread action at the local level (Toulmin et al, 1992). It follows the enactment of land tenure reform in 1984, to ensure fair access to land and resources and to encourage greater local involvement in managing and restoring degraded land. The programme is carried out in four stages and now involves about 380 villages. A village level committee is established after village discussions and training, which then works with programme staff to define and demarcate village boundaries. After a resource inventory has been made, the last two stages involve the negotiation and finalizing of a contract between government and the village committee about the investments needed for better village productivity and management of village resources.

AGRICULTURAL RESEARCH, EXTENSION, EDUCATIONAL AND PLANNING SERVICES

Although both policy and legal reforms are essential for the transition to a more sustainable agriculture, they are not sufficient conditions. What governments also need to do is to take action to reform both the internal workings of their agricultural institutions and the way they interact with other organizations. This will need changes in the provision of research, extension, educational and planning services.

National Agricultural Research and Extension Systems

The first place governments can have an impact is on their own national research and extension systems. Many of these systems are under financial pressure, with governments short of resources and often unable to find adequate facilities or ensure that they hold on to quality professionals. Most are also oriented towards modernized agriculture, and there are deeply ingrained structures that bias research and extension against sustainable and participatory agriculture.

As has been illustrated in Chapter 6, there is growing acceptance that participatory approaches are essential for the development of a more sustainable agriculture. But most government and state institutions are currently limited in their capacity to conduct such research and extension. This is partly because they are organized according to the principles of the positivist paradigm (see Chapter 1). Researchers have the prestigious role of being the source of new technologies; and extensionists pass knowledge from centres of learning to farmers, who are assumed to be passive recipients.

Another problem is that excessive centralization and inflexible management tend to suffocate new initiatives. Reward systems for researchers are usually based on scientific publications, so discouraging them from working in the field, where research is less 'controllable' or on topics that may be seen as less scientific. In Nepal, where users' forestry has expanded rapidly in recent years (see Chapter 5), there are considerable constraints within existing institutions preventing professionals from adopting new approaches. As two forest rangers put it, their new roles are *'not part of our job description, thus only those with a personal interest do the work and they get no recognition for it... There has to be a firm and clear long-term policy .... Just as for user forestry to work in a village everyone in that village has to agree to support it, so it is the same for us in the Forest Department'* (Deo and Yadav, 1990).

As a result, intervention in agricultural research and extension systems greatly favours Green Revolution lands and better-off farmers. Worldwide research expenditure by the private sector alone is some US$1.9 billion per year, which is nearly half of the entire public agricultural research expenditure of developing countries (Vorley, 1993). The entire expenditure of the CGIAR system is just 15 per cent of this total spent on pesticides. In the USA, some $4.45 million is allocated for research and education programmes for sustainable agriculture, but this is just 0.5 per cent of the total USDA research and education budget (Reganold et al, 1990).

In India, there are overwhelming biases towards modern agriculture in postgraduate research. Researchers seeking a position in a university or government research establishment must select topics to investigate that reflect existing interests and expertise. Anil Gupta and colleagues' analysis (1989) shows how this translates into a repeated focus on 'modern' methods of farming and widespread ignorance of alternative resource-conserving technologies and practices. They studied the abstracts of research and extension theses completed in 32 agricultural colleges and universities between 1974–84 (Table 8.4). These illustrate a highly unbalanced agenda for postgraduate research. These problems are

Table 8.4 The biases in Indian agricultural postgraduate research

Of all 1128 theses on all topics:
- 4.5% dealt with drought-prone areas
- 22% dealt with rainfed agriculture
- 73.5% dealt with irrigated agriculture

Of 900 extension theses:
- 0.02% dealt with issues involving agricultural professionals and methods of science

Of the 376 agronomy theses:
- 2% dealt with organic fertilizers and green manuring
- 27% dealt with inorganic fertilizers
- 11% dealt with irrigation
- 0.8% dealt with salinity
- 16% dealt with intercropping
- 18% dealt with millet and sorghum
- 30% dealt with cereals

Of the 329 sociology and extension theses:
- 33% dealt with dryland regions
- 1.2% dealt with livestock
- 1% dealt with millets and sorghums
- None dealt with pulses, fodder and forestry

Source: Gupta et al, 1989

compounded by deep gender biases in institutions. In India, more than 70 per cent of the farm work is done by women, yet the Ministry of Agriculture report that only 0.5 per cent of all extension officers are female – and this includes Kerala where 25 per cent of extension workers are women (Antholt, 1992). The long run answer has to lie in reforms of both recruitment procedures and in incentives for training in higher education.

Another problem is that many state institutions have narrow mandates. Often crop production is the responsibility of one department in a ministry, while livestock may be in another department or ministry altogether; tree crops fall into the responsibility of a forestry department; water resources into another; and soil and water conservation in perhaps yet another. This is in stark contrast to the diversity needed on farms and in communities for sustainability. Division of responsibility among external agencies hinders the ability of their professionals to support mixed livelihood systems if they are working alone. However, if individuals and institutions work more closely together in alliances and networks, including with non-government organizations, then this need not be an insurmountable problem. Better alliances and linkages are widely recognized to be a better option than putting all services into one single, multi-sectoral institution (Pretty and Chambers, 1993a). It encourages the development of a better capacity for learning, and for responding to the changing needs of farmers and rural people (see Chapter 6).

Despite the widespread problems, there are a growing number of successful innovations in national agricultural research and extension systems (see Box 6.5). There are some similarities. In many there was a recognition that past approaches had failed both farmers and researchers. Pressure for change came in certain cases from external sources, such as from donors and NGOs. In some cases, support came from senior staff, who created the space for innovators who, in turn, were often charismatic individuals able to promote and achieve change. In some cases, small and autonomous groups within larger bureaucracies became a model for the rest, leading to wider systemic change.

The International Agricultural Research Institutions

The international agricultural research centres of the CGIAR (Consultative Group for International Agricultural Research) are a special case in the provision of agricultural services to farmers. Although they are largely independent of the national policies of governments, save for those donors that fund the CG system, they do have a professional influence out of all proportion to their size and budgets. In the early 1990s, the CG expenditure was some 6 per cent of the global expenditure on agricultural research. Nevertheless, agricultural scientists worldwide see the CG institutions as centres which embody professional excellence. Through their training of national scientists, their international networking of research programmes, their publications and their prestige, the centres have long spread and sustained the dominant concepts and values of the transfer of technology paradigm (Pretty and Chambers, 1993a).

Despite recent critiques of the CG system, particularly in relation to poverty alleviation and sustainable agriculture (Gibbon, 1992; Ravnborg, 1992; Scoones and Thompson, 1994), there are signs of change. Partly this has come about through pressure on funds. In 1993–4, donor support to the CG system fell by 30 per cent, largely because of the sudden removal of support by some bilateral donors. This provoked a period of hectic evaluation of the impacts of past research. Curiously, little was known within many international centres of the impact in rural communities of their research. What this soul-searching and impact analysis may have done is to bring into sharper focus just how research institutions do learn about the effect of their work and how they can improve performance (Conway et al, 1994).

The other factor is that groups of professionals within some centres have already been conducting successful participatory research in partnership with other organizations and groups of farmers (Box 8.3). These have received some publicity and are demonstrating to colleagues what can be achieved. However, they are not the norm. Those individuals who have succeeded in developing and using participatory approaches have tended to be isolated and marginalized within their institutions. At least until recently, they have been more recognized and respected in the world of national agricultural systems and NGOs than by their colleagues.

Box 8.3 A selection of groups within some international centres of the CGIAR already conducting successful participatory research in partnership with other organizations and groups of farmers

Participatory research efforts include:

- post-harvest potato research with Peruvian farmers, from CIP;
- bean research with Bolivian, Colombian and Rwandan farmers and NGOs, from CIAT;
- aquaculture systems research and development with Malawian and Filipino farmers, from ICLARM;
- women in rice systems programme, from IRRI;
- upland conservation research and development in the Philippines and elsewhere, from IRRI;
- pigeonpea research with women farmers in Andhra Pradesh and pearl millet research in Rajasthan from ICRISAT;
- soil and water conservation research with Indian NGOs and farmers, from ICRISAT;
- countrywide network for potato research in Philippines, UPWARD (User's Perspective with Agricultural Research and Development) at CIP;
- participatory research by national scientists in West, central and East Africa, working with AFRENA and AFNETA, from IITA.

Note: this is not intended to be comprehensive list; it is to illustrate the range of activities and institutions involved

Sources: Rhoades and Booth, 1982; Ashby et al, 1989; Bebbington and Farrington, 1992; Sperling et al, 1993; Lightfoot and Noble, 1992; Paris and Del Rosario, 1993; Fujisaka, 1989, 1991a; Pimbert, 1991; Kerr and Sanghi, 1992; UPWARD, 1990

Non-Governmental Organizations (NGOs) as Partners

Although government agencies have not been good at working with NGOs, it is increasingly being recognized that they cannot go it alone when it comes to agricultural and rural development. There is a wealth of skills and knowledge in the non-government sector, and governments are increasingly recognizing the benefits of policies that support closer working with NGOs.

The scale, scope and influence of NGOs concerned with development has grown enormously in recent years (Korten, 1990; Edwards and Hulme, 1992; Fowler, 1992; Pretty and Chambers, 1993a; Farrington and Bebbington, 1993; Bebbington and Thiele, 1993). In the South, there are perhaps some 10–20,000 development NGOs and in the OECD countries a further 4000. Their activities are now very diverse and, in some of the poorest areas and countries, they perform many of the roles carried out elsewhere by government. This includes not only relief, welfare, community development, and agricultural research and extension, but also advocacy and lobbying, development education, legal reform, training, alliance building, and national and international networking.

In some locations, the coverage by farmers' groups and NGOs in extension, training and input supply is better than that provided by the public sector, such as in eastern Bolivia, where there are 130 agricultural staff employed by farmers and NGOs compared with 55 by the public sectors; and in northern Ghana, where there are three times as many extension agents operating from church-based agricultural stations as from the public sector (Farrington and Biggs, 1990).

These varied functions and roles mean that NGOs are both critical actors in their own right, as well as potential partners for government and international institutions. A number of factors particular to NGOs make them successful (Pretty and Chambers, 1993a). They have the flexibility to choose the subject area and sources of information. They have the freedom to develop their own incentives for professionals. They have the capacity to struggle to get things right, and so more ability at the local level to question, puzzle, change and learn. They have the strength in supporting community level initiatives, and helping to organize federations and caucuses. And they can work on longer time horizons, as they are less affected by the time and target-bound 'project' culture.

Some of these factors make them very different to government agencies, who cannot 'cherry-pick' where they will work. Some NGOs also choose to work alone, such as when, in their opinion, there is little of relevance in the public sector programmes for their clientele. Government institutions may be bypassed, because they are weak, a trap for human capital, or simply repressive, and funds channelled to NGOs to create parallel structures. Creating parallel structures, though, is likely to be both inefficient and non-persistent. The alternative is to work with governments, so that NGOs *'identify how best they might support but not substitute for what exists'* (Roche, 1991). The principal objective must now be to foster change from within, not to threaten power but to pressurize, and to support innovative individuals. None the less, there are problems for developmental NGOs and it tends to be their virtues rather than their vices that have been emphasized (Zadek and Mayo, 1994).

NGOs are successful at small-scale initiatives and, as they are locally based, may be a better defence against repressive states. But where there have been transitions to elected democracies, *'NGOs are presented with the difficult fact that governments are to some extent popularly elected whilst NGOs are not'* (Bebbington, 1991). Many NGOs, at least non-membership organizations, are not accountable, and just because they are NGOs does not mean they are not subject to corruption.

There is a strong case for encouraging better collaboration between NGOs and the public sector. Many types of relationships have developed between NGOs and governments. Some collaborate in agricultural research and information dissemination; some provide support for marginalized regional administrations, such as by ACORD in Mali (Roche, 1991); some train government staff, such as in India, where they have been at the forefront of training government officials in participatory approaches (Mascarenhas et al, 1991; Shah, 1994a).

Others federate into higher level organizations, so as to influence policies. This has led to the creation of consortia of government, NGO and farmers' organizations for joint planning and coordination. In Ecuador, such collaboration has been planned to lead to regional agricultural technology development committees, in which government organizations, NGOs and farmers' organizations all have voting power (Bebbington and Farrington, 1994a).

These links between government agencies and NGOs have significant implications for new state–society relations (Curtis, 1991). There are benefits, from synergism, from greater efficiency of resource use, and from NGOs and farmer organizations becoming more accountable. There are also costs and dangers. The state's capabilities may be weakened in two ways: through NGOs substituting for government activities; and through a brain drain to NGOs, as increasingly NGOs are able to attract skilled people away from the public sector.

Formal Educational Institutions

Governments also provide support for agriculture through the educational establishment. This helps in the development of human resource capital, and plays a critical role in the training of agricultural professionals for both government and non-government sectors.

But universities and their agricultural faculties are often the most conservative of agricultural organizations. They have generally been slow to adopt innovative ideas, methods and staff development activities. They remain in the conceptual strait-jacket of positivism and modernization, arising partly out of the functional and practical demarcation of research and teaching, and the focus on 'teaching' rather than learning (Pearson and Ison, 1990; Pretty and Chambers, 1993a; Scoones and Thompson, 1994). These structures create biases hugely in favour of the modernist paradigm, with significant implications for sustainable agriculture (Box 8.4).

Most have developed structures that reflect the proliferation of

Box 8.4 The biased structures of agricultural universities and faculties

The biases are that:

- they are frequently organized along authoritarian rather than participatory management lines;
- management positions are often held on basis of seniority rather than management skills;
- creative and eccentric innovation is rarely tolerated;
- institutional rewards, particularly senior authorship of papers, promotes individual and isolated research – making many institutions lonely places;
- organizations become introspective and resistant to new ideas, processes and changing environmental circumstances;
- staff development, if it exists, is frequently in the form of refresher training, where content (new facts) is the primary input, rather than a balance between content and the development of new management or learning skills;
- explicit or implicit status divisions become set in stone, eg researcher versus extensionist, natural versus social scientist, so increasing the difficulties of integrating disciplines.

Source: Ison, 1990

disciplines which have emerged over the past 30 years. An innovative field or area of study is usually accommodated by adding on a new sector, without basic restructuring. The adoption of a focus on Farming Systems Research, for example, has commonly resulted in the creation of new courses or departments, so treating it as another discipline rather than as something that integrates across and between sectors (Gibbon, 1992). New ideas have rarely stimulated radical rethinking or restructuring. Agricultural universities, thus, have a very poor record in training professionals to be real-world problem solvers.

The fundamental requirement for sustainable agriculture is for universities to evolve into communities of participatory learners. Academics must become involved in learning, learning about learning, facilitating the development of learners, and exploring new ways of understanding their own and others' realities. Participatory learning implies mutual learning – from farmers, from students' own learning and from colleagues. Although there are some in agricultural universities who work closely with farmers in a participatory mode, this has had only a slight influence on the style of teaching and learning in universities with students. More radical change is required. The education system does not need patching and repairing; it needs transformation.

Such changes are very rare in universities, an exception being Hawkesbury College, which is now part of the University of Western Sydney, Australia (Bawden, 1992). It is more common in training institutions linked less to the mass production of graduates and more to the development of capable professionals (Lynton and Pareek, 1990); and in some adult education institutions (Rogers, 1985). However, a regional consortium of NGOs in Latin America concerned with agroecology and low input agriculture recently signed an agreement with 11 colleges of agriculture from Argentina, Bolivia, Chile, Mexico, Peru and Uruguay, to help in the joint reorientation of curriculum and research agendas towards sustainability and poverty concerns (Altieri and Yuryevic, 1992; Yuryevic, 1994). The agreement defines collaboration to develop a more systemic and integrated curriculum, professional training programmes, internship programmes, collaborative research efforts and the development of training materials.

Because of the widespread failure of the formal educational sector to provide the necessary learning environments for the development of new professionals, it has chiefly been NGOs from both the South and the North which have led the way. Enlightened individuals in government organizations, NARSs, and CG institutes, and farmers have also played their part. The investment is not so much in knowledge, in the formal sense, but in the capacity to learn, reflect and know. Training is centred on learning by doing and bringing scientists, extensionists and farmers together to negotiate and learn from each other on a personal level. The farmer field schools of the IPM programmes in Indonesia and Honduras, and the organic training centres in Dominican Republic, are fine examples of this new focus on human resource development for a more sustainable agriculture (Kenmore, 1991; Bentley, 1994; Röling and van der Fliert, 1992; Ornes, 1988). The key principles of farmer field schools are described in Box 8.5.

National Planning, Decentralization and Participation

The final area where government can lower the long-term costs to farmers and communities is through appropriate planning processes. There are

Box 8.5 The key principles of farmer field schools

1. What is relevant and meaningful is decided by the learner, and must be discovered by the learner. Learning flourishes in a situation in which teaching is seen as a facilitating process that assists people to explore and discover the personal meaning of events for them.
2. Learning is a consequence of experience. People become responsible when they have assumed responsibility and experienced success.
3. Cooperative approaches are enabling. As people invest in collaborative group approaches, they develop a better sense of their own worth.
4. Learning is an evolutionary process, and is characterized by free and open communication, confrontation, acceptance, respect and the right to make mistakes.
5. Each person's experience of reality is unique. As they become more aware of how they learn and solve problems, they can refine and modify their own styles of learning and action.

Sources: adapted from Kingsley and Musante, 1994; Kenmore, 1991; Stock 1994

many different systems of planning within the government machinery. National development plans are usually prepared by a central ministry with multi-sectoral responsibilities, such as a ministry of planning or finance. As planning horizons tend to be short, typically five years, environmental and social concerns have generally been subordinated to crude measures of economic performance, such as gross domestic product, employment generation and foreign exchange earnings. Sectoral plans are more limited in scope, reflecting the mandate of the responsible agency, such as agriculture, forestry, fisheries, mining health and so on. Programme plans are more specific and project plans still more so. Regional and local plans relate to specific areas, and may be sectoral, such as forestry plans, or cross-sectoral, such as land-use plans. Within these planning frameworks, national environmental policies tend to be concerned with specific goals, such as establishment of protected areas, water quality and pesticides regulation.

But most normal planning is deeply flawed (Dalal-Clayton and Dent 1994; Pretty and Scoones, 1994). It tends to focus on a narrow technical view, rather than considering the social and economic complexities of farming and livelihood systems. It is usually data hungry, with information needs being partly defined by sophisticated technologies, such as satellite imagery. These measure too few factors, become the domain of technically skilled outsiders, claim accuracy and are rarely ground-truthed. Outsiders tend to define local needs, and develop technologies and innovations on research stations, so there is little use of local expertise, knowledge and skills. Once land use classifications are completed, there is no room for adjustment in the face of local environmental, economic and social change.

However, many national governments, are increasingly attempting to integrate environmental, economic and social concerns in national planning processes. These strategies include national conservation strategies (NCSs), national environmental action plans (NEAPs), and more recently the national sustainable development strategies (NSDSs) (IIED/IUCN, 1993). The concept of the NSDS was highlighted in Agenda 21, the action plan of the UN Conference on Environment and Development (or 'Earth Summit') held in Rio de Janeiro in 1992. At the core of these new planning processes is the notion that local people must be fully involved in developing and implementing strategies, including contributing to design, information exchange and sharing in decision making. This implies that planning itself will have to become much more adaptive and capable of responding to diverse needs and opportunities.

Agenda 21 indicates that governments *'should implement programmes with a focus on empowerment of local and community groups by delegating authority, accountability and resources to most appropriate level'*. A variety of measures can be taken to encourage such institutionalization of adaptive planning (Pretty and Scoones, 1994). It is generally felt that financial accountability, in the form of successful cost-recovery or cost-contribution, is a measure of the value that people put on an intervention or change. Support for the local level can encourage local autonomy and independence, but this may depend on the degree to which these revenue-earning technologies are supported. Who has a stake is also important. Local people could have an increased stake if they are empowered to make decisions; local governments could more effectively achieve developmental goals; donors could see a more efficient use of funds; but state-wide institutions, with competing interests, may be threatened.

It will be essential to sensitize bureaucrats to the needs of adaptive planning. In some cases they poorly understand the skills and knowledge of villagers. Many have a poor understanding of informal approaches to participatory data gathering. They may lack local credibility and may be restricted to establishing dialogue with 'traditional' authorities. Planners must be trained in the use of local level information, which will require linkages with the formal government planning system, methods of articulating local responses with sectoral concerns of line ministries/ agencies, as well as integrating conventional and new approaches to planning. Where this has been done in Britain, there have been significant changes in attitudes among both local people and planners (Box 8.6).

Box 8.6 Planning for Real: adaptive planning in urban Britain

At public meetings and consultations, local planners and other outsiders sit on a platform, behind a table, maintaining their superiority. When only a few people turn up, and only a few of them speak up, they blame local indifference. Planning for Real attempts to bridge this gap by focusing on a model of the neighbourhood. Unlike an architect's model, this should be touched, played with, dropped, changed around. At the first meeting the neighbourhood model is constructed, using houses and apartment blocks made from card and paper on a polystyrene base. The model then goes into the community, to the launderette, the school foyer, the fish and chip shop, so that people see it and get to hear of the second consultation.

At the second meeting the objective is to find out whether the planners have got it right. There is no room for passivity, not many chairs, no platform, with the model in the middle of the room. People spot the landmarks, discuss, identify problems and glimpse solutions. They are permitted to put more than one solution on the same place – so allowing for conflicts to surface. Often people who put down an idea wait for others to talk first about it. The process permits people to have first, second and third thoughts – they can change their minds. The model allows people to address conflicts without needing to identify themselves. It depersonalises conflicts and introduces informality where consensus is more easily reached.

The professionals attend too. These local planners, engineers, transport officials, police, social workers, wear a badge identifying themselves, but can only talk when they are spoken to. The result is they are drawn in, and begin to like this new role. The 'us and them' barriers begin to break down.

Priorities are assessed and local people involved in local skills surveys. The human resources are documented and planning can then capitalize on these hitherto hidden resources.

Source: Gibson, 1991

In Kenya, the district approach to planning has decentralized decision making and allocation of resources to District Development Committees, on which sit representatives of various line agencies. Decisions are inevitably framed as if there is little variability in needs from community to community, particularly within well-defined agro-ecological zones. But emerging evidence from the community-based planning by the Ministry of Agriculture is showing clearly that there is a huge diversity in needs within these zones (Muya et al, 1992; Mucai et al, 1992). As extension officers are now working more closely with local communities, they are able to respond to these specific needs and, as a result, local planning has become more effective (Pretty et al, 1994).

It is also critical that planning processes develop the mechanisms for conflict resolution and social mediation. It is inevitable that differences already exist or will arise over what constitutes the best route at any given time towards sustainability (see Chapter 1). As the successes in Chapter 7 spread or become stronger, so they too will be brought into conflict with existing power structures. What planning will need to do is find the processes for making these complex trade offs. These will need to bring together individuals and institutions with different interests and values, and then mediate between them. What is needed is *'a structure and procedure of negotiation, founded on a common information... to assist governments and groups of land users to come to a maximum degree of agreement on land use decisions'* (Brinkman, 1993). Where this has been done successfully, such as by the US Corps of Engineers over water resources in the USA and in India in the Joint Forest Management programmes where there are conflicts between communities over regenerating resources, then there can be significant benefits for all actors (Delli Priscoli, 1989; USACE, 1990; SPWD, 1992; Röling, 1994).

In terms of agricultural development efforts, the benefits of adaptive planning processes will be seen if there is an evolution away from the 'project culture'. Planning is often thought to be synonymous with intervention and the starting of 'projects', implying the involvement of outsiders and external funding. The development aid business reinforces this with its concentration on the conventional project cycle with discrete project identification and funding (Korten, 1980; Uphoff, 1990; Chambers, 1993; Carley, 1994). The project process has been described by Robert Chambers (1993) as a 'pathology' of development, and embodies all the certainties of the modernist paradigm (see Chapters 1 and 2). The convention is that projects are subject to systematic and rigorous procedures, but in the real world this cannot be the case, particularly if participation is taken seriously. The principal problem is that project cycles set out future actions in the form of blueprints. These do not easily permit change and adaptations.

If agricultural projects are to become adaptive and participatory, this will imply significant reform for the way the 'project' is conceived and organized. As Chris Roche (1994) put it: *'it demands questioning of the whole concept of the 'project'. In participatory projects, elements of the project cycle come in new and different sequences and often occur simultaneously. There is a complex and permanent interaction between action, identification, monitoring and evaluation, which are not divisible into neat boxes beloved of planners'.*

Support is needed for learning- or para-projects (Chambers, 1993; Uphoff, 1990), which have several features: they are more labour-intensive than capital-intensive; they mobilize local resources, including knowledge and management skills; they are more flexible and so focus on a learning process approach; and their goals are qualitative shifts in the ways people and institutions and individuals interact and work together. All of these are important for the development of a more sustainable agriculture.

POLICIES THAT SUPPORT SUSTAINABLE AGRICULTURE

There have been an increasing number of policy initiatives in recent years specifically oriented towards improving the sustainability of agriculture. Most of these have focused on input reduction strategies, because of concerns over foreign exchange expenditure or environmental damage. Only a few as yet represent coherent plans and processes that clearly demonstrate the value of integrating policy goals. A thriving and sustainable agricultural sector requires both integrated action by farmers and communities, and integrated action by policy makers and planners. This implies both horizontal integration with better linkages between sectors and vertical integration with better linkages from the micro to macro level.

Sustainable agricultural sector policy reforms will also have to be synchronized with macro, industrial sector and trade reforms. The removal of agricultural input subsidies, for example, can be very costly to farmers in the short term if they are not matched by the development of systems for farmer training and learning, the realignment of exchange rates and reductions in industrial sector protection.

Some Recent Policy Initiatives in the North

The environment ministers of the OECD countries, meeting in January 1991, identified agriculture as one sector in which improved policy integration offered major returns (OECD, 1993c). They noted that both environmental and agricultural goals could be pursued within the context of agricultural reform, with a view to moving towards a more sustainable agriculture. Setting appropriate prices for agricultural inputs and outputs to reflect better their full environmental and social costs has been identified as one way to achieve better policy integration. Market failure, such as non-payment or under-payment for resource degradation, or intervention failure, such as under-supply of public goods, are important barriers to assigning real prices to raw materials, goods and services. Several countries have found that increasing the costs of inputs is an effective way to reduce their use, as well as in cutting on-farm and off-farm pollution.

The most common approach has been to introduce taxes and input levies on fertilizers and pesticides (Table 8.5). Some of this revenue is used to subsidize exports (such as in Finland); to support further the input reduction programme (such as in Sweden, where $3–3.5 million are raised

Table 8.5 Current taxes and charges on fertilizers and pesticides in selected countries

Fertilizer taxes		*Pesticide taxes*	
Austria	$0.42 / kg N $0.25/kg P $0.14/kg K	Finland	2.5% on retail price
Finland	$0.25 / kg P $0.03/kg N and K	Norway	11% on retail price
Norway	15% tax on N, P, K	Sweden	20% 'price regulation' charge, + $5.65 per hectare for each application of pesticides; + 10% environmental tax
Sweden	$0.08 / kg N $0.15/kg P + additional 20% levy on retail price		
USA:			
Iowa State	$0.00096 / kg N		
Wisconsin	$0.00020 / kg N		

Source: OECD, 1989, 1992, 1993c; Conway and Pretty, 1991

each year); to support research into alternative agriculture (such as in Iowa and Wisconsin) or to return to farmers' resources in the form of income support (such as in Norway). There are also proposals to introduce similar taxes in Belgium, Denmark, The Netherlands and Switzerland.

It is generally felt, however, that these levels have been set too low significantly to affect consumption (OECD, 1989, 1992; Baldock, 1990). Studies in the USA, for example, comparing alternative agriculture farms with conventional neighbours conclude that a 25 per cent tax on fertilizers and herbicides would not be sufficient to encourage farmers to convert from conventional to sustainable systems (Dobbs et al, 1991). Even though it would reduce net income by $10 per ha on conventional farms compared with just $1.25 on alternatives, taxes would have to be much higher to lead to big reductions in input use (Reichelderfer, 1990).

But other factors may be important, including revised advice from the agricultural ministry, growing public concern over high rates of application and general changes in cropping practices. Supplemented by other policy measures, such as new regulations, training programmes, provision of alternative control measures and reduced price support, there have been some substantial reductions in input use in recent years. In Sweden, pesticide consumption was cut by half between 1985 and 1990; in Austria there has been a decline in consumption of fertilizers, especially potassium; and in Denmark there has been a fall in consumption of pesticides by 18 per cent between 1986–91 (OECD, 1992, 1993c; Beaumont, 1993).

Levies have also been used to limit pollution in the livestock sector. Levy payments are used in The Netherlands to penalize those farmers producing more livestock waste than their land can absorb. The Netherlands has also introduced a levy on manufactured feed to help pay for research and advisory services dealing with pollution from livestock waste.

Several countries have now set ambitious national targets for the reduction of input use (Beaumont, 1993). Sweden aims to reduce nitrogen consumption by 20 per cent by the year 2000. The Netherlands is seeking a cut in pesticide use of 50 per cent by the year 2000 as part of its 'Multi-Year Plan for Crop Protection'. The cost of this reduction programme has been estimated at $1.3 billion, most of which will be raised by levies on sales. Denmark is aiming for a 50 per cent cut in its pesticide use by 1997, a plan which relies on advice, research and training, with no taxes or levies. In the USA, the Clinton administration announced in 1993 a programme to reduce pesticide use while promoting sustainable agriculture. The aim is to see IPM programmes on 75 per cent of the total area of farmland by the year 2000. There have been no equivalent policies within the Common Agricultural Policy (CAP) of the EU, though it is expected that there will be reductions in input use as a result of decreasing crop prices arising out of the last round of CAP reform.

The alternative to penalizing farmers is to encourage them to adopt alternative low or non-polluting or degrading technologies by acting on subsidies, grants, credit or low-interest loans. These could be in the form of direct subsidies for low input systems or the removal of subsidies and other interventions that currently work against alternative systems. Acting on either would have the effect of removing distortions and making the low input options more attractive. Such initiatives are much rarer, and tend to be fragmented, with support to farmers for maintaining landscapes or habitats. An exception is the MEKA project in Baden Würtemberg, Germany, where the principle is to pay farmers not to damage the environment. The scheme is voluntary and open to all farmers, who are able to choose the aspects of the scheme with which they wish to comply. They then receive payments on a points system (Box 8.7).

Box 8.7 The MEKA grant scheme, Baden Würtemberg, Germany

The MEKA scheme aims to reduce over-production and promote more integrated and environmentally sensitive farming. Points, worth a cash payment of around $10 each, are awarded on a per hectare basis for specific agricultural practices. For example, using no growth regulator attracts 10 points; sowing a green manure crop in the autumn earns 6 points, applying no herbicides and using mechanical weeding gets 5 points; cutting back livestock to 1.2–1.8 adult units per hectare brings 3 points; and direct drilling on erosive soils earns 6 points. Direct environment protection measures include up to 15 points for reduced stocking on areas designated as of special scientific interest and points can also be earned for keeping rare breeds.

The total cost of the scheme in the first year was $48 m and was split between the Federal government and the regional government. *'By encouraging care of the environment and the traditional landscape with grant aid, the scheme is helping family farming businesses like ours to survive'*, says George Mayer, who farms a 60 ha mixed farm. More than 43,000 farmers had joined this scheme by 1992.

Source: FW, 1992

Most policy initiatives are still piecemeal. They affect a small part of a individual farmer's practices, but do not necessarily lead to substantial shifts towards sustainable agriculture. However, one of the first nations to convert the principles of sustainable development into a series of clear steps in a national strategy is The Netherlands. This is the National Environmental Policy Plan (NEPP) (VROM, 1989, 1990; WRI, 1994). It is probably the best example of an action-oriented plan, with 220 prescribed steps towards clear targets. It integrates land use, transport and energy plans with agricultural, industrial and economic planning. It declares that environmental problems are connected and that society must end its practice of making others pay for the cost of degradation. Its governing principle is that polluters must pay and, despite reservations in some sectors, there remains widespread support for the NEPP.

Some Recent Policy Initiatives in the South

Probably the best example of an integrated National Conservation Strategy (NCS) comes from Pakistan (EUAD, 1992). This covers an impressive range and depth of issues relating to sustainable development, as well as making prescriptions about a wide range of institutional and societal processes. These include administrative matters, communication, education, legislation, grassroots organization, and research and development roles. It also sets out indicators against which national and local progress can be measured. What is important is that the NCS recognizes that it is not the plan or document that is important. Rather, it is the enhanced communication and linkages between institutions and actors, the increased mutual understanding and the development of a national network of people directly committed to continuing the process. These are all valuable resources for sustainable development.

The Group Farming initiative of the Kerala state government in India is a good example of how coordinated action within the agricultural sector can have a significant impact on farming practice (Sherief, 1991). Land reform in the 1970s led to the formation of a new class of small farmers, with some 70 per cent owning less than 2 ha. But as the costs of inputs, pest and disease control spiralled, so the area under rice fell from 810,000 to 570,000 ha between 1975 and 1988. Small farmers had been unable to adopt the whole technological package.

In 1989, the group farming for rice programme was launched as a collaborative effort between the Kerala Agricultural University and the State Department of Agriculture, with the primary objective of reducing the costs of farming. Agricultural offices were opened in every panchayat, each with two to three local extensionists. Local committees comprising all rice farmers were formed and these chart out a detailed plan of farming. In this group activities, such as for water management and labour operations, are jointly agreed. Costs are reduced through community nursery raising of rice; fertilizer applications on the basis of soil testing; the introduction of IPM and minimum use of pesticides; and the formation of plant protection squads. The average cost reduction to farmers has been Rs1000 per ha and rice yields have improved by 500 kg/ha. As Sherief put it *'instead of pampering the cultivators with subsidies, stress was given to self-reliance and timely action for all agricultural operations'.*

In China, agricultural policy is encouraging farmers to grow green manures in the rice fields (Yixian, 1991). During the 1980s, the continuous and monocropping of rice was widely observed to have caused soil fertility and pest problems. The Agricultural Ministry set up multiplication bases for green manures on some 1200 ha, which are expected to produce 5.5 million kg seed each year. In some regions, farmers selling green manure seed to state-run farm cooperatives receive fertilizers at lower prices. Green manures and plant residues are now used on 68 per cent of the 22 million ha of rice fields.

Similar successes have been observed in Indonesia where, during the late 1980s, pesticide subsidies were cut from 85 per cent to zero by January 1989 and farmer field schools (FFSs) established for IPM (see Chapters 2 and 7). The country saves some $130–160 million each year and pesticide production fell by nearly 60 per cent between 1985–90. Rice yields have continued to improve, despite the cut in inputs. This raises a crucial policy issue when it comes to cutting inputs. Farmers who are dependent on external inputs need support to make the transition to a more sustainable agriculture. As these FFSs demonstrate, if this support is in the form of increasing farmers' capacity to learn and act on their own farms, then this is a more than adequate substitute.

The FAO is supporting similar rice–IPM programmes in eight countries in addition to Indonesia: Bangladesh, China, India, Malaysia, the Philippines, Sri Lanka, Thailand and Vietnam. Together, these programmes are training many thousands of farmers and have saved many millions of dollars in pesticides. For many years, there have been concerns that Third World countries have been unable to ban the use of hazardous pesticide products that are banned in the north (ADB, 1987; Conway and Pretty, 1991; Beaumont, 1993). However, there are increasing cases of countries now banning hazardous products, such as by the Dominican Republic and Philippines in 1991 and 1992, even though this action affects a substantial proportion of the domestic pesticide markets (Siedenberg, 1992; Thomen, 1992).

One of the most remarkable coordinated policy efforts by a southern country had been in Cuba, where a national alternative agricultural strategy has been in development since 1990 (Rossert and Benjamin, 1993). Up to this time, Cuba's agricultural and food sector was heavily dependent on external support from the Soviet bloc. It imported 100 per cent of wheat, 90 per cent of beans, 57 per cent of all calories consumed, 94 per cent of fertilizer, 82 per cent of pesticides and 97 per cent of animal feed. It was also paid three times the world price for its sugar. At this time, Cuba also had the most scientists per head of population in Latin America, the most tractors per ha, the second highest grain yields, the greatest increase in per capita food production in the 1980s, the lowest infant mortality, the highest number of doctors per head population, the highest secondary school enrolment and lowest teacher:pupil ratios.

But in 1990, trade with the Soviet bloc collapsed, leading to severe shortages in all imported goods. Within two years, petroleum imports fell to half of the pre-1990 level, fertilizers to a quarter, pesticides to a third, and food imports to less than half. Over a very short period, a modern and industrialized agriculture was faced with the dual challenge of having to double food production on less than half the inputs.

The response of the Ministry of Agriculture was to declare an 'alternative model' as the official policy for agriculture. This has been set against the former 'classical model', in that it seeks to focus on resource-conserving technologies that substitute local knowledge, skills and resources for the former external inputs. It also emphasizes the diversification of agriculture; the breeding of oxen to replace tractors; the use of IPM to replace pesticides; the introduction of new practices in science; the need for widespread training; the promotion of better cooperation among farmers both within and between communities; and reversal of the rural exodus by encouraging people to remain in rural areas.

Fortunately, there was a good base in research institutions for this rapid conversion, as some younger scientists had been conducting research into alternative agriculture, often against the wishes of senior colleagues, during the 1980s. The impact of the new policy has already been remarkable. There are now more than 200 village-based Centres for the Reproduction of Entomophages and Entomopathogens, and 90 per cent of agricultural land uses monitoring for pests and diseases. Many biological control methods are proving more efficient than pesticides. The use of cut banana stems baited with honey to attract ants, which are then placed in sweet potato fields, has led to the complete control of sweet potato borer by the predatory ants. There are 173 vermicompost centres, the production from which grew from 3000 to 93,000 tonnes in 4 years. Crop rotations, green manuring, intercropping and soil conservation are all more common. Planners have also sought to encourage urban people to move to the countryside, as labour needs for alternative agriculture are now a constraint. Programmes are now aiming to create more attractive housing in the countryside supplemented with services and to encourage urban people to work on farms for periods of two weeks to two years.

None the less, these policies that are supporting a more sustainable agriculture will not necessarily succeed without action in other sectors. They will quickly be undermined by inappropriate macro- and micro-economic policies, and by systems of governance that are controlling and enforcing rather than democratic.

SUMMARY

Most successes in sustainable agriculture are still only on a relatively small scale. This is largely because an enabling policy environment is missing in almost every country. Although there are rare exceptions, most policy frameworks still actively encourage agriculture that is dependent on external inputs, technologies and knowledge.

Governments have long provided public support to their domestic agricultural sectors. Through a wide range of direct and indirect monetary transfers from consumers and taxpayers to farmers and other producers, they have sought to ensure that agriculture provides the food and other products needed by the non-farming population. Support measures have included market price support, direct payments, input cost reductions and provision of general services. The levels of support vary hugely between countries. Most types of support, however, have encouraged the greater use of external inputs, such as pesticides, fertilizers, credit and water.

Most, if not all, of the policy measures used to support agriculture currently act as powerful disincentives against sustainability. In the short term, this means that farmers switching from high input to resource-conserving technologies can rarely do so without incurring some transition costs. The principal problem is that policies simply do not reflect the long-term social and environmental costs of resource use. The external costs of modern farming, such as soil erosion, health damage or polluted ecosystems, are not incorporated into individual decision making by farmers. For the full benefits of sustainable agriculture to accrue, policies must be more integrated and more directed towards alternative practices.

The first place that such integration for sustainable agriculture can occur is in the provision of appropriate regulatory frameworks. These can be used either to encourage or to discourage particular types of farming practice or technology, and can act from local to national levels. A particularly important area is policies that affect farm size and land tenure, as these have a crucial bearing on sustainability.

Although both policy and legal reforms are essential for the transition to a more sustainable agriculture, they are not sufficient conditions. What governments also need to do is to take action to reform both the internal workings of their agricultural institutions and the way they interact with other organizations. This will need changes in the provision of research, extension, educational and planning services. This will need also changes in the provision of national and international research, with internal reforms combined with new relationships with other actors and institutions. NGOs will have to be seen much more as potential partners in research and extension. Formal educational institutions need reform too and much can be learned about learning from innovations with farmer field schools. Planning also has to become more adaptive and responsive to local needs and changing circumstances.

Policy reform has been under way in many countries, with some new initiatives supporting elements of a more sustainable agriculture. Most of these have focused on input reduction strategies, because of concerns over foreign exchange expenditure or environmental damage. Only a few as yet represent coherent plans and processes that clearly demonstrate the value of integrating policy goals. None the less, it is clear that many policy reforms are leading to changes in the sustainability of agriculture.

9

POLICIES THAT WORK FOR SUSTAINABLE AGRICULTURE

POLICIES THAT WORK

As has been illustrated throughout this book, sustainable agriculture can be economically, environmentally and socially viable. It does, however, need coordinated action by national governments to encourage and nurture the transition from modernized systems towards more sustainable alternatives. Without appropriate policy support, it will remain at best localized in extent and at worst wither away.

There is much, however, that governments can do with existing resources. The intention of this chapter is to set out 25 tested policies that are known to work. These are intended to address existing constraints in the three essential areas of action (see Figure 1.2). These are the encouraging of resource-conserving technologies and practices; the fostering of local group and community action; and the reforming of external institutions. In all three areas there are economic, regulatory and institutional approaches that can be marshalled to achieve progress towards a more sustainable agriculture.

Before setting out these policies, it is important to be clear about just how policies should be trying to address issues of sustainability. As has been suggested in Chapter 1, precise and absolute definitions of sustainability, and therefore of sustainable agriculture, are impossible. Sustainability itself is a complex and contested concept. In any discussions of sustainability, it is important to clarify what it is being sustained, for how long, for whose benefit, over what area and measured by what criteria. Answering these questions is difficult, as it means assessing and trading off values and beliefs.

Farming and rural problems are always open to interpretation. As all actors have uniquely different perspectives on what is a problem and what constitutes improvement in agriculture, what is important is the focus on sharing these perspectives and insights. The question of defining what we are trying to achieve with sustainable agriculture is part of the problem, as each individual has different values and objectives.

Sustainable agriculture should not, therefore, be seen as a set of practices to be fixed in time and space. It implies the capacity to adapt and change as external and internal conditions change. Yet there is a danger that policy, as it has tended to do in the past, will prescribe the practices that farmers should use rather than create the enabling conditions for locally generated and adapted technologies.

Throughout the world, environmental policy has tended to take the view that rural people are mismanagers of natural resources. The history of soil and water conservation, rangeland management, protected area management, irrigation development and modern crop dissemination shows a common pattern: technical prescriptions are derived from controlled and uniform conditions, supported by limited cases of success, and then applied widely with little or no regard for diverse local needs and conditions. Differences in receiving environments and livelihoods then often make the technologies unworkable and unacceptable. When they are rejected locally, policies shift to seeking success through the manipulation of social, economic and ecological environments, and eventually through outright enforcement (see Chapter 2).

For sustainable agriculture to succeed, policy formulation must not repeat these mistakes. Policies must arise in a new way. They must be enabling, creating the conditions for sustainable development based more on locally available resources, and local skills and knowledge. Achieving this will be difficult. In practice, policy is the net result of the actions of different interest groups pulling in complementary and opposing directions. It is not just the normative expression of governments. Effective policy will have to recognize this, and seek to bring together a range of actors and institutions for creative interaction.

The problem with traditional policy making is that it is appropriate only for non-complex systems. In coming to terms with ecological and social systems, policy processes will have to address multiple realities, chaotic systems, non-linearities and unpredictability. As Thompson and Trisoglio (1994) put it: '*And in then reaching, uncritically, for our familiar Newtonian tools – neo-classical economics in particular... – we are committing ourselves to a most unwise path. We are aspiring to manage the unmanageable*'.

What will be required will be the development of new processes for participation and mediation that could lead to the formulation of policies with much wider positive benefits.

Policy 1: Declare a National Policy for Sustainable Agriculture

The first action that governments can take is to coordinate policies and institutions more clearly. Policies have long focused on generating external solutions to farmers' needs. This has encouraged dependencies on external inputs, even when they are financially more costly, environmentally damaging and therefore economically inefficient when compared with resource-conserving options.

New policies must be enabling, creating the conditions for development based more on locally available resources, and local skills and knowledge. Policy makers will have to find ways of establishing dialogues and alliances with other actors, and farmers' own analyses could be facilitated and their organized needs articulated. Dialogue and interaction would give rapid feedback, allowing policies to be adapted iteratively. Agricultural policies could then focus on enabling people and professionals to make the most of available social and biological resources.

Declaring a national policy for sustainable agriculture helps to raise the profile of these processes and needs, as well as giving explicit value to alternative societal goals. It would also establish the necessary framework within which the more specific actions (Nos 2–25) listed below can fit and be supported.

ENCOURAGING RESOURCE-CONSERVING TECHNOLOGIES AND PRACTICES

Although some resource-conserving technologies and practices are currently being used, the total number of farmers using them is still small. This is because the adoption of these technologies is not a costless process for farmers. They cannot simply cut their existing use of fertilizer or pesticides and hope to maintain outputs, so making their operations more profitable. They will need to substitute something in return. They cannot simply introduce a new productive element into their farming systems and hope it succeeds. They will need to invest labour, management skills and knowledge. But these costs do not necessarily go on for ever, and much can be done to support this transition.

Many of the actions in this section (Nos 2–12) seek to reduce these transition costs. In particular, they relate to increasing research to demonstrate the wide applicability of resource-conserving technologies, increasing the dissemination of information to farmers and reducing the perceived short-term risks.

Policy 2: Establish a National Strategy for IPM

Integrated pest management (IPM) is the integrated use of a range of pest control strategies in a way that not only reduces pest populations to satisfactory levels but is sustainable and non-polluting. Inevitably IPM is a more complex process than, say, relying on regular calendar spraying of pesticides. It requires a level of analytical skill and certain basic training in crop monitoring and ecological principles. Large-scale IPM for rice programmes are now demonstrating that ordinary farmers can rapidly acquire new principles and approaches, and these are producing substantial reductions in insecticide use, while maintaining yields and increasing profits.

What is needed is for governments to establish a national strategy or manifesto for IPM, linking together research and extension institutions to support the development of appropriate pest and predator management technologies, and putting farmers at the centre as IPM-experts. This is then combined with policies to discourage the overuse or inappropriate use of pesticides.

Policy 3: Prioritize Research into Sustainable Agriculture

There is a need for increased research by agricultural departments and colleges into resource-conserving technologies. Current practices are heavily biased towards modern agricultural practices. At the present, too little is known about the economic and environmental benefits of resource-conserving agriculture because of a lack of professional and scientific incentives, including central and local funds, for research. Governments can raise funds by imposing levies on external inputs.

Where possible farmers should be involved closely in research design and implementation, as it is they who know their local conditions best. Research institutions must find ways of working closely with farmers and rural communities. Research should, therefore, constitute both more basic research into resource-conserving technologies in a wide variety of biophysical and socioeconomic contexts, and more analysis and understanding of what farmers are already doing through case studies and participatory analysis. Indigenous knowledge and management systems form an important focus for such research.

Policy 4: Grant Farmers Appropriate Property Rights

Sustainable agriculture incorporates the notion of giving value to the future availability of resources. But where there is lack of secure tenure and clear property rights, this discriminates against the long-term investment necessary for sustainable agriculture. If farmers are uncertain how long they will be permitted to farm a piece of land, then they will have few incentives to invest in practices that only pay off in the long term, such as soil and water conservation, agroforestry, planting hedgerows and building up soil fertility. In some places, tenants risk eviction if they improve the land they farm – if the land becomes too productive, landlords may claim it and farm it for themselves. Where land reform has occurred, there have been substantial impacts on agricultural growth, poverty alleviation and investments in resource-conserving technologies.

The best option is to grant property and titling rights through national programmes for land reform. This can be supported by the innovative use of tenancy laws that encourage action by landlords to set lease conditions that specify the use of regenerative technologies and that also ensure tenants receive the full economic value of any resource-conserving investments they have made during the course of their tenancy.

Policy 5: Promote Farmer-to-Farmer Exchanges

Farmers are the best educators of other farmers, and so farmer-to-farmer extension, visits and peer training can greatly help in information exchange and dissemination. External agencies can help in several ways. Most common are farmer exchange visits, in which farmers are brought to the site of a successful innovation or useful practice, where they can discuss and observe benefits and costs with adopting farmers. Professionals play the role of bringing interested groups together and facilitating the process of information exchange. During the visits, participants are stimulated by the discussions and observations, and many will be provoked into trying the technologies for themselves.

But one of the greatest constraints for promoting wider use of farmer-to-farmer exchanges lies in the quality of available facilitators. They must be well acquainted with the farmers; they must know about the different systems and practices present in the various communities; they must be able to facilitate discussions, interjecting where necessary to guide the conversation; and they must be able to stimulate the discussion while not dominating it. They must, therefore, have all the qualities of a new agricultural professionalism.

Policy 6: Offer Direct Transitionary Support to Farmers

Farmers face real adjustment costs when converting to a more sustainable agriculture and so a particularly important element is the transitionary period. One policy option is to offer subsidies for just this transitionary period. These would have so-called 'sunset provisions'. Such transitionary subsidies would be supported by codes of good management practice to illustrate how transition costs can be reduced or shortened, together with grants, subsidies or low-interest loans for resource-conserving machinery, equipment and so on. The adoption of transitionary payments, or conversion grants, has been used for encouraging farmers to convert entire holdings to organic practices in several European countries.

Policy 7: Direct Subsidies and Grants Towards Sustainable Technologies

Many OECD countries have a wide range of conservation and environmental schemes currently available for farmers. These offer direct financial support in return for the adoption by farmers of conservation-oriented practices and technologies. This conservation focus is increasingly being tied through existing support payments, rather than implying additional financial resources. But many of these schemes are still fragmented, being restricted to designated areas or specific landscapes. What are required are nationwide initiatives available to all farmers. Policies should be relevant to local conditions and requirements, but not restricted to those farmers who happen to farm within a particular designated area. Such integration is essential for the development of a more sustainable agriculture. This directing of subsidies also implies the elimination of any subsidies that encourage resource degradation or depletion.

Policy 8: Link Support Payments to Resource-Conserving Practices

In Europe and North America, 'set-aside' and the conservation reserve programmes have been imposed as the solution to overproduction. Farmers receive direct support only if they comply with the conditions of the payments, which include taking out of production fixed proportions of their land. But at present, these payments are not closely tied to resource-conserving practices. One option would be to link whole farm management agreements with area payments, so ensuring integrated and productive farms. This may in the end mean that set-aside is no longer needed, as production will fall with the wide adoption of a more sustainable agriculture.

Policy 9: Set Appropriate Prices (Penalize Polluters) with Taxes and Levies

Current policies tolerate external environmental and public health costs because of lack of markets for public goods, such as landscapes, soil, biodiversity and groundwater quality. These external costs are not accounted for by farmers. However, there are policies that seek to ensure that polluters pay some or all of these external costs. These include the imposing of taxes or levies on external inputs to reduce their use; and the adoption of transferable rights or permits systems, such as irrigation entitlements and transferable permits for nitrogen. These can be supplemented by establishing regulations to enforce compliance, such as groundwater protection zones, nitrate vulnerable zones, well-field protection, riverine protection zones, and wetland and erodible land protection.

The attraction for countries in the North of imposing taxes on inputs, particularly on inorganic nitrogen, is that it should reduce surpluses and decrease environmental impacts, such as groundwater contamination, at the same time. But the high returns available from external inputs make it likely that taxes would have to be set at very high levels if they were to achieve a 'desired' reduction in chemical use. Taxes below a certain threshold could simply result in a net flow of income from the agricultural sector, with consequent impacts on rural communities. At the moment, these schemes simply provide a convenient means of raising revenue for the government, though this is usually used to support research or other activities for sustainable agriculture.

Policy 10: Provide Better Information for Consumers and the Public

There is imperfect information for consumers to select food products according to whether resource-conserving or degrading practices have been used. The opportunity exists for policies to couple food markets to the environment. There are many options including new cosmetic standards and publicity campaigns to demonstrate to consumers that poor appearance does not necessarily mean poor quality. 'Eco-labelling' of foods can also help, so that consumers may exercise greater choice. But this would need to be linked to increased publicity to demonstrate the difference between products from regenerative and organic agriculture. The establishment and publication of maximum residue limits for pesticides in foods also allows consumers to make more choices.

Some of this is beginning to happen in some countries in the North, with some supermarket chains increasingly selecting for quality assurance production methods. But these foods still tend to be considerably more expensive than conventional products. For them to be available to all sectors of society, then there should be no difference in price between conventionally and alternatively grown food.

Policy 11: Encourage the Adoption of Natural Resource Accounting

Current methods for determining national and sectoral income are very misleading indicators of sustainable economic development. By convention, national income accounts ignore natural assets, assuming that the productivity of these resources is not relevant to national economic health. When conventional and natural resource accounting methods of calculating national income are compared, they show that what has been counted as income is actually losses in the form of natural resource depletion. Because resource depletion is equivalent to capital consumption, conventional accounting methods inevitably overstate income and so give inappropriate signals to planners. Conventional accounting effectively values natural capital at zero.

When natural resource accounting is used, national income accounts are adjusted to reflect the depreciation of natural capital and the direct costs of environmental degradation, either by developing monetary measures, which produce a figure of net national welfare, or by physical accounting systems, which present material and environmental resource accounts in biophysical terms. Natural resource accounting practices should also become standard practice in agricultural projects. Billions of dollars of agricultural loans and grants are made each year by bilateral donors, regional development banks and the World Bank, yet few economic analyses take account of natural capital.

Policy 12: Establish Appropriate Standards and Regulations for Pesticides

A primary line of control of pesticides is licensing and registration. Exhaustive testing is required of all new products, and they are then registered for use on specific crops or animals, provided they are applied according to certain conditions. These registrations vary hugely from country to country, with some governments permitting the use of products widely banned elsewhere. None the less, some have taken the step of banning large numbers of products widely used elsewhere because of local concerns over pest resurgences and health impacts.

Little can be done practically at the stage of application of pesticides, save for making recommendations on operator protection, spray equipment and regularity of spraying. Reliance is usually placed on the promotion of 'good agricultural practice'. Towards the end of the food production process, however, the level of regulation increases again, as maximum residue standards are set to protect consumers. These are usually national systems backed by the international Codex Committee on Pesticide Residues sponsored by WHO and FAO.

SUPPORTING LOCAL GROUPS FOR COMMUNITY ACTION

As has been illustrated in Chapter 8, coordinated action is necessary for sustainable agriculture to have any significant impact on local and national economies and environments. Such coordinated action would have the benefit of reversing some of the breakdown of social and economic structure in rural communities. The principal current constraints relate to the lack of local institutions and groups that help to ensure regular contacts between farmers and farmers, and farmers with other sectors of society, particularly local people in their community, but also with urban consumers. In highly modernized agriculture, external inputs have substituted for labour and many farmers now work alone on their enterprises.

The options in this section (Nos 13–18) relate to finding ways of increasing the social linkages between farmers and communities.

Policy 13: Encourage the Formation of Local Groups

The first action is to encourage more coordinated local action through better linkages between farmers. The success of sustainable agriculture depends not just on the motivations, skills and knowledge of individual farmers, but on action taken by groups or communities as a whole. Simply extending the message that sustainable agriculture can match conventional agriculture for profits, as well as producing extra benefits for society as a whole, will not suffice. What is also required will be increased attention to community-based action through local institutions and users' groups.

Six types of local group or institution are directly relevant to the needs for a sustainable agriculture: community organizations; natural resource management groups; farmer research groups; farmer-to-farmer extension groups; credit management groups; and consumer groups.

Policy 14: Foster Rural Partnerships

The success of sustainable agriculture in many communities and different countries has illustrated how people and agencies with apparently conflicting interests can be brought together in effective partnerships for action. There is a need for a coordinated national approach to rural development that puts community action and social cohesion as the primary goal. These should emphasize the need for local diversity, community involvement in decisions, local added value for agricultural produce, provision of services, and good networks and communications to achieve sustainable development. Such partnerships need a facilitating agency or a policy that makes nationwide provisions for local facilitators.

Policy 15: Support for Farmers' Training and Farmer Field Schools

Many farmers have neither the time nor opportunity to attend meetings or courses to learn about new technologies or practices for a more sustainable agriculture. There is a need for schemes that encourage farmer training in their own communities and on their own fields. Agricultural systems are complex and constantly changing, and what works in one village may not work in another. Farmer field schools place an emphasis on facilitating knowledge processes, continuous observation and feedback from local environments, enhancing local decision-making capacity, and group learning helping to cement local linkages and understanding.

Research and educational institutions need to be encouraged to give support for these approaches to farmer training and learning, and for the training of facilitators to conduct such training.

Policy 16: Incentives for On-Farm Employment

It is clear that resource-conserving technologies require more labour than their modern equivalents. Labour, knowledge and management skills have to be substituted back for the former higher use of external inputs. Most of this labour will have to be skilled, such as for pest monitoring, tree management or hedge laying. Some will require fewer skills, such as for weeding. But there are two problems. Most farmers in industrialized countries balk at the notion of increased costs incurred by employing more staff. They have come to believe that the best way of increasing efficiency is to cut labour. In many Third World countries, the problem is more likely to be a shortage of labour, particularly during key seasonal peaks. Many young people no longer want to work on farms, preferring to be unemployed or to seek opportunities in urban centres.

What are needed are incentive schemes to encourage the employment of local people on farms. This may require additional financial support, either directly to farmers or indirectly for rural services to improve the rural environment and economy.

Policy 17: Assign Local Responsibilities for Landscape Conservation

Most local people value in both aesthetic and recreational terms elements of their landscape, yet they are rarely involved in decisions and processes that shape it. As land managers and owners, farmers clearly should be making decisions about how best to farm their land. But if responsibility is assigned to both farmers and communities for landscape conserving activities, where local people are encouraged to become involved in local farming in an indirect way, then again more understanding would be created among different interests.

This involves the devolution of planning and monitoring to local people, who are no longer seen simply as informants, but as teachers, extension agents, activists and monitors of change. These local paraprofessionals include village energy workers, villager extension agents, pest control experts, village game wardens, women veterinarians and so on. Recognizing that village specialists come from all sectors and classes of the community facilitates the integration of marginalized groups, so allowing their skills and knowledge to influence development priorities. Given the chance, local people are able to monitor environmental change and so take action when required.

Policy 18: Permit Groups to Have Access to Credit

It has long been assumed that poor people cannot save money. Because they are poor and have little or no collateral, they are too high a risk for banks and so have to turn to traditional money lenders. These inevitably charge extortionate rates of interest and very often people get locked into even greater poverty while trying to pay off debts.

Recent evidence is emerging, however, to show that when local groups are trusted to manage financial resources, they can be more efficient and effective than external bodies, such as banks. They are more likely to be able to make loans to poorer people. They also recover a much greater proportion of loans. In a wide range of countries, local credit groups are directly helping poorer families both to stay out of debt and reap productive returns on small investments on their farms.

What is needed to support these efforts is for banks to change their rules about lending. The convention has been that they only lend to individuals with collateral. But where banks have been instructed or have chosen to lend to groups as an institution, with the groups taking collective responsibility for the loans, many more poor and needy people have access to credit. They are also better at paying back loans.

REFORMING EXTERNAL INSTITUTIONS AND PROFESSIONAL APPROACHES

As has been illustrated throughout this book, sustainable agriculture implies the integrated use of resources at local level so as to meet both productivity and sustainability goals. Yet, most external agencies, both government and non-government, are organized along sectoral lines, so making it difficult for farmers and professionals to engage in meaningful debate and action. Agricultural research and extension professionals tend to be too narrowly trained, tend not to work in a multi-disciplinary fashion and tend not to work closely with farmers.

The actions for this section (Nos 19–25) relate to the wider coordination of policy processes, better working linkages between professionals and farmers, and changes to training and teaching programmes. These imply the development of a new professionalism for agriculture, in which responsibility is placed on individuals as well as institutions.

Policy 19: Encourage the Formal Adoption of Participatory Methods and Processes

Many organizations have a poor record when it comes to participation with farmers. Yet there is good evidence that participation can lead to significant changes in economic and environmental status of communities. A wide range of participatory methods have been shown to be effective over recent years in almost every country of the world. However, professionals familiar with these methods and, more importantly, the attitudes required for sensitive working with local people, are still in the minority. What is needed is wider support for the use of participatory methods and processes, and the establishment of appropriate incentives to encourage their institutionalization by researchers, extensionists and planners.

None the less, it will be important to ensure that a too rapid institutionalization does not occur. If training manuals and guidelines convert participatory methods into simplistic steps, then methodologies will become too rigid and unable to change to suit diverse institutional and ecological conditions.

Policy 20: Support for Information Systems to Link Research, Extension and Farmers

Sustainable agriculture farmers need more specialized and interactive information systems. The poor linkages between agencies of different sectors and professionals of different disciplines means that farmers and communities are rarely involved in research and extension activities. The problem lies with a critical lack of multi-disciplinary and communication skills in professionals, and an adherence to sectoral rather than systemic approaches. This means that external institutions tend to miss the complexities perceived by people at the local level.

There are several options for strengthening linkages between agencies and sectors. The best are to ensure that training and capacity building in the use of community development and participatory methods occur in the field; and the explicit adoption of farmer-to-farmer extension methods. As farmers develop their own expertise, so they are more capable of making demands on research and extension systems.

Policy 21: Rethink the Project Culture

For sustainable agriculture to succeed, projects must choose a learning process rather than relying on blueprints. These projects should start small and cheaply. They should have uncomplicated design and not try to over-innovate. At the outset, the approach should focus on what people articulate is most important to them. This may mean starting with activities that are not central to project remits, for which funding flexibility from donors is necessary. The best introduced technologies are low risk, easy to teach, tested under local conditions and offer the prospect of clear, on-site benefits in the coming season or year. A common feature of successful projects has thus been an early period for experimentation and building local capacity. Continual dialogue allows outsiders to learn, plan with the local people and replan. These linkages reduce lags in information flows, as feedback occurs during the project cycle.

Policy 22: Strengthen the Capacities of NGOs to Scale Up

Most NGOs are quite small, though often conspicuous. Coverage by NGOs as a whole is usually patchy and small compared with that of government organizations. If governments adopt an alternative approach to agriculture, even if quality is diluted, the wider impact can be enormous. At the same time, it is important to find ways to widen the impact of the successful actions of NGOs. There are three different approaches to this scaling up.

The additive strategy, in which NGOs increase their size and expand operations, is currently widespread as donors have tended to foster operational and organizational expansion. But it has dangers. Some of the comparative advantage of NGOs is liable to be lost when they expand. Close relationships with farmers, the capacity to experiment and the ability to be flexible to local contexts may all be weakened. Internal organizational objectives may come to dominate and displace development objectives.

The multiplicative strategy, in which NGOs achieve impact through deliberate influence, networking, training and policy reform, can take many forms. Intermediary NGOs can provide stimulus, resources and technical assistance for the formation and functioning of community-based organizations. Such NGOs can channel financial and technical resources from other agencies to community-based organizations instead of using those resources themselves.

The diffusive strategy, in which NGOs achieve impact through informal and spontaneous spread of ideas, approaches and methods, entails developing and spreading ideas, approaches and methods which others pick up, and then spread on their own.

Policy 23: Foster Stronger NGO–Government Partnerships

There are significant benefits in encouraging NGOs and governments to work together. The size of human capital and resources locked up in government institutions usually represents a huge under-utilized potential. Opportunities for innovative work to catalyse change within governments do exist, particularly under conditions of increased decentralization. Collaboration between NGOs and governments to realize more of the potential and exploit more of the opportunities means working together in a mutually independent fashion.

The primary objective can, therefore, be to foster change from within government. This can often be pursued best through supporting and working with innovative individuals and programmes. Personalities and relationships are a vital element in successful partnerships. NGOs working in this mode try to enlist support at all levels of the system, particularly among higher level administrators and politicians. All need to understand and appreciate the demands of the new approaches, and to be aware of the potential benefits.

Policy 24: Reform Teaching and Training Establishments

Sustainable agriculture implies new thinking about teaching and learning. The central concept of sustainable agriculture is that it enshrines new ways of learning about the world. Teaching and learning, though, are not the same thing. Learning does not necessarily result from teaching. The preoccupation with teaching has constrained the effectiveness of higher education and limited its abilities to meet society's demands. Professionals who are to work with local complexity, diversity and uncertainty need to engage in sensitive learning about the particular conditions of rapid change.

A move from a teaching to a learning style has profound implications. Everyone involved in agriculture, including farmers, trainers, educators, researchers, extensionists and administrators becomes important, as do the relationships between them. The focus is then less on what we learn and more on how we learn. Institutions will need to provide creative learning environments, conditions in which learning can take place through experience, and through personal exploration and experimentation. The pedagogic goals become self-strengthening for people and groups through self-learning and self-teaching. This is a huge agenda for the educational establishment.

Policy 25: Develop Capacity in Planning for Conflict Resolution and Mediation

Policy conflicts are inevitable in debates over what represents sustainability and what steps should be taken towards it. Trade-offs have to be made and the key policy challenge will be to find productive ways of doing this that ensure some consensus for constructive action. The process of social mediation requires wisely managed participation among all actors.

As community-led initiatives become stronger and more capable of drawing on external resources, and even of influencing higher level decision makers, this may bring local people into conflict with existing power structures. One problem is that there generally is a lack of mediation processes between the macro- and micro-levels. Establishing mechanisms for meso-level dispute resolution will be an important component of successful policies for sustainable agriculture. These require facilitation by 'neutral' actors, who seek to establish consensus and joint ownership among the various actors and interests in agriculture and rural development.

In practice, of course, most trade-off decisions are still made by politicians, representing to a lesser or greater extent the interests of the public. They are likely to see increasing participation, and the increasing democracy it implies, as a threat to their power base. Perhaps the greatest challenge for sustainable agriculture will, therefore, be to persuade politicians and high-level decision makers to adopt participatory approaches and to demonstrate just how they, and their constituencies and countries, will benefit from so doing.

SUMMARY

Sustainable agriculture can be economically, environmentally and socially viable. There are resource conserving technologies, local institutional structures and enabling external institutions that are all known to work. But, until recently, there have been few policies known to be effective. This chapter has indicated that there are at least 25 policies that can be used to support the transition to greater sustainability and self-reliance.

But as few have been applied in an integrated way, the full benefits of a more sustainable agriculture have yet to be achieved. The major new challenge lies in the development of new approaches to policy formulation. The problem with traditional policy making is that it is appropriate only for non-complex systems. For a more sustainable agriculture to succeed, policy formulation must not repeat the past mistakes of coercion and control. Rather, it will have to find ways of being enabling, so creating the conditions for sustainable development based more on local resources, skills and knowledge.

The greatest challenge, therefore, will be reform of policy processes themselves. These will have to focus more on participation and social mediation if the contested complexities and uncertainties of sustainability are to be continually addressed.

REFERENCES

AAN. 1993a. Use of conservation tillage on the increase survey finds. *Alternative Agriculture News* 11 (12), 2.

AAN. 1993b. Census bureau drops survey of farm residents. *Alternative Agriculture News* 11(11): 2.

Abedin Z and Haque H. 1989. Innovator workshops in Bangladesh. In: Chambers R, Pacey A and Thrupp, L-A (eds). *Farmer First: Farmer Innovation and Agricultural Research.* Intermediate Technology Publications Ltd., London.

Abrams P. 1982. *Historical Sociology*. Open Books, Shepton Mallet.

Acaba M, Apura, D, Cabiling J, De Pedor R and Lightfoot C. 1987. *A Study of Farmers' Evaluation of Camote Varieties*. Farming Systems Development Project, Eastern Visayas, Visayas State College of Agriculture, Leyte, Philippines.

ACORA. 1990. *Faith in the Countryside*. The Archbishops' Commission on Rural Areas, London.

ADB. 1987. *Handbook on the Use of Pesticides in the Asia-Pacific Region*. Asian Development Bank, Manila.

Adnan S, Barrett A, Nurul Alam S M and Brustinow A. 1992. *People's Participation. NGOs and the Flood Action Plan*. Research and Advisory Services, Dhaka.

AED. 1991. *Communication for Technology Transfer in Agriculture*. Academy for Educational Development, Washington DC.

Agarwal A and Narain S. 1989. *Towards Green Villages*. Centre for Science and the Environment, New Delhi.

Agarwal B. 1984. Rural women and high yielding variety rice technology. *Economic and Political Weekly* 19(13): A39–A52.

Agarwal B. 1985. Women and technological change in agriculture: the Asian and African experience. In: Ahmed I (ed) *Technology and Rural Women: Conceptual and Empirical Issues*. George Allen and Unwin, London.

Aggarwal P K and Garrity D P. 1987. Intercropping of Legumes to Contribute Nitrogen in Low-Input Upland Rice Based Cropping Systems. International Symposium Nutrient Management for Food Crop Production in Tropical Farming Systems. Malang, Indonesia.

Agresti A. 1979. *Analysis of Association Between 2,4,5-T Exposure and Hospitalised Spontaneous Abortions*. Environmental Health Sciences Centre, Oregon State University, Corvallis.

AKRSP. 1994. *Aga Khan Rural Support Programme Annual Report*. Gilgit, Pakistan.

Alcorn J B. 1984. Development policy, forests and peasant farms: reflections on Haustec-managed forest contribution to commercial and resource conservation. *Economic Botany* 38(4): 389–406.

Alexander A. 1993. Modern trends in special fertilisation practices. *World Agriculture* 35–38.

Alexander H. 1993. *Lessons in Landcare. Australia's model for a better farming future*. Victoria, Australia.

Allen W A, Kazmeirczak R F, Lambur M T, Norton G W and Rajotte E G. 1987. The national evaluation of extension's integrated pest management programs. *Virginia Cooperative Extension Services Publication* 491-010, Blacksburg.

Altieri M A and Liebman M. 1986. Insect, weed and plant disease management in multiple cropping systems. In: Francis C A (ed) *Multiple Cropping Systems*. Macmillan, New York.

Altieri M A and Schmidt L L. 1986. The dynamics of colonising arthropod communities at the interface of abandoned, organic and commercial apple orchards and adjacent woodland habitats. *Agric. Ecosystems and Environm.* 16, 29–43.

Altieri M A and Yurjevic A. 1992. Changing the agenda of the universities. *ILEIA Newsletter* 2/92, 39.
Altieri M A. 1990. Agroecology and rural development in Latin America. In: Altieri M and Hecht S (eds) *Agroecology and Small Farm Development*. CRC Press, Boca Raton.
Altieri M. 1992. Farmers' initiatives to maintain diversity. In: *The Gene Traders: Security or Profit in Food Production*. Proceedings of an International Conference, 14–15 April 1992. ITDG, Rugby and New Economics Foundation, London.
Anderson D. 1984. Depression, dust bowl, demography, and drought: the colonial state and soil conservation in East Africa during the 1930s. *African Affairs* 321–43.
Antholt C H. 1991. Agricultural extension in the 21st century. Lessons from South Asia. In: Rivera, W and Gustafsen, D (eds) *Agricultural Extension: Worldwide International Evolution and Forces for Change*. Elsevier, New York.
Antholt C H. 1992. *Relevancy, Responsiveness and Cost-Effectiveness: Issues for Agricultural Extension in the 21st Century*. Asia Region, Technical Dept, World Bank, Washington DC.
Antholt C H. 1994. *Getting Ready for the Twenty-First Century. Technical Change and Institutional Modernisation in Agriculture*. World Bank Technical Paper 217. Asia Technical Department Series, World Bank, Washington DC.
Argyris C and Schön D. 1978. *Organisational Learning*. Addison-Wesley, Reading, Mass.
Argyris C, Putnam R and Smith D M. 1985. *Action Science*. Jossey-Bass Publishers, San Francisco and London.
Arthur B. 1989. Competing technologies, increasing returns and lock-in by historical events: the dynamics of allocation order increasing returns. *Economic Journal* 99, 116–31.
Ashby J, Quiros C and Rivers Y. 1987. Farmer participation in on-farm varietal trials. *ODI Agricultural Administration (Research and Extension) Network Discussion Paper*, 22. Overseas Development Institute, London.
Ashby J, Quiros C and Rivers Y. 1989. Experiences with group techniques in Colombia, pp. 127–31. In: Chambers R, Pacey A and Thrupp L A (eds) *Farmer First: Farmer Innovation and Agricultural Research*. Intermediate Technology Publications Ltd, London.
ASOCON. 1991. Small watershed planning case study. ASOCON Workshop on Conservation Project Design. Asia Soil Network Conservation Network, Jakarta, Indonesia.
Attah-Krah A N and Francis PA. 1987. The role of on-farm trials in the evaluation of composite echnologies: the case of alley farming in Southern Nigeria. *Agric. Systems* 23: 133–52.
Attaviroj P. 1991. Soil erosion and land degradation in the northern Thai uplands. An economic case study. *Contour* III (1), 2–7.
Atuma S S. 1985. Residues of organochlorine pesticides in some Nigerian food materials. *Bull. Environ. Contam. Toxicol.* 35: 735–8.
Atuma S S and Okor D I. 1985. Pesticide usage in Nigeria – need for a baseline study. *Ambio* 14: 340–1.
Ault, W O. 1965. Open field husbandry and the village community – a study of agrarian by-laws in Medieval England. *Trans. Amer. Philos. Soc.* 55.
Axinn G H. 1988. T & V (Tragic and Vain) Extension. *INTERPAKS, INTERCHANGE, International Agriculture* 5(3). College of Agriculture, University of Illinois, Urbana.
Ayeh E. 1990. Global 2000: what's in it for the farmer? *ILEIA Newsletter* 6(3), 10–11.
Bagadion B U and Korten F F. 1991. Developing irrigators' organisations; a learning process approach. In: Cernea M M (ed) *Putting People First*. Oxford University Press, Oxford, 2nd edn.
Bahuguna V K. 1992. *Collective Resource Management. An Experience of Harda Forest Division*. Regional Centre for Wastelands Development, Bhopal.
Bailey C H. 1914. The composition and quality of wheat grown in mixtures with oats. *Agron J*. 6: 204–10.
Baker G, Knipscheer H C and de Souza Neto J. 1988. The impact of regular research field hearings (RRFH) in on-farm trials in northeast Brazil. *Expl. Agric.* 24, 281–8.
Balbarino E A and Alcober D L. 1994. Participatory watershed management in Leyte, Philippines: experiences and impacts after 3 years. Paper for *IIED New Horizons* conference, Bangalore, India, Nov 1994. IIED, London.
Balbarino E A, Tung L and Obusa A P. 1992. FARMI's experiences on PTD: the case of contour hedgerows in the calcareous uplands of Matalom. *On-Farm Research Notes* No 12, Visayas College of Agriculture, Baybay, Leyte, Philippines.

Baldock D. 1990. Economic instruments for reducing agricultural pollution. Working paper for IEEP Programme on Economics Incentives in Agriculture. IEEP, London.

Bannister M E and Nair P K R. 1990. Alley cropping as a sustainable agricultural technology for the hillsides of Haiti: experience of an agrofirestry outreach project. *Am. J. Altern. Agric.* 5(2): 51–9.

Barbier E B and Burgess J. 1990. *Malawi. Land Degradation in Agriculture.* Report of World Bank Economic Mission on Environmental Policy. IIED, London.

Barbier E B. 1991. Environmental Management and Development in the South. Prerequisites for sustainable development. Paper prepared for UNCED. IIED, London.

Bartra R. 1990. Agrarian crisis and political crisis in Mexico. In: Bernstein M, Crow B, Mackintosh M and Martin C (eds) *The Food Question: Profits versus People.* Earthscan Publications Ltd, London.

Bastian E and Gräfe W. 1989. Afforestation with multipurpose trees in *media lunas:*a case study from the Tarija basin, Bolivia. *Agroforestry Systems* 9, 93–126.

Batie S S and Taylor D B. 1989. Widespread adoption of non-conventional agriculture: profitability and impacts. *Amer. J. Alternative Agriculture* 4 (3 and 4), 128–34.

Bawden R. 1991. Systems thinking and practice in agriculture. *Journal of Dairy Science* 74: 2362–73.

Bawden R. 1992. Creating learning systems: a metaphor for institutional reform for development. Paper for joint *IIED/IDS Beyond Farmer First: Rural People's Knowledge, Agricultural Research and Extension Practice* Conference, 27–9 October, Institute of Development Studies, University of Sussex, UK. IIED, London.

Bawden R. 1994 A learning approach to sustainable agriculture and rural development: reflections from Hawkesbury. Hawkesbury College, Australia. *Mimeo.*

Bax J E and Fisher G E. 1993. The Viability Grass/White Clover Swards For Dairy Production In The UK. Paper presented at the *XVII International Grassland Congress* Massey University, New Zealand, 1993.

Bax J E. 1992. SAC clover-based dairy systems. In: MMB/SAC. *Extensification of Milk Production Using White Clover.* Milk Marketing Board Research and Development and Scottish Agricultural College, Dumfries.

Beaumont P. 1993. *Pesticides, Policies and People.* The Pesticides Trust, London.

Bebbington A. 1991. *Farmer Organisations in Ecuador: Contributions to Farmer First Research and Development.* Sustainable Agriculture Programme Gatekeeper Series SA26. IIED, London.

Bebbington A and Farrington J. 1992. The Scope for NGO-Government Interactions in Agricultural Technology Development: An International Overview. *Agric. Admin. (R&E) Network Paper* 33, ODI, London.

Bebbington A and Thiele G, with Davies P, Prager M and Riveros H. 1993. *NGOs and the State in Latin America: Rethinking Roles in Sustainable Agricultural Development.* Routledge, London.

Beckett J V. 1990. *The Agricultural Revolution.* Basil Blackwell, Oxford.

Beckjord P. 1991. Paulownia, the tree of choice in China. *International Ag-Sieve* IV(i), 1–2.

Bedford G O. 1980. Biology, ecology and control of palm rhinoceros beetles. *Ann. Rev. Entomol.* 25: 309–39.

Beinart W. 1984. Soil erosion, conservationism and ideas about development: a southern African exploration, 1900–1960. *J. Southern African Studies II*, 52–83.

Beingolea J, Beingolea K and Ruddell E D. 1992. *Report on field trials to determine appropriate potato varieties and methods of fertilization for potatoes produced between 2800 and 3900 masl: North Potosi, Bolivia.* World Neighbors, Santiago, Chile.

Belbin M. 1992. *Building the Perfect Team.* Industrial Training Research Unit, Cambridge. Video.

Benhke R and Scoones I. 1992. *Rethinking Range Ecology: Implications for Rangeland Management in Africa.* Drylands Programme Issues Paper No 33. IIED, London.

Bennett H H. 1939. *Soil Conservation.* McGraw-Hill, New York.

Benor D, Harrison J Q and Baxter M. 1984. *Agricultural Extension: The Training and Visit System.* World Bank, Washington DC.

Bentley J W, Rodríguez G and González A. 1993. Science and the people: Honduran campesinos and natural pest control inventions. In: Buckles D (ed) *Gorras y Sombreros: Caminos hacia la Colaboración entre Técnicos y Campesionosia.* Dept of Crop Protection, El Zamarano, Honduras.

Bentley J. 1992. Promoting farmer experiments in non-chemical pest control. Paper for joint *IIED/IDS Beyond Farmer First: Rural People's Knowledge, Agricultural Research and Extension Practice* Conference, 27–29 October, Institute of Development Studies, University of Sussex, UK. IIED, London.
Bentley J. 1994. Stimulating farmer experiments in non-chemical pest control in Central America. In: Scoones I and Thompson J (eds) *Beyond Farmer First: Rural People's Knowledge, Agricultural Research and Extension Practice*. IT Publications, London.
Bentley J and Melara W. 1991. Experimenting with Honduran farmer-experimenters. *ODI Agricultural Administration (Research and Extension) Network Discussion Paper* 24: 31–48.
Berardi G M. 1978. Organic and conventional wheat production: examination of energy and economics. *Agro-Ecosystems* 4: 367–76.
Berman M. 1982. *All That is Solid Melts into Air*. New York.
Bernstein R. 1985. *Habermas and Modernity*. Oxford University Press, Oxford.
Berry W. 1977. *The Unsettling of America: Cultures and Agriculture*. Sierra Club Book, San Francisco.
Bhatnagar B and Williams A. 1992. *Participatory Development and the World Bank: Potential Directions for Change*. World Bank, Washington DC.
Biggs S. 1989. Resource-poor farmer participation in research: a synthesis of experience from nine national agricultural research systems. *OFCOR Project Study No 3*. ISNAR, The Hague.
Billings M and Singh A. 1970. Mechanisation and the wheat revolution: effects on female labour in Punjab. *Economic and Political Weekly* 5(52).
Bimbao M P, Cruz A V and Smith I R. 1992. An economic assessment of rice–fish culture in the Philippines. In: Hiemstra W, Reijntjes C and Van der Werf E (eds). *Let Farmers Judge*. IT Publications, London.
Bishop J and Allen J. 1989. *The On-Site Costs of Soil Erosion in Mali*. Environment Department Working Paper No 21, World Bank, Washington DC.
Bishop J. 1990. *The Cost of Soil Erosion in Malawi*. Malawi Country Operations Division, The World Bank, Lome, Togo.
Blackler A. 1994. Indigenous versus imposed: soil management in the Mixteca Alta, Oaxaca, Mexico. Paper presented to *Rural History Centre Conference*, 10 May, University of Reading.
Boardman J and Evans R. 1991. *Flooding at Steepdown*. A report to Adur District Council, West Sussex.
Boardman J. 1990. Soil erosion on the South Downs: a review. In: Boardman J, Foster I D L and Dearing J A (eds). *Soil Erosion on Agricultural Land*. John Wiley and Sons, Chichester.
Boardman J. 1991. The Canadian experience of soil conservation: a way forward for Britain? *Intern J. Environmental Studies* 37, 263–9.
Bonsu M. 1983. Organic residues for less erosion and more grain in Ghana. In: El-Swaify M, Molderhauer W and Lo A (eds) *Soil Erosion and Conservation*. Soil Conservation Service, Ankery, Iowa.
Boonkerd N, Kesawapitak P and Chaiwanakupt S. 1991. Successful integrated plant nutrition in prevailing crop mixes in Thailand. In: *Asian Experiences in Integrated Plant Nutrition: Report of the Expert Consultation of the Asia Network of Bio and Organic Fertilizers*. Regional Office for Asia and the Pacific (RAPA), FAO, Bangkok.
Borlaug N. 1958. The impace of agricultural research on Mexican wheat production. *Trans. of New York Academy of Sciences*, Series II xx(3), 279–95.
Borlaug N. 1992. Small-scale agriculture in Africa: the myths and realities. *Feeding the Future* (Newsletter of the Sasakawa Africa Association) 4: 2.
Borowitz S. 1989. Lessons from a traditional agroecosystem. *The Cultivar* 7(1): 1–4.
Bossert T J. 1990. Can they get along without us? Sustainability of donor-supported health projects in central America and Africa. *Soc. Sci. Med.* 30(9), 1015–23.
Bowers J K and Cheshire P. 1983. *Agriculture, the Countryside and Land Use*. Methuen and Co, London.
Bowler I. 1979. *Government and Agriculture. A Spatial Perspective*. Longman, London.
Brinkman R. 1993. Recent developments in land use planning. Paper presented to *Conference on the Future of the Land*. Wageningen Agricultural University, August 1993.
Bromley D W. 1993. Common property as metaphor: systems of knowledge, resources and the decline of individualism. *The Common Property Resource Digest* 27, 1–8. IASCP, Winrock and ICRISAT, Hyderabad.

Bromley D W and Cernea M M. 1988. *The Management of Common Property Natural Resources and Some Conceptual and Operational Fallacies.* World Bank Discussion Papers, No 57. The World Bank, Washington, DC.
Brouwers J H A M. 1993. *Rural People's Response to Soil Fertility Decline. The Adja Case (Benin).* Wageningen Agricultural University Papers 93.4. Wageningen Agricultural University, The Netherlands.
Bryden J and Watson D. 1991. *Local Initiatives and Sustainable Development in Rural Scotland.* A report for the Scottish Development Department. The Arkleton Trust (Research) Ltd and Landwise Scotland.
Bunch R. 1983. *Two Ears of Corn: A Guide to People-Centred Agricultural Improvement*. World Neighbors, Oklahoma City.
Bunch R. 1987. Case study of the Guinope Integrated Development Program, Guinope, Honduras. Paper presented at *IIED Conference on Sustainable Development*, London 28–30 April 1987.
Bunch R. 1989. Encouraging farmer's experiments. In: Chambers R, Pacey A and Thrupp L-A (eds) *Farmer First: Farmer Innovation and Agricultural Research.* Intermediate Technology Publications Ltd, London.
Bunch R. 1990. *Low Input Soil Restoration in Honduras: The Cantarranas Farmer-to-Farmer Extension Programme.* Sustainable Agriculture Programme Gatekeeper Series SA23. IIED, London.
Bunch R. 1991. People-centred agricultural improvement. In: Haverkort et al (eds) *Joining Farmers' Experiments.* IT Publications, London.
Bunch R. 1993. EPAGRI's work in the State of Santa Catarine, Brazil: Major New Possibilities for Resource-Poor Farmers. COSECHA, Tegucigalpa, Honduras.
Bunch R and López G. 1994. Soil recuperation in central America: measuring impact 4 to 40 years after intervention. Paper for *IIED New Horizons* conference, Bangalore, India, Nov 1994. IIED, London.
Burrell, E D R. 1960. *An Historical Geography of the Sandlings of Suffolk, 1600 to 1850.* MSc Thesis, University of London.
Bussell F P. 1937. Oats and barley on New York farms. *Cornell Extension Bulletin* 376.
Caird J. 1852. In: Mingay G E (ed) 1968. *English Agriculture in 1850–51.* Frank Cass and Co., London.
Calva J L (ed).1988. Crisis Agricola y Alimentaria en Mexico : 1982–88. *Fontamara* 54, Mexico.
Campbell A. 1992. Community First – Landcare in Australia. Paper for *IIED/IDS Beyond Farmer First: Rural People's Knowledge, Agricultural Research and Extension Practice* Conference, 27–29 October, Institute of Development Studies, University of Sussex, IIED, London.
Campbell A. 1994a. Participatory inquiry: beyond research and extension in the sustainability era. Paper for *International Symposium Systems-Oriented Research in Agriculture and Rural Development*, Montpelier, France 21–5 November 1994.
Campbell A. 1994b. *Landcare : Communities Shaping the Land and the Future.* Allen and Unwin, Sydney.
Campbell A. 1994c. *Community First: Landcare in Australia.* Gatekeeper Series 42. IIED Sustainable Agriculture Programme. IIED, London.
Campbell J. 1992. *Joint Forest Management.* The Ford Foundation, New Delhi.
Campion D, Hall D R and Prevett P F. 1987. Use of pheromones in crop and stored products pest management: control and monitoring. *Insect Sci. Applic.* 8, 803–5.
Campion D G and Hosny M M. 1987. Biological, cultural and selective methods for control of cotton pests in Egypt. *Insect Sci Applic* 8: 803–5.
Carley M. 1994. *Policy Management Systems and Methods of Analysis for Sustainable Agriculture and Rural Development.* FAO, Rome and IIED, London.
Carruthers I and Chambers R. 1981. Rapid Rural Appraisal: rationale and repertoire. *IDS Discussion Paper No 155.* Institute of Development Studies, University of Sussex, UK.
Casteñeda C P and Rola A C. 1990. Regional pesticide review: Philippines. A country report. Paper at *IDRC Regional Pesticide Review Meeting*, 24 March, Genting Highlands, Malaysia.
Castillo G T. 1992. Sustainable agriculture: in concept and in deed. *ODI Agric Admin (R&E) Network Paper* 36: 1–32. ODI, London.
Cernea M M. 1987. Farmer organisations and institution building for sustainable development. *Regional Development Dialogue* 8, 1–24.

Cernea M M. 1991. *Putting People First*. Oxford University Press, Oxford. 2nd edn.
Cernea M M. 1993. Culture and organisation. The social sustainability of induced development. *Sustainable Development* 1(2), 18–29.
CFDA. *passim. Summary of Illnesses and Injuries Reported by California Physicians as Potentially Related to Pesticides 1972–1990*. California Department of Food and Agriculture, Sacramento, California.
Chambers R. 1983. *Rural Development: Putting the Last First*. Longman, London.
Chambers R. 1985. Normal professionalism, new paradigms and development. Paper for *Seminar on Poverty, Development and Food: Towards the 21st Century*. IDS, Sussex, December 1985.
Chambers R. 1991. Farmers' practices, professionals and participation: challenges for soil and water management. Paper for the 1991 workshop on *Farmers' Practices and Soil and Water Conservation Programmes*. ICRISAT, June 1991.
Chambers R. 1992a. The self-deceiving state: psychosis and therapy. *IDS Bulletin* 23(4): 31–42.
Chambers R. 1992b. Methods for analysis by farmers: the professional challenge. Paper for 12th Annual Symposium of Association for FSR/E, Michigan State University, 13–18 September.
Chambers R. 1992c. *Rural Appraisal: Rapid, Relaxed and Participatory*. IDS Discussion Paper 311. IDS, Brighton.
Chambers R. 1993. *Challenging the Professions: Frontiers for Rural Development;* Intermediate Technology Publications, London.
Chambers R and Ghildyal B. 1985. Agricultural research for resource poor farmers – the farmer first and last model. *Agricultural Administration* 20: 1–30.
Chambers R, Pacey A and Thrupp L A (eds). 1989. *Farmer First: Farmer Innovation and Agricultural Research*. Intermediate Technology Publications Ltd, London.
Chand R, Sindhu D S. and Kaul J L. 1985. Impact of agricultural modernization on labour use pattern in Punjab with special reference to women's labour. *Indian Journal of Agric. Econ*. XL(3).
Chand S and Gurung B. 1991. Informal research with farmers: the practice and prospects in the hills of Nepal. *Journal of Farming Systems Res–Extension* 2(2): 69–79.
Chapman N. 1988. The impact of T&V extension in Somalia. In: Howell J (ed) *Training and Visit Extension in Practice*. Overseas Development Institute, London.
Chartres J. 1985. The marketing of agricultural produce. In: Thirsk J (ed) *The Agrarian History of England and Wales Vol V. 1640–1750. II. Agrarian Change*. Cambridge University Press, Cambridge.
Chaudhri D P. 1992. Employment consequences of the Green Revolution: some emerging trends. *Indian J. Labour Econ* 35(i), 23–36.
Checkland P and Scholes J. 1990 *Soft Systems Methodology in Action*. John Wiley and Sons, Chichester.
Checkland P B. 1981. *Systems Thinking, Systems Practice*. John Wiley, Chichester.
Chen D F, Meier P G and Hilbert M S. 1987. Organochlorine pesticide residues in paddy fish in Malaysia and the associated health risk to farmers. *Bull. WHO* 62: 251–3.
Chen T P and Yenpin L I. 1986. Integrated agriculture – aquaculture studies in Taiwan. In: *Proc of the ICLARM – SEARCA Conference on Integrated Agriculture–Aquaculture Farming Systems*. Manila, Philippines.
Christiansson C. 1988. Degradation and rehabilitation of agropastoral land. Perspectives on environmental change in semi-arid Tanzania. *Ambio.*
CIAT. 1987. Biological control halts cassava hornworm. *CIAT Report* 1987: 34–6. CIAT, Cali, Colombia.
CIAT. 1989. Farmer organsiations in technology adaptation and transfer. *CIAT Report* 1989: 11–14. CIAT, Cali, Colombia.
CIDICCO. *passim. Cover Crops News*. CIDICCO Tegucigalpa, Honduras.
Clarke A C. 1973. Profiles of the Future. In: *A Dictionary of Contemporary Quotations*. Pan Books, London.
Clunies-Ross T and Hildyard N. 1992. *The Politics of Industrial Agriculture*. SAFE Alliance and Earthscan Publications Ltd, London.
CNPPA. 1993. *Parks for Life*. Report of the IVth World Congress on National Parks and Protected Areas. Ed by McNeely J. IUCN, Geneva.
Collier W L, Soentoro, Wiradi G, Basandaran E, Sontoso K and Stepanek J F. 1982. The acceleration of rural development on Java: from village studies to a national perspective. *Agro-Economic Survey Occasional Paper No 6*, Bogor, Indonesia.

Collier W L, Wiradi G and Soentoro. 1973. Recent changes in rice harvesting methods. Some serious social implications. *Bulletin of Indonesian Economic Studies* 9(2): 36–45.

Conroy A. 1990. Fertilizer and maize seed supply and distribution in Malawi. Lilongwe, Malawi.

Conroy C and Litvinoff M. 1989. *The Greening of Aid*. Earthscan Publications Ltd, London.

Conway G R. 1971. Better methods of pest control. In: Murdoch W W (ed) *Environment, Resources, Pollution and Society*. Sinauer Assoc Inc, Stanford.

Conway G R 1987. The properties of agroecosystems. *Agric Systems* 24, 95–117.

Conway G R and Barbier E B 1990. *After the Green Revolution*. Earthscan Publications Ltd, London.

Conway G R and Pretty J N. 1991. *Unwelcome Harvest. Agriculture and Pollution*. Earthscan Publications Ltd, London. 645 pp.

Conway G R, Lele U, Peacock J and Piñero M. 1994. *Sustainable Agriculture for a Food Secure World: A Vision for the Consultative Group on International Agricultural Research (CGIAR)*. CGIAR, Washington DC.Cook T and Campbell D. 1979. *Quasi-Experimentation: Design and Analysis Issues for Field Settings*. Rand McNally, Chicago.

Cordova V, Herdt R W, Gascon F B and Yambao L. 1981. Changes in rice production technology and their impact on rice farm earnings on Central Luzon, Philippines 1966–1979. *Dept of Agricultural Economics Paper* 18–19. IRRI, Los Banos, Philippines.

Cornelius J. 1993. Stress and the family farm. Paper presented to the *Centre for Agricultural Strategy/Small Farmers Association Symposium*. 30–31 March, University of Reading.

Craig I A and Pisone U. 1988. *A Survey of NERAD Promising Processes, Methodologies and Technologies for Rainfed Agriculture in N E Thailand*. NERAD Project, Tha Phra, Khon Kaen.

Craig I A. 1987. *Pre-rice crop green manuring: a technology for soil improvement under rainfed conditions in N E Thailand*. NERAD Project, Tha Phra, Khon Kaen.

Critchley W. 1991. *Looking After Our Land: New Approaches to Soil and Water Conservation in Dryland Africa*. Oxfam and IIED, London.

Croft B A and Strickler K. 1983. Natural enemy resistance to pesticides. In: Georghiou G and Saito T (eds) *Pest Resistance to Pesticides*. Plenum, New York.

Crosson P and Ostrov J E. 1988. Alternative agriculture: sorting out its environmental benefits. *Resources* 92, 13–16.

Cunningham A B. 1990. People and medicines: the exploitation and conservation of traditional Zulu medicinal plants. Proceedings of the 12th Plenary Meeting of aetfat, Symposium VIII, pp 979–90, Mitt Inst Allg Bot, Hamburg.

Curtis D. 1991. *Beyond Government: Organisations for Common Benefit*. Macmillan Education Ltd, London.

D'Souza E R and Palghadmal T J. 1990. *Sustainable Water-use System: A Case Study of Sase-Gandhalewadi Lift Irrigation Project*. Social Centre, Ahmednagar, India.

Dabbert S. 1990. Zur Optimalen Organisation Alternativer Landwirtschaftlicher Betriebe. *Agrarwirtschaft Sonderheft* 124. Verlag Alfred Strothe, Frankfurt.

Dahlberg K A. 1990. The industrial model and its impacts on small farmers: the green revolution as a cane. In: Altieri M and Hecht S B (eds) *Agroecology and Small Farm Development*. CRC press. Boca Raton, Florida.

Dalal-Clayton B and Dent D. 1994. Surveys, plans and people: a review of land resource information and its use in developing countries. *Environment Planning Group Issues* 2. IIED, London.

Daly H E and Cobb J B. 1989. *For the Common Good: Redirecting the Economy Towards Community, the Environment and a Sustainable Future*. Beacon Press, Boston.

Darma G. 1984. Residu Pesticidas dalam Sayuran-Sayuran Tanah Air. Wahana Link Kungan Hidup, Jakarta, Indonesia.

Darwin C. 1859. *Origin of Species*. London.

Datta K K and Joshi P K. 1993. Problems and prospects of cooperatives in managing degraded lands. *Econ. and Political Weekly* 28 (12–13) A16–A24.

Datta K K and de Jong C. 1991. The effort of substance drainage on farm economy. International Institute for Land Reclamation and Improvement, Wageningen, The Netherlands.

Davies R. 1992. Brave new era beckons. *Farmers Weekly*. 20 November: 62–5.

de la Cruz, Lightfoot C and Sevilleja. 1992. A user perspective on rice–fish culture in the Philippines. In: Hiemstra W, Reijntjes C and Van der Werf E (eds) *Let Farmers Judge*. IT Publications, London.

De Datta S K. 1986. Improving notrogen fertiliser efficiency in lowland rice in tropical Asia. *Fert. Res.* 9, 171–86.

De los Reyes R and Jopillo S G. 1986. *An Evaluation of the Philippines Participatory Communal Irrigation Program*. Institute of Philippine Culture, Quezon City.

Dei G J S. 1989. Hunting and gathering in a Ghanaian rain forest community. *Ecology of Food and Nutrition* 22: 225–43.

Delli Priscoli J. 1989. Public involvement, conflict management: means to EQ and social objectives. *Journal of Water Resource Planning and Management* 115(1), 31–42.

Dempster J P and Coaker T H. 1974. Diversification of crop ecosystems as a means of controlling pests. In: Price-Jones D and Solomon M E (eds) *Biology in Pest and Disease Control*. Blackwell, Oxford.

Denevan W M. 1970. Aboriginal drained-field cultivation in the Americas. *Science* 169, 647–54.

Denzin N K. 1984. *Interpretive Interactionalism*. Sage Publications, London.

Deo O P and Yadav N P. 1990. Why Nepal needs user forestry in the hills. *AERDD Bulletin* 30, p 23, University of Reading.

Devavaram J. 1994. Paraikulum Watershed, Tamil Nadu. Case study for IIED collaborative research project *New Horizons: The Economic and Environmental Benefits of Participatory Watershed Development*. IIED, London.

Dewey K G. 1981. Nutritional consequences of the transformation from subsistence to commercial agriculture in Tabasco, Mexico. *Human Ecology* 9(2): 151–87.

Dhar S K, Gupta J R and Sarin M. *Participatory Management in the Shivalik Hills: Experience of the Haryana Forest Dept*. Sustainable Forest Management Working Paper No 5. Ford Foundation, New Delhi.

Dinham B. 1993. *The Pesticide Hazard*. Zed Books, London.

Diop A. 1992. Farmer-extensionist-research partnerships: Rodale International's experience. Paper for joint IIED/IDS *Beyond Farmer First: Rural People's Knowledge, Agricultural Research and Extension Practice*. Conference, 27–29 October 1992.

Dobbs T L, Becker D L and Taylor. 1991. Sustainable agriculture policy analyses: South Dakota on-farm case studies. *Journ Farming Systems Research-Extension* 2(2), 109–24.

Donkin R A. 1979. *Agricultural Terracing in the Aboriginal New World*. University of Arizona Press, Tucson.

Doubleday O. 1992. Role of crop protection agents in farming systems: protecting the apple. In: BCPC Monographic No 49. *Food Quality and Crop Protection Agents*, p 69–76.

Doubleday O and Wise C J C. 1993. Achieving quality – a grower's viewpoint. In: Tyson D (ed) *Crop Protection for UK Horticulture?* British Crop Protection Council, London.

Douthwaite R. 1992. *The Growth Illusion*. Routledge, London.

Dovring F. 1985. Energy use in United States agriculture: a critque of recent research. *Energy in Agriculture* 4: 79–86.

Dowswell C and Russell N C. 1991. Workshop summary. In: *Africa's Agricultural Development in the 1990s: Can it be Sustained?* CASIN/SAA/Global 2000. Sasakawa Africa Association, Tokyo.

Dowswell C. 1993. Strengthening the institutional foundations for modern agriculture in Africa. In: *Developing African Agriculture: New Initiatives for Institutional Co-operation*. SAA/Global 2000/CASIN. Sasakawa Africa Association, Tokyo.

Drinkwater M. 1992. Knowledge, consciousness and prejudice: developing a methodology for achieving a farmer-researcher dialogue in adaptive agricultural research in Zambia. Paper for joint *IIED/IDS Beyond Farmer First: Rural People's Knowledge, Agricultural Research and Extension Practice* Conference, 27–9 October 1992.

Duke of Westminster (DoW). 1992. *The Problems in Rural Areas*. A report of recommendations arising from an inquiry chaired by His Grace the Duke of Westminster DL. Brecon, Powys.

Easterby-Smith. 1992. Creating a learning organisation. *Personnel Review* 19(5), 24–28.

Eaton D. 1993. *Soil Erosion and Farmer Decision-Making: Some Evidence from Malawi*. Dissertation for MSc in Environmental and Resource Economics, University College, London.

Eckbom A. 1992. *Economic Impact Assessment of Implementation Strategies for Soil Conservation. A comparative analysis of the on-farm and catchment approach in Trans Nzoia, Kenya*. Unit for Environmental Economics, Dept of Economics, Gothenburg University, Sweden.

Edwards C A and Lofty J R. 1977. *The Biology of the Earthworm*. Chapman and Hall, London.
Edwards M and Hulme D. 1992. *Making a Difference? NGOs and Development in a Changing World*. Earthscan Publications Ltd, London.
Eisner E W. 1990. The meaning of alternative paradigms for practice. In: Guba E G (ed) *The Paradigm Dialog*. Sage Publications, Newbury Park.
Ekins P. 1992. *Real-Life Economics*. Routledge, London.
El Titi A and Landes H. 1990. Integrated farming system of Lautenbach: A practical contribution toward sustainable agriculture. In: Edwards C A, Lal R, Madden P, Miller R H and House G (eds) *Sustainable Agricultural Systems*. Soil and Water Conservation Society, Ankeny.
El-Swaify S A, Arsyad S and Krishnarajah P. Soil loss and conservation planning in ten plantations of Sri Lanka. In: Carpenter R A (ed) *Natural Systems for Development: What Planners Need to Know*. Macmillan, New York.
Ellis J E and Swift D M 1988. Stability of African pastoral ecosystems: alternative paradigms and implications for development. *J. of Range Management* 41(6): 450–59.
Enshayan K. 1991. Some implied assumptions of industrial agriculture. Ohio State University, Columbus, Ohio.
ERCS/IIED. 1988. *Rainbow Over Wollo*. Ethiopian Red Cross Society, Addis Ababa and IIED, London.
Erickson C L and Candler K L. 1989. Revised fields and sustainable agriculture in the lake Titicaca Basin of Peru. In: Browder (ed) *Fragile Lands of Latin America: Strategies for Sustainable Development*. Westview Press, Boulder, Colorada.
Ernle, Lord. 1912. *English Farming. Past and Present*. London.
EUAD. 1992. *The Pakistan National Conservation Strategy*. Environment and Urban Affairs Division, Government of Pakistan, Islamabad.
Evans G E. 1960. *The Horse in the Furrow*. Faber and Faber, London.
Evans G E. 1966. *The Pattern Under the Plough*. Faber and Faber, London.
Evans R. 1990a. Water erosion in British farmers' fields: some causes, impacts, predictions. *Progress in Physical Geog*. 14(2), 199–219.
Evans R. 1990b. Soils at risk of accelerated erosion in England and Wales. *Soil Use and Management* 6(3), 125–31.
Faeth P (ed). 1993. *Agricultural Policy and Sustainability: Case Studies from India, Chile, the Philippines and the United States*. WRI, Washington DC.
Faeth P, Repetto R, Kroll K, Dai Q and Helmers G. 1990. *Paying the Farm Bill: US Agricultural Policy and the Transistion to Sustainable Agriculture*. World Resources Institute, Washington DC.
Fagi A. 1993. Fishing in the rice paddy. *IDRC Reports* July 1993, 19–20.
Family F and Vicsek T (eds). 1991. *Dynamics of Fractal Systems*. World Scientific Publ Co, Singapore.
FAO. 1976. *Energy for Agriculture in the Developing Countries*. FAO, Rome.
FAO. 1991. *Issues and Perspectives in Sustainable Agriculture and Rural Development*. Main Document. FAO Newsletter, Conference on Agriculture and the Environment. 's-Hertogenbosch, Netherlands 15–19 April 1991.
FAO. 1992. *The Keita Integrated Development Project*. FAO, Rome.
FAO. 1993a. *Strategies for Sustainable Agriculture and Rural Development (SARD): The Role of Agriculture, Forestry and Fisheries*. FAO, Rome.
FAO. 1993b. *Harvesting Nature's Diversity*. FAO, Rome.
FAO. 1993c. *The State of Food and Agriculture*. FAO, Rome.
FAO. *passim*. *Agricultural Production Indices*. FAO, Rome.
Farrington J and Bebbington A. 1993. *Reluctant Partners? Non-governmental organisations, the state and sustainable agriculture development*. Routledge, London.
Farrington J and Biggs S. 1990. NGOs, agricultural technology and the rural poor. *Food Policy* December: 479–91.
Fernandez A. 1992. *The MYRADA Experience: Alternative Management Systems for Savings and Credit of the Rural Poor*. MYRADA, Bangalore.
Fernandez A. 1993. *The Interventions of a Voluntary Agency in the Process and Growth of People's Institutions for Sustained and Equitable Management of Micro-Watersheds*. MYRADA Rural Management Systems paper 18, Bangalore.
Fernandez A. 1994. The Myrada experience: towards a sustainable impact analysis in participatory micro watershed management. Paper for *IIED New Horizons* conference, Bangalore, India,, Nov 1994. IIED, London.

Feyeraband P. 1975. *Against Method: Outline of Anarchistic Theory of Knowledge*. Verso, London.
Figueiredo P. 1986. *The Yield of Crops on Terraced and Non-Terraced Land. A Field Survey in Kenya*. Swedish University of Agricultural Sciences, Uppsala.
Finsterbusch K and van Wicklen W A. 1989. Beneficiary participation in development projects: empirical tests of popular theories. *Econ. Development and Cultural Change* 37(3), 573–93.
Fish S K and Paul R. 1992. Prehistoric landscapes of the Sonoran desert Hohokam. *Population and Environment: A Journal of Interdisciplinary Studies* 13(4).
Flinn J C and De Datta S K. 1984. Trends in irrigated rice yields under intensive cropping at Philippine research stations. *Field Crops Research* 9, 1–15.
Flinn J C, De Datta S K and Labadan E. 1981. An analysis of long-term rice yields at IRRI farm. *Agricultural Economics Development Paper* 81–04. IRRI, Los Banos, Philippines.
Flores, M. 1989. Velvetbeans: an alternative to improve small farmers' agriculture. *ILEIA Newsletter* 5, 8–9.
Fowler A. 1992. *Prioritizing Institutional Development: A New Role for NGO Centres for Study and Development*. Sustainable Agriculture Programme Gatekeeper Series SA35. IIED, London.
Fowler C and Mooney P. 1990. *The Threatened Gene: Food, Policies and the Loss of Genetic Diversity*. The Lutterworth Press, Cambridge.
Fowler F J and Mangione T W. 1990. *Standardized Survey Interviewing: Minimising Interviewer-Related Error*. Applied Social Research Methods Series Volume 18. Sage Publications, Inc, Newbury Park, California.
Fox J J. 1992. Managing the ecology of rice production in Indonesia. In: Hardjono J (ed) *Indonesia: Resources, Ecology and the Environment*. Oxford University Press, Singapore.
Francis C A. 1986. *Multiple Cropping Systems*. John Wiley and Sons.
Francis C A and Hildebrand P F. 1989. Farming systems research-extension and the concepts of sustainability. *FSRE Newsletter* 3: 6–11. University of Florida, Gainsville.
Fre, Z. 1993. Ethnoveterinary knowledge among pastoralists in eastern Sudan and Eritrea: implications for animal health, participatory extension and future policy. *IIED Sustainable Agriculture Programme Research Series* 1(2): 1–23. IIED, London.
Freire P. 1968. *Pedagogy of the Oppressed*. Penguin Books, London.
Fujisaka S, Mar M, Swe A, Wah L, Mordy K, Thein C, Lwin T and Palis R K. 1992. Rice in Myanmar: a diagnostic survey. *Myanmar J. Agric. Science* 4(1), 1–13.
Fujisaka S. 1989. *Participation by Farmers, Researchers and Extension Workers in Soil Conservation*. Sustainable Agriculture Programme Gatekeeper Series SA16. IIED, London.
Fujisaka S. 1990. Rainfed lowland rice: building research on farmer practice and technical knowledge. *Agric, Ecosystems and Environ*. 33, 57–74.
Fujisaka S. 1991a. Thirteen reasons why farmers do not adopt innovations intended to improve the sustainability of agriculture. In: *Evaluation for Sustainable Land Management in the Developing World. Volume 2: Technical Papers*. IBSRAM, Thailand. IBSRAM Proceedings 12(2): 509–22.
Fujisaka S. 1991b. A diagnostic survey of shifting cultivation in northern Laos: targeting research to improve sustainability and productivity. *Agroforestry Systems* 13, 95–109.
Furusawa K. 1988. Agricultural crisis: Japan and the World. *RONGEAD (European NGO's Network on Agriculture, Food and Development, Lyon)* 88(2–3): 7–9.
Furusawa K. 1991. Life rooted in the rice plant *Resurgence* 137, 20–3.
Furusawa K. 1992. Sustainable agriculture and the new paradigm – changes in production, distribution and consumption. Mejiro Gakuen Women's College, Tokyo.
Furusawa, K. 1994. Co-operative alternatives in Japan. In: Conford P (ed) *A Future for the Land, Organic Practice from a Global Perspective*. Resurgence Books, Bideford.
FW. 1991a. USA farm suicides twice national rate. *Farmers Weekly* 18 October 1991.
FW. 1991b. Value of yield mapping is already proven. *Farmers Weekly* 6 September, p 31.
FW. 1992. Germans embrace novel environmental scheme. *Farmers Weekly* 9 October, p 8.
FW. 1993a. Low-level attack on mildew. *Farmers Weekly* 4 July, p 44.
FW. 1993b. Wild oat treatment really hits the spot. *Farmers Weekly* 5 February, p 54.
FW. 1993c. GPS tailors fertilizer use. *Farmers Weekly* 9 April, p 64.
FW. 1993d. Pest control by predator to take off? *Farmers Weekly* 15 January, p 31.
FW. 1993e. Nature aids pest control ... as value of predators grows in bank. *Farmers Weekly* 21 May, p 46.

FW. 1993f. Organic beef beats a conventional system. *Farmers Weekly* 26 March, p 61.
Game Conservancy. 1993. *The Game Conservancy Review of 1992*. The Game Conservancy, Fordingbridge, Hampshire.
GAO. 1989. Export of unregistered pesticides is not adequately monitored by EPA. General Accounting Office, Washington.
García-Barrios R and García-Barrios L. 1990. Environmental and technological degradation in peasant agriculture; a consequence of development in Mexico. *World Development* 18(11): 1569–85.
Garrity D P, Kummer D M and Guiang E S. 1993. The Philippines. In: NRC. *Sustainable Agriculture and the Environment in the Humid Tropics*. National Academy Press, Washington DC.
Gartrell M J, Craun J C, Podrebarac D S and Gunderson E L. 1986a. Pesticides, selected elements and other chemicals in infant and toddler total diet samples, October 1980–March 1982. *J. Assoc. Off. Anal. Chem.* 69: 123–45.
Gartrell M J, Craun J C, Podrebarac D S and Gunderson E L. 1986b. Pesticides, selected elements and other chemicals in adult total diet samples, October 1980–March 1982. *J Assoc. Off. Anal. Chem.* 69: 146–59.
Gaur A C and Verma L N. 1991. Integrated plant nutrition in prevailing crop mixes in India. In: *Asian Experiences in Integrated Plant Nutrition: Report of the Expert Consultation of the Asia Network of Bio and Organic Fertilizers*. Regional Office for Asia and the Pacific (RAPA), FAO, Bangkok.
Georghiou G P. 1986. The magnitude of the problem. In: NRC. *Pesticide Resistance: Strategies and Tactics for Management*. National Academy Press, Washington DC.
Gibbon D. 1992. The future of farming systems research in developing countries. In: Raman K and Balaguru T (eds) *Farming Systems Research in India: Strategies for Implementation*. NAARM, Rajendranagar, Hyderabad.
Gibson T. 1991. Planning for Real. *RRA Notes* 11, 29–30.
Gichuki F N. 1991. Conservation Profile. In: *Environmental Change and Dryland Management in Machakos District, Kenya 1930–90*. ODI Working Paper 56. ODI, London.
Giddens A. 1987. *Social Theory and Modern Society*. Blackwell Oxford.
Gide A. 1925. *Les Faux-Monnayeurs*. Editions Gallimard, Paris.
Gill G. 1991. But how does it compare with the 'real' data? *RRA Notes* 14, 5–13
Gill G. 1993. *OK, the Data's Lousy, But It's All We've Got (being a critique of conventional methods)*. Sustainable Agriculture Programme Gatekeeper Series SA38. IIED, London.
Gladwin C H. 1979. Cognitive strategies and adoption decisions; a case study of non-adoption of an agronomic recommendation. *Econ. Dev. and Cultural Change* 28: 155–73.
Gleick J. 1987. *Chaos: Making a New Science*. Heinemann, London.
Gleissman S R. 1990. Understanding the basis of sustainability for agriculture in the tropics: experiences in Latin America. In: Edwards et al (eds) *Sustainable Agricultural Systems*. Soil and Water Conservation Society, Ankeny, Iowa.
Glewwe P and van der Gaag J. 1990. Identifying the poor in developing countries: do different definitions matter? *World Development* 18(6), 803–14.
Goethert R and Hamdi N. 1988. *Making Microplans: A Community-Based Process in Programming and Development*. IT Publications, London.
GoI. 1992. *Farmers as Experts. The Indonesian National IPM Program*. Government of Indonesia.
Goldschmidt W. 1978. *As You Sow; Three Studies in the Social Consequences of Agribusiness*. Allanheld, Osmun & Co, Montclain, New Jersey.
Gómez-Pompa A, Morales H L, Ávilla E J and Ávilla J J. 1982. Experiences in traditional hydraulic agriculture. In: Flannery K V (ed) *Maya Subsistence*. Academic Press, New York.
Gómez-Pompa A and Jiménez-Orsonio J J. 1989. Some reflections in intensive traditional agriculture. In: Gladwin C and Truman K (eds) *Food and Forum. Current Debates and Policies*. University Press of America.
Gómez-Pompa A and Kaus A. 1992. Taming the wilderness myth. *Bioscience* 42(4): 271–9.
Gómez-Pompa A, Kaus A, Jiménez-Orsiono J J, Bainbridge D and Rorive V M. 1993. In: NRC. *Sustainable Agriculture and the Environment in the Humid Tropics*. National Academy Press, Washington DC.
Gould S J. 1989. *Wonderful Life: The Burgess Shale and the Nature of History*. Penguin Books, London.

Grandin B. 1987. *Wealth Ranking*. IT Publications, London.
Gravert H O, Pabst K, Ordoloff D and Trietel U. 1992. Milk production in alternative agriculture. *Ecology and Farming (IFOAM)*, 4, 8–9.
Greig-Smith P, Frampton G and Hardy T (eds). 1993. *Pesticides, Cereal Farming and the Environment: The Boxworth Project*. HMSO, London.
Griffin E and Dennis H. 1969. A Mexican corporate campaign in conservation. *The Professional Geographer*, 21, 358–9.
Grönvall M. 1987. *A Study of Land Use and Soil Conservation on a Farm in Mukurweini Division, Central Kenya*. Dept of Physical Geography, University of Stockholm.
GTZ. 1992. *The Spark Has Jumped the Gap*. Deutsche Gessellschaft für Techniscle Zusammenarbeit (GTZ), Eschborn.
Guba E G and Lincoln Y. 1989. *Fourth Generation Evaluation*. Sage, London.
Guba E G (ed). 1990. *The Paradigm Dialog*. Sage Publications, Newbury Park.
Guba E. 1981. Criteria for assessing the trustworthiness of naturalistic inquiries. *Educational Communication of Technology Journal* 29: 75–92.
Gubbels P. 1993. *Peasant farmer organization in farmer-first agricultural development in West Africa: new opportunities and continuing constraints*. Agric Admin (R and E) Network Paper 40. ODI, London.
Gubbels P. 1994. Farmer-driven research and the Project Agro-Forestier in Burkina Faso. In: Scoones I and Thompson J (eds) *Beyond Farmer First*. Intermediate Publications Ltd, London.
Guijt I. 1991. *Perspectives on Participation. An Inventory of Institutions in Africa*. IIED, London.
Guijt I. 1992. Diagrams for village land use planning: how MARP can help to understand local resource use. *Haramata: Bulletin of the Drylands Programme* No 18, 18–21, IIED London.
Guijt I and Pretty J N (eds). 1992. *Participatory Rural Appraisal for Farmer Participatory Research in Punjab, Pakistan*. Pakistan-Swiss Potato Development Project, PARC, Islamabad and IIED, London.
Guijt I and Thompson J. 1994. Landscapes and livelihoods. Environmental and socioeconomic dimensions of small-scale irrigation. *Land Use Policy* 11(1). Forthcoming.
Gulati A. 1990. Fertilizer subsidy: is the cultivator 'net-subsidised'? *Indian Journal of Agri Econ*. 45, (1) 1–11.
Gupta A K, Patel N T and Shah R N. 1989. Review of post-graduate research in agriculture (1973–1984): are we building appropriate skills for tomorrow? Centre for Management in Agriculture, IIM, Ahmedabad, India.
Haagsma B. 1990. *Erosion and Conservation on Santao Antao. No Shortcuts to Simple Answers*. Working document 2, Santao Antao Rural Development Project, Republic of Cape Verde.
Habermas J. 1987. *The Philosophical Discourse of Modernity*. Oxford University Press, Oxford.
Helmers G A, Langemeier M R and Atwood J. 1986. An economic analysis of alternative cropping systems for east-central Nebraska. *Am. J. Altern. Agric*. 1, 153–8.
Hammond W N O, Neuenschwander P and Herren H R. 1987. Impact of the exotic parasitoid *Epidinocarsis lopezi* on cassava mealybug populations. *Insect Sci. Applic*. 8, 887–91.
Handy C. 1985. *Understanding Organisations*. Penguin Books Ltd, Harmondsworth.
Handy C. 1989. *The Age of Unreason*. Business Books Ltd, London.
Hansen G E. 1978. Bureaucratic linkages and policy-making in Indonesia. In: Jackson K D and Pye L W (eds) *Political Power and Communications in Indonesia*. University of California Press, Berkeley.
Hanson J C, Johnson D M, Peters S E and Janke R R. 1990. The profitability of sustainable agriculture on a representative grain farm in the mid-Atlantic region, 1981–1989. *Northeastern J. Agric. and Resource Econ*. 19(2), 90–8.
Harder M. 1991. *Launching a Self-Evaluation Process in Farmer Organisations*. Intercooperation Self Help Support Programme and National Development Foundation, Sri Lanka.
Harlan J R. 1989. Wild-grass seed harvesting in the Sahara and Sub-Sahara of Africa. In: Harris D R and Hillman G C. *Foraging and Farming: The Evolution of Plant Exploitation, One World Archaeology-B*, pp 79–98, Unwin Hyman, London.
Harrison P. 1987. *The Greening of Africa*. Penguin Books Ltd, London.
Harvey D. 1989. *The Condition of Postmodernity*. Basil Blackwell Ltd, Oxford.
Hassan I. 1985. The culture of postmodernism. *Theory, Culture and Society* 2(3): 119–32.

Haverkort B, van der Kamp and Waters-Bayer A (eds). 1991. *Joining Farmers' Experiments: Experiences in Participatory Development*. IT Publications, London.
Hazell P and Ramasamy C. 1991. *The Green Revolution Reconsidered: the Impact of High-Yielding Rice Varieties in South India*. The Johns Hopkins University Press, Baltimore and London.
Headley J C. 1985. Soil conservation and cooperative extension. *Agricultural History* 59, 290–306.
Heermans J.1988. The Guesselbodi experiment: bushland management in Niger. In: Conroy C and Litvinoff M (eds) *The Greening of Aid*. Earthscan Publications, London.
Heinrich G, Worman F and Koketso C. 1991. Integrating FPR with conventional on-farm research programmes: an example from Botswana. *J. Farming Systems Res-Extension* 2(2): 1–15.
Helmers G A, Langemeier M R and Atwood J. 1986. An economic analysis of alternative cropping systems for east-central Nebraskas. *Am. J. Altern. Agric.* 1: 153–8.
Herzog D C and Funderbank J E. 1986. Ecological bases for habitat management and pest control. In: Kogan M (ed) *Ecological Theory and Integrated Pest Management Practice*. John Wiley and Sons, New York.
Hewitt de Alcantara C. 1976. *Modernizing Mexican Agriculture: Socioeconomic Implications of Technological Change 1940–1970*. United Nations Research Institute for Social Development Report No 76.5. UNRSID, Geneva.
HL. 1990. *The Future of Rural Society*. House of Lords Select Committee on the European Communities. HMSO, London.
Hoar S K, Blair A, Holmes F F, Boysen C D, Robel R J, Hoover R and Fraumeni J F. 1986. Agricultural herbicide use and risk of lymphoma and soft-tissue sarcoma. *J. Am. Med. Assn.* 256: 1141–7.
Hoar S K, Weisenberger D D, Babbitt P A, Saal R C, Cantor K P and Blair A. 1988. A case-control study of non-Hodgkin's lymphoma and agricultural factors in eastern Nebraska. *Am. J. Epidimiol.* 128: 901.
Hobbelink H, Vellve R and Abraham M. 1990. *Inside the Bio Revolution*. IOCU & GRAIN, Penang and Barcelona.
Holderness B A. 1989. Prices, productivity and output. In: Mingay G E (ed) *The Agrarian History of England and Wales. Vol. VI. 1750–1850*. Cambridge University Press, Cambridge.
Holland J H, Holyoak KJ, Nisbett R E and Thagard P R. 1986. *Induction: Processes of Inference, Learning and Discovery*. MIT Press, Cambridge, Mass.
Holling C S. 1978. *Adaptive Environmental Assessment and Management*. John Wiley and Sons, Chichester.
Holmes B. 1993. Can sustainable farming win the battle of the bottom line? *Science* 260, 1893–5.
Horwith B J, Windle P N, MacDonald E F, Parker J K, Ruby A M and Elfring C. 1989. The role of technology in enhancing low resource agriculture in Africa. *Agric. Human. Values, VI (3)*
Howell J (ed). 1988. (ed). *Training and Visit Extension in Practice*. Overseas Development Institute, London.
Hruska A J. 1993. Care policy promotes IPM. *International Ag-Sieve* V(6): 4–5.
HSE. 1993. Health and Safety Executive. *Pesticide Incidents Investigated in 1992/1993*.
Huang H T and Pei Yang. 1987. Ancient cultural citrus art as a biological agent. *BioScience* 37, 665–71.
Huby M. 1990. *Where You Can't See the Wood for Trees*. Kenya Woodfuel Development Programme Series, Beijer Institute, Stockholm.
Hudson N. 1991. *A Study of the Reasons for Success or Failure of Soil Conservation Projects*. FAO Soils Bulletin 64. FAO, Rome.
Hudson N and Cheatle R J. 1993. *Working with Farmers for Better Land Husbandry*. IT Publications, London.
Humphries J. 1990. Enclosures, common rights and women: the proletarianization of families in the late eighteenth and early nineteenth centuries. *J. Econ. History* 50: 17–42.
Hunegnaw T. 1987. *Technical Evaluation of Soil Conservation Measures in Embu District, Kenya*. Report of a minor field study. IRDC, Swedish University of Agricultural Sciences, Uppsala.
Hunger Project. 1991. The Hunger Project Update: The Alan Shawn Feinstein World Hunger Project, Brown University, Rhode Island, USA.

Hunter J. 1991. *The Claim of Crofting*. Mainstream.
Hussain S S, Byerlee D and Heisey P W. 1994. Impacts of the training and visit extension system on farmers' knowledge and adoption of technology: evidence from Pakistan. *Agric. Econ.* 10: 39–47.
Huxley E. 1960. *A New Earth. An Experiment in Colonialism*. Chatto and Windus, London.
Huyssens A. 1984. Mapping the post-modern. *New German Critique* 33: 5–52.
IARC. 1991. *Occupational Exposure in Insecticides Application and Save Pesticides*. IARC Monographs on the Evaluation of Carcinogenic Risks to Humans, Vol 53. IARC, Lyon.
ICAITI. 1977. *An Environmental and Economic Study of the Consequences of Pesticide Use in Central American Cotton Production*. Instituto Centroamericano de Investigacion y Technologia Industrial, Guatemala City, Guatemala.
ICRISAT. 1993. Will the pod-borer become the farmer's pall bearer? *SAT News* 7–10.
IDS/IIED. 1994. *PRA and PM&E Annotated Bibliography*. IDS, Sussex and IIED, London.
IFAD. 1992. *Soil and Water Conservation in Sub-Saharan Africa*. IFAD, Rome.
IFOAM. *passim*. *Ecology and Farming*. IFOAM, Tholey-Theley, Germany.
IIED/IUCN. 1994. *Strategies for National Sustainable Development*. IIED, London and IUCN, Geneva.
IIED. 1988–present. *RRA Notes*. Sustainable Agriculture Programme, IIED, London.
IITA/ABCP. 1988. *Annual Report and Research Highlights, April 1988*. IITA, Ibadan, Nigeria.
Ikerd J, Monson S and Dyne D V. 1992. *Potential Impacts of Sustainable Agriculture*. University of Missouri, Agricultural Economics Dept, Columbia.
International Ag-Sieve. 1988. Biological control halts canava hornworm. *Int. Ag-Sieve* 1(2), 1–2.
International Federation of Agricultural Producers. 1992. *Toward Self-supporting Farmers' Organisations*. IFAD, Paris.
International Fertilizer Development Center. 1992. *Annual Report: Colors of Sustainability*. Muscle Shoals, Alabama.
IPCC. 1990. Intergovernmental Panel on Climate Change. *Scientific Assessment of Climate Change*, report of Working Group 1, and accompanying Policymakers Summary. World Meteorological Organization, Geneva.
IRRI. 1981. *Consequences of Small Rice Farm Mechanization Project*. Dept of Agricultural Engineering, IRRI, Philippines.
Ison R. 1990. *Teaching Threatens Sustainable Agriculture*. Gatekeeper Series SA21, IIED, London.
Iwamoto Y 1994. Paper presented to *Rural History Centre Conference*, 10 May, University of Reading
Jacks G V and Whyte R O. 1939. *The Rape of the Earth: A World Survey of Soil Erosion*. Faber and Faber, London.
Jackson M C. 1991. The origins and nature of critical systems thinking. *Systems Practice* 4(2), 131–49.
Jaggard K. 1993. Are big crop yields equivalent to big profits? *Arable Research Institute Association Newsletter*, 22–4.
Jain L C, Krishnamurthy B V and Tripathi P M. 1985. *Grass Without Roots: Rural Development Under Government Auspices*. Sage Publications, New Delhi.
Jeyeratnam J. 1990. Acute pesticide poisoning: a major global health problem. *World Health Statistics Quarterly* 43, 139–43.
Jiggins J. and de Zeeuw H. 1992. Participatory technology development in practice: process and methods. In: Reijntjes C, Haverkort B and Waters-Bayer A (eds) *Farming for the Future*. Macmillan and ILEIA, Netherlands.
Jin W. 1991 Production and evaluation of organic manure in China. In : *Asian Experiences in Integrated Plant Nutrition: Report of the Expert Consultation of the Asia Network of Bio and Organic Fertilizers*. Regional Office for Asia and the Pacific (RAPA), FAO, Bangkok.
Jintrawet, A, Smutkupt S, Wongsamun C, Katawetin R and Kerdsuk V. 1987. *Extension Activities for Peanuts after Rice in Ban Sum Jan, Northeast Thailand: A Case Study in Farmer-to-Farmer Extension Methodology*. Khon Kaen University, Khon Kaen, Thailand.
Jodha N S. 1990. *Rural Common Property Resources: A Growing Crisis*. Sustainable Agriculture Programme Gatekeeper Series SA24. IIED, London.
Johansen C. 1993. Two legumes unbind phosphate. *International Ag-Sieve* V(5) 1–3.
Johnson K. 1979. *Rain and Stormwater Harvesting in the USA and Latin America*. Special report to UNEP, Nairobi.

Jonjuabsong L and Hawi-khen A. 1991. A summary of some Thai experiences in sustainable agriculture in collaboration with government agencies. *Agric Admin (R&E) Network* 28, 24–35. ODI, London.
Jordan V W L, Hutcheon J A and Glen D M. 1993. *Studies in Technology Transfer of Integrated Farming Systems. Considerations and Principles for Development*. AFRC Institute of Arable Crops Research, Long Ashton Research Station, Bristol.
Jordan V. 1993. The integrated farming systems research. The 'LIFE' project – future directions. *Arable Research Institute Association Newsletter* 20–1.
Joshy D. 1991. Successful integrated plant nutrition in prevailing crop mixes – Nepal. In: *Asian Experiences in Integrated Plant Nutrition: Report of the Expert Consultation of the Asia Network of Bio and Organic Fertilizers*. Regional Office for Asia and the Pacific (RAPA), FAO, Bangkok.
Juma C. 1989. *Biological Diversity and Innovation: Conserving and Utilizing Genetic Resources in Kenya*. African Centre for Technology Studies, Nairobi, Kenya.
Jutsum A R. 1988. Commercial application of biological control: status and prospects. *Phil. Trans. R. Soc. Lond.* B 318, 357–71.
Kamp K, Gregory R and Chowhan G. 1993. Fish cutting pesticide use. *ILEIA Newsletter* 2/93, 22–3.
Kang B T, Wilson G F and Lawson T L. 1984. *Alley Cropping: A Stable Alternative to Shifting Agriculture*. IITA, Ibadan.
Kaphalia B S, Siddiqui F S and Setu T D. 1985. Contamination levels in different food items and dietary intake of organochlorine residues in India. *Ind. J. Med. Res.* 81: 71–8.
Karch-Türler C. 1983. Betriebsw. Untersuchungen emf dem Gebiet des Alternatives Landbaus. Agarwirt Studies No 18, Swiss Federal Institute of Technology (ETH), Zurich.
Kartasubrata J. 1993. Indonesia. In: *Sustainable Agriculture and the Environment in the Humid Tropics*. National Academy Press, Washington DC.
Kassogue A with Dolo J and Ponsioen T. 1990. *Traditional Soil and Water Conservation on the Dogon Plateau, Mali*. Issues Paper 23, Drylands Programme. IIED, London.
Kaul Shah M. 1993. *Impact of Technological Change in Agriculture: Women's Voices from a Tribal Village in South Gujarat, India*. IDS, University of Sussex, April 1993.
Kelly L C. 1985. Anthropology in the Soil Conservation Service. *Agricultural History*, 59, 136–47.
Kenmore P. 1991. *How Rice Farmers Clean up the Environment, Conserve Biodiversity, Raise More Food, Make Higher Profits. Indonesia's IPM – A Model for Asia*. FAO, Manila, Philippines.
Kenmore P E, Carino F O, Perez C A, Dyck V A and Gutierrez A P. 1984. Population regulation of the brown planthopper within rice fields in the Philippines. *Journ Plant Protection in the Tropics* 1(1), 19–37.
KEPAS. 1984. *The Sustainability of Agricultural Intensification in Indonesia*. KEPAS, Jakarta.
Kerkhof P. 1990. *Agroforestry in Africa. A Survey of Project Experience*. Panos Institute, London.
Kerr J. 1994. How subsidies distort incentives and undermine watershed development projects in India. Paper for IIED *New Horizons* conference, Bangalore, India, Nov 1994. IIED, London.
Kerr J and Sanghi N K. 1992. *Soil and Water Conservation in India's Semi Arid Tropics*. Sustainable Agriculture Programme Gatekeeper Series SA34, IIED, London.
Kesseba A M. 1992. Strategies for developing a viable and sustainable agricultural sector in sub-saharan Africa: some issues and options. *IFAD Staff Working Paper 6*. IFAD, Rome.
Khush G S. 1990. Multiple disease and insect resistance for increased yield stability in rice. In: IRRI. *Progress in Irrigated Rice Research*. IRRI, Los Baños, Philippines.
Kiara J, Segerros M, Pretty J and McCracken J. 1990. *Rapid Catchment Analysis in Murang'a District, Kenya*. Ministry of Agriculture, Kenya.
Kingsley M A and Musante P. 1994. Activities for developing linkages and cooperative exchange mong farmers' organisations, NGOs, GOs and researchers: case study of IPM in Indonesia. Paper presented at IIED *In Local Hands: Community-Based Sustainable Development* Symposium, 4–8 July, Brighton.
Kirk V and Miller M L. 1986. *Reliability and Validity in Qualitative Research*. Qualitative Research Series 1. Sage Publications, Beverly Hills.
Kiss A and Meerman F. 1991. *Integrated Pest Management in African Agriculture*. World Bank Technical Paper 142. African Technical Dept Series. World Bank, Washington DC.

KKU. 1987. *Rapid Rural Appraisal*. Proceedings of an International Conference. Rural Systems Research Project, Khon Kaen University, Thailand.

Knipling E F. 1960. The eradication of the screw-worm fly. *Scientific American* 203, 54–61.

Kolhe S S and Mitra B N. 1987. Effects of azolla as an organic source of nitrogen in rice-wheat cropping system. *J. Agron. Crop. Sci.* 159, 212–15.

Kopke U. 1984. *The Development of Cropping Systems Under Varied Ecological Conditions*. GTZ, Eschborn.

Korten D. 1980. Community organisation and rural development – a learning process approach. *Public Administration Review* 40(5): 480–511.

Korten D. 1990. *Getting to the 21st Century: Voluntary Action and the Global Agenda*. Kumarian Press, West Hartford, Connecticut.

Kothari A, Pande P, Singh S and Dilnavaz. 1989. *Management of National Parks and Sanctuaries in India. Status Report*. Indian Institute of Public Administration.

Kotschi J, Waters-Bayer A, Adelhelm R and Hoesle U. 1989. *Ecofarming in Agricultural Development*. GTZ, Eschborn.

Kottak C P. 1991. When people don't come first: some sociological lessons from completed projects. In: Cernea M (ed) *Putting People First*. Oxford University Press, Oxford. 2nd edn.

Krishna A. 1994. Large-scale government programmes: watershed development in Rajasthan, India. Paper for *IIED New Horizons* conference, Bangalore, India, Nov 1994. IIED, London.

Krishnarajah P. 1985. Soil erosion control measures for tea land in Sri Lanka. *Sri Lanka Journal of Tea Science* 54 (2), 91–100.

Kroese R and Butler Flora C. 1992. Stewards of the land. *ILEIA Newsletter* 2/92, 5–6.

Kuhn T. 1962. *The Structure of Scientific Revolutions*. Chicago University Press, Chicago.

Kurokawa K. 1991. *Intercultural Architecture. The Philosophy of Symbiosis*. Academy Editions, London.

Kydd J. 1989. Maize research in Malawi: lessons from failure. *Journal of International Development* 1: 112–44.

Laird J. 1992. Ecofarming in China brings green profits. *Our Planet* 4(1): 12–13.

Lal R. 1989. Agroforestry systems and soil surface management of a Tropical Alfisol. In Soil moisture and crop yields. *Agroforestry Systems* 8: 7–29.

Lampkin N (ed). 1992. *Collected Papers on Organic Farming*. 2nd reviewed and updated edn. Centre for Organic Husbandry and Agroecology, University College of Wales, Aberystwyth.

Lane C. 1990 Barabaig natural resource management: sustainable land use under threat of destruction. *Discussion Paper* 12. UNRISD, Geneva.

Lane C. 1993. The state strikes back: extinguishing customary rights to land in Tanzania. In: *Never Drink from the Same Cup*. Proceedings of the Conference on Indigenous Peoples in Africa, Denmark. CDR/IWIGIA Doc 72.

Lane C 1994. The Barabaig/NAFCO conflict in Tanzania: on whose terms can it be resolved? *Forest, Trees and People Newsletter* 20.

Lane C and Pretty J N. 1990. *Displaced Pastoralists and Transferred Wheat Technology in Tanzania*. Sustainable Agriculture Programme Gatekeeper Series SA20. IIED, London.

Lathwell D. 1990. Legume green manures: a potential substitute for fertilizer in maize. *International Ag-Sieve* 111(3), 7.

Lauraya F M, Sala A L R and Wijayaratna C M. 1991. Self-assessment of performance by irrigation associations. Paper presented at *International Workshop on Farmer-managed Irrigation Systems*. Mendoza, Argentina.

Leach G. 1976. *Energy and Food Production*. IPC Science and Technology Press, Guildford and IIED, London.

Leach G. 1985. Energy and agriculture. Paper for USAID meeting on *Agriculture and Rural Development and Energy, IRRI, Philippines*, 24–26 April 1985.

Leach G and Mearns R. 1988. *Beyond the Woodfuel Crisis*. Earthscan Publications Ltd, London.

Lewis T. 1969. The diversity of insect fauna in a hedgerow and neighbouring fields. *J. Appl. Ecol.* 6, 453–8.

Liebhardt W, Andrews R W, Culik M N, Harwood R R, Janke R R, Radke J K and Rieger-Schwartz S L. 1989. Crop production during conversion from conventional to low-input methods. *Agronomy Journal* 81(2), 150–9.

Lightfoot C and Noble R. 1992. Sustainability and on-farm experiments: ways to exploit participatory and systems concepts. Paper for *12th Annual Farming Systems Symposium, Michigan State University, 13–18 September*.
Lincoln Y and Guba E. 1985. *Naturalistic Inquiry*. Sage Publications, Newbury Park.
Lincoln Y S. 1990. The making of a constructivist. A remembrance of transformations past. In: Guba, E G (ed) *The Paradigm Dialog*. Sage Publications, Newbury Park.
Lipton M with Longhurst R. 1989. *New Seeds and Poor People*. Unwin Hyman, London.
Litsinger J A, Canapi B L, Bandong J P, Dela Cruz C M Apostol R F, Pantua P C, Lumaban M D, Alviola A L, Raymundo F, Libertario EM, Loevinsohn M E and Joshi RC. 1987. Rice crop loss from insect pests in wetland and dryland environmentds of Asia with emphasis on the Philippines. *Insect Sci Applic* 8: 677–92.
Liu C C and Weng B Q. 1991. The function and potential of biofertilizer and organic manure in agricultural production – the new models and research in Fujia, China. In: *Asian Experiences in Integrated Plant Nutrition: Report of the Expert Consultation of the Asia Network of Bio and Organic Fertilizers*. Regional Office for Asia and the Pacific (RAPA), FAO, Bangkok.
Lobao L. 1990. *Locality and Inequality: Farm and Industry Structure and Socio-Economic Conditions*. State University of New York Press, New York.
Lobley M. 1993. Small farms and agricultural policy: a conservationist perspective. Paper presented to the *Centre for Agricultural Strategy/Small Farmers Association Symposium*, 30–31 March, University of Reading.
Lobo C and Kochendörfer-Lucius G. 1992. *The Rain Decided to Help Us. An Experience in Participatory Watershed Development in Maharashtra State, India*. Social Centre, Ahmednagar.
Lockeretz W, Shearer G and Kohl D H. 1984. Organic farming in the corn belt. *Science* 211, 540–7.
Loevinsohn M E. 1987. Insecticide use and increased mortality in rural central Luzon, Philippines. *The Lancet* i: 1359–62.
Long N and Long A (eds). 1992. *Battlefields of Knowledge: the Interlocking of Theory and Practice in Social Research and Development*. Routledge, London.
Lorenz E N. 1993. *The Essence of Chaos*. UCL Press Ltd, London.
Lovelock J. 1979. *Gaia: A New Look at Life on Earth*. Oxford University Press, Oxford.
Luo S M and Lin R J. 1991. High bed-low ditch system in the Pearl River Delta, South China. *Agric. Ecosystems and Environ*. 36, 101–9.
Luo S M and Han C R. 1990. Ecological agriculutre in China. In: NRC. *Sustainable Agriculture and the Environment in the Humid Tropics*. National Academy Press, Washington DC.
Lynton R P and Pareek U. 1990. *Training for Development*. Kumarian Press, West Hartford, Conn. 2nd edn.
MacDonald S. 1977. The diffusion of knowledge among Northumberland farmers, 1780–1815. *Agric. Hist. Rev.* 29: 30–9.
MacKinnon J, MacKinnon K, Child G and Thorsell J 1986. *Managing Protected Areas in the Tropics*. IUCN, Gland.
MacRae R J, Hill S B, Henning J and Mehuys G R. 1989. Agricultural science and sustainable agriculture: a review of the existing scientific barriers to sustainable food production and potential solutions. *Biol. Agric. J. Hortic.* 6, 173–217.
Macrory R. 1990. The legal control of pollution. In: Harrison R (ed) *Pollution Cause, Effects and Controls*. 2nd edn. Royal Society of Chemistry, Nottingham.
Madden J P and Dobbs T. 1990. The role of economics in achieving low-input farming systems. In: Edwards et al (eds) *Sustainable Agricultural Systems*. Soil and Water Conservation Society, Ankeny, Iowa.
MAFF. *passim*. *Agricultural and Horticultural Census Statistics*. Ministry of Agriculture, Fisheries and Food, London.
Magrath W B and Arens P. 1987. *The Costs of Soil Erosion on Java: A Natural Resource Accounting Approach*. World Resources Institute, Washington DC.
Magrath W B and Doolette J B. 1990. Strategic issues in watershed development. In: Doolette and Magrath (eds) *Watershed Development in Asia*. World Bank Technical Paper No 127. Washington DC: The World Bank, 1990.
Makwete J. 1993. Contribution of the Sasakawa Global 2000 Project in Tanzania. In: *Developing African Agriculture: New Initiatives for Institutional Cooperation*. SAA/Global 2000/CASIN. Sasakawa Africa Association, Tokyo.

Manning R E. 1989. The nature of America: visions and revisions of wilderness. *Nat. Res. J.* 29, 25–40.

Manoharan M, Velayudham K and Shanmugavalli N. 1993. Farmers' preferences for red rice. Tamil Nadu Agricultural University, Killikulum, Tamil Nadu, India.

Marchal J-Y. 1978. L'espace des techniciens et celui des paysans histoire d'un périmètre antiérosif en Haut-Volta. In: ORSTOM. *Maîtrice de L'Espace Agrarian et Développement en Afrique Tropicale.* ORSTOM, Paris.

Marchal J-Y. 1986. Vingt ans de lutte antiérosive au nord du Burkina Faso. *Cahiers ORSTOM, Série Pédalologique* XXII (2), 173–80.

Marquez C B, Pingali P Z and Palis F G. 1992. *Farmer Health Impacts of Long-term Pesticide Exposure – a medical and economic analysis in the Philippines.* IRRI, Los Baños, Philippines.

Marsden T, Munton R and Ward N. 1992. Incorporating social trajectories into uneven agrarian development: farm businesses in upland and lowland Britain. *Sociologica Ruralis* 32, 408–30.

Marshall C. 1990. Goodness criteria. Are they objective or judgement calls? In: Guba E G (ed) *The Paradigm Dialog.* Sage Publications, Newbury Park.

Martineau J Revd. 1993. The Church of England and the crisis on the small family farm. Paper presented to the *Centre for Agricultural Strategy/Small Farmers Association Symposium,* 30–31 March, University of Reading.

Mascarenhas J, Shah P, Joseph S, Jayakaran R, Devararam J, Ramachandran V, Fernandez A, Chambers R and Pretty J N. 1991. Participatory Rural Appraisal. *RRA Notes* 13. IIED, London.

Mathema S B and Galt D. 1989. Appraisal by group trek. In: Chambers et al (eds). *Farmer First.* IT Publications, London.

Matose F and Mukamuri B. 1992. Trees, people and communities in Zimbabwe's Communal Lands: Local knowledge and extension practice. Case Study presented at the *IIED/IDS Beyond Farmer First: Rural People's Knowledge, Agricultural Research and Extension Practice Workshop, 27–29 October , Institute of Development Studies, University of Sussex, UK.* IIED, London.

Matteson P C. 1992. 'Farmer First' for establishing IPM. *Bulletin of Entomological Research* 82, 293–6.

Matteson P C, Gallagher K D and Kenmore P E. 1992. Extension of integrated pest management for planthoppers in Asian irrigated rice: empowering the user. In: Denno RF and Perfect T J (eds) *Ecology and Management of Planthoppers.* Chapman and Hall, London.

Matthias-Mundy E. 1989. Of herbs and healers. *ILEIA Newsletter* 3, 20–2.

Maturana H and Varela F. 1987. *The Tree of Knowledge. The Biological Roots of Human Understanding.* Shambala Publications, Boston.

Maurya D. 1989. The innovative approach of Indian farmers. In: Chambers et al (eds) *Farmer First.* Intermediate Technology Publications Ltd, London.

McCorkle C M, Brandsletter R H and McClure D. 1988. *A Case Study on Farmer Innovation and Communication in Niger.* Communication for Technology Transfer in Africa, Academy of Educational Development, Washington.

McCorkle C M. 1989. Veterinary anthropology. *Human Organisation* 48(2), 156–62.

McCown R, Haaland G and de Haan C. 1979. The interaction between cultivation and livestock production. *Ecological Studies* 34, 297–332.

McCracken J. 1987. Conservation priorities and local communities. Introduction. In: Anderson D and Grove R (eds) *Conservation in Africa. People, Policies and Practice.* Cambridge University Press, Cambridge.

McCracken J A, Pretty J N and Conway G R. 1987. *An Introduction to Rapid Rural Appraisal for Agriculutural Development.* IIED, London.

Mearns R. 1991. *Environmental Implications of Structural Adjustment: Reflections on Scientific Method.* IDS Discussion Paper 284. IDS, Brighton.

MELU. 1977. *Auswertung drei-jähringer Erhebungen in neun biologisch-dynamisch bewirtschafteten Betrieben.* Baden Würtemberg. Ministrum für Ernährung, Landwirtschaft und Umwelt, Stuttgart.

Merrill-Sands D and Collion M-H. 1992. Making the farmers' voice count: issues and opportunities for promoting farmer-responsive research. Paper for *12th Annual Farming Systems Symposium,* Michigan State University, 13–18 September.

Merrill-Sands D, Biggs S, Bingen R J, Ewell P, McAllister J and Poats S. 1991. Institutional considerations in strengthening on-farm client-oriented research in National Agricultural

Research Systems: Lessons from a nine-country study. *Experimental Agriculture* 27, 343–73.
Michael Y G. 1992. The effects of conservation on production in the Audit-Tid area, Ethiopia. In: Tato K and Hurni H (eds) *Soil Conservation for Survival*. Soil and Water Conservation Society, Ankeny, Iowa.
Millar D. 1993. The relevance of rural people's knowledge for re-orienting extension, research and training: a case study from Northern Ghana. In: Rural People's Knowledge, Agricultural Research and Extension Practice. Africa Papers. *Sustainable Agriculture Research Series* 1(2): 41–61. IIED, London.
Millington S. 1992. In *Farmers Weekly*. Red route to green manure, 12 June, p 54.
Mingay G E. 1989. *The Agrarian History of England and Wales. Vol VI. 1750–1850*. Cambridge University Press, Cambridge.
Mishra P R and Sarin M. 1988. Social security through social fencing: Sukhomajri and Nada, North India. In: Conroy C and Litvinoff M (eds) *The Greening of Aid*. Earthscan Publications Ltd, London.
Mitchell P. 1987. Letter to the *Herald Tribune*, 6 January, from Chief of Information, World Food Programme.
MMB/SAC. 1992. *Extensification of Milk Production Using White Clover*. Milk Marketing Board Research and Development Committee and Scottish Agricultural College, Dumfries.
MMB. *passim*. Milk Marketing Board, Britain.
Mndeme K C H. 1992. Combating soil erosion in Tanzania: The HADO experience. In Tato K and Hurni H (eds). *Soil Conservation for Survival*. SCS, Ankeny, Iowa.
MOA/MALDM. *passim*. Reports of Catchment Approach Planning and Rapid Catchment Analyses. Soil and Water Conservation Branch, Ministry of Agriculture, Livestock Development and Marketing, Nairobi, 1988–94.
MOA. 1981. *Soil Conservation in Kenya*. Especially in small-scale farming in high potential areas using labour intensive methods. SWCB, Ministry of Agriculture, Nairobi. 7th edn.
MOA. 1992. *Soil and Water Conservation: National Strategy Formulation Workshop*. Ministry of Agriculture, Nairobi, Kenya.
Montemayor L. 1992. A farmer-based approach is crucial. *ILEIA Newsletter* 2/92, 7.
Montgomery J D. 1983. When local participation helps. *Journ. Policy Analysis and Management* 3 (1), 90–105.
Moris J. 1990. *Extension Alternatives in Africa*. ODI, London.
Morrison J. 1993. Protected areas and aboriginal interests in Canada. WWF Canada Discussion paper.
Morse D. 1988. Policies for sustainable agriculture; getting the balance right. Paper for International Consultation on Environment, Sustainable Development and the Role of Small Farmers. IFAD, Rome.
Moss J. 1993. Pluriactivity and survival? A study of family farms in N Ireland. Paper presented to the *Centre for Agricultural Strategy/Small Farmers Association Symposium*, 30–31 March. University of Reading.
Mosse D. 1992. Community management and rehabilitation of tank irrigation systems in Tamil Nadu: a research agenda. Paper for *GAPP Conference in Participatory Development*, 9–10 July. London.
Mucai G M, Ndungu M, Wanjiku J, Thompson J, Mwaniki J M, Odeny A O and Msumai J M. 1992. *Rapid Catchment Analysis of Ringuti Catchment, Kiambu, November 1991*. SWCB, Ministry of Agriculture, Nairobi.
Mukherjee D, Ghosh B N, Chakraborty J and Roy B R. 1980. Pesticide residues in human tissues. *Indian J. Med. Res.* 72: 583–7.
Mukherjee N. 1992. Villagers' perceptions of rural poverty through the mapping methods of PRA. *RRA Notes* 15, 21–6. IIED, London.
Mullen J. 1989. Training and visit system in Somalia: contradictions and anomalies. *Journal of International Development* 1, 145–67.
Munton R and Marsden T. 1991. Occupancy change and the farmed landscape: an analysis of farm-level trends. *Environment and Planning* A 23, 499–510.
Murphree M. 1993. Communities as resource management institutions. IIED Sustainable Agriculture Programme *Gatekeeper Series* 36. IIED, London.
Muya F S, Njoroge M, Mwarasomba L I, Sillah P K, Kwedilima I and Pretty J N. 1992. *Rapid Catchment Analysis of Thigio Catchment, Kiambu, November 1991*. SWCB, Ministry of Agriculture, Nairobi.

Mwenda E. 1991. *Soil and Water Conservation. Field Guide Notes for Catchment Planning.* SWCB, Ministry of Agriculture, Nairobi.

Nabhan G P, Rea A M, Reichhardt K L, Mellink E and Hutchinson C F. 1982. Papago influences on habitat and biotic diversity: Quitovac oasis ethnoecology. *J. Ethnobiol.* 2, 124–43.

Nair P K R (ed). 1989. *Agroforestry Systems in the Tropics.* Kluwer Academic Publishers, Dordrecht, Netherlands.

Narayan D. 1993. *Focus on Participation: Evidence from 121 Rural Water Supply Projects.* UNDP-World Bank Water Supply and Sanitation Program, World Bank, Washington DC

NEDCO. 1984. Sediment transport in the Mahaweli Ganga. Report funded by Kingdom of Netherlands to Ministry of Land and Land Development. Hydrology Division, Irrigation Dept, Colombo.

Netting R McC, Stone M P and Stone G. 1989. Kofyar cash cropping: choice and change in indigenous agricultural development. *Human Ecology* 17, 299–399.

Neuenschwander P, Hammond W N O, Guiterrez A P, Cudjoe A R, Baumgartner J U, Regev U and Adjaklow R. 1989. Impact assessment of the biological control of the cassava mealybug by the introduced parasitoid *Epidinocarsis lopezi. Bull. Entomol. Res.* 79, 579–94.

Neuenschwander P and Herren H R. 1988. Biological control of the cassava mealybug, *Phenacoccus manihoti,* by the exotic parasitoid *Epidinocarsis lopezi* in Africa. *Phil. Trans. R. Soc. Lond. B..* 318: 319–33.

Newby H. 1980. *Green and Pleasant Land? Social Change in Rural England.* Hutchinson, England

Nonaka I. 1988. Towards middle-up-down management: accelerating information creation. *Sloan Management Review* Spring 1988, 9–18.

Norbu C. 1991. Integrated nutrition management studies – Bhutan. In: *Asian Experiences in Integrated Plant Nutrition: Report of the Expert Consultation of the Asia Network of Bio and Organic Fertilizers.* Regional Office for Asia and the Pacific (RAPA), FAO, Bangkok

Norgaard R. 1988. The biological control of cassava mealybug in Africa. *Am. J. Agric. Econ.* 70, 366–371.

Norgaard R. 1989. The case for methodological pluralism. *Ecol. Econ. 1: 37–57.*

Norman D, Baker D, Heinrich G, Jonas C, Maskiara S and Worman F. 1989. Farmer groups for technology development: experience in Botswana, pp. 136–46. In: Chambers R, Pacey A and Thrupp L-A. (eds). *Farmer First. Farmer Innovation and Agricultural Research.* Intermediate Technology Publications, London.

Norton A. 1992. Analysis and action in local institutional development. Paper for *GAPP Conference on Participatory Development,* 9–10 July, 1992 London.

NRC. 1989. *Alternative Agriculture.* National Research Council. National Academy Press, Washington DC.

NRI. 1994. Integrated pest management. *Natural Resources Institute Annual Review* 8–11. NRI, Chatham.

NRSP. 1994. First Quarterly Report. National Rural Support Programme, Islamabad, Pakistan.

OECD/International Energy Agency. 1992. *Energy Balances of OECD Countries.* OECD, Paris.

OECD. 1989. *Agricultural and Environmental Policies.* OECD, Paris.

OECD. 1992. *Agents for Change.* Summary report from the OECD workshop on Sustainable Agriculture Technologies and Practices. OECD, Paris.

OECD. 1993a. *Agricultural Policies, Markets and Trade. Monitoring and Outlook 1993.* OECD, Paris.

OECD. 1993b. *World Energy Outlook.* OECD, Paris.

OECD. 1993c. *Agricultural and Environmental Policy Integration. Recent Progress and New Directions.* OECD, Paris.

Ogilvy S. 1993. Talisman: Assessing the consequences of reduced inputs. In: *HGCA Proceedings of the Cereals R & D Conference.* Robinson College, Cambridge. 5–6 January 1993.

Ogle B M and Grivetti L E. 1985. Legacy of the chameleon: edible wild plants in the Kingdom of Swaziland, Southern Africa. A cultural, ecological, nutritional study. Part II–Demographics, species availability and dietary use, analysis by ecological zone. *Ecology of Food and Nutrition* 17, 1–30.

Ohnox K. 1988. The decline of the food self-sufficiency rate and the deterioration of agriculture in Japan. *Rongead (European NGO's Network on Agriculture, Food and Development, Lyon)* 88 (2–3), 9–11.
Okafor J C. 1989. *Agroforestry Aspects*. World Wide Fund for Nature, Surrey, UK.
Omohundro J T. 1985. Efficiency, sufficiency and recent change in Newfoundland subsistence horticulture. *Human Ecology* 13(3), 291–308.
Ornes F. 1988. Community training centres for organic agriculture and appropriate technology, Dominion Republic. In: Conroy C and Litvinoff M (eds) *The Greening of Aid*. Earthscan Publications Ltd, London.
Orstrom E. 1990. *Governing the Commons: The Evolution of Institutions for Collective Action*. Cambridge Univesrity Press, Cambridge.
Östberg W and Christiansson C. 1993. *Of Lands and People*. Working Paper No 25 from the Environment and Development Studies Unit, Stockholm University, Stockholm
OTA. 1988. *Enhancing Agriculture in Africa: A Role for US Development Assistance*. US Office of Technology Assessment. US Government Printing Office, Washington DC.
Ouedraogo S. 1992. *Evaluation du Projet Agro-Forestier: Volet Information Generale*. Oxfam Impact Study, Burkina Faso.
OXFAM. 1987. Soil and water conservation activities in Hararghe region, Ethiopia. Oxfam, Oxford. mimeo.
P T. 1990. *King Cotton and the Pest*. The Pesticides Trust, London
P T. 1993. Pesticide threat to *Bacillus thuringiensis*. *Pesticides News* 21 (September 1993), 14. The Pesticides Trust, London.
Paavo A. 1989. *Land to the Stealer. An Open Letter to the Canadian People*. By Baha N, Gidabuyokt G, Gihuja B, Gembut A and Hesod G. On behalf of the Barabaig people, Hanagn District, Arusha Region, Tanzania. Briarpatch, September. 1989: 23–25.
Padilla H. 1992. High and stable crop yields: the Bontoc rice terraces. In: Hiemstra W, Reijntjes C and Van der Werf E (eds) *Let Farmers Judge*. IT Publications, London.
Palmer I. 1976. *The New Rice in Asia: Conclusions from Four Country Studies*. UNRISD, Geneva.
Palmer I. 1977. *The New Rice in Indonesia*. UNRISD, Geneva.
Palmer I. 1981. Seasonal dimensions of women's roles. In: Chambers R, Longhurst R and Pacey A (eds) *Seasonal Dimensions to Rural Poverty*. IDS, Sussex.
Palmer J. 1992. The sloping agricultural land technology. In: Hiemstra et al (eds) *op cit*
Pandit S. 1991. Participatory management of forests in West Bengal. *The Indian Forester* 117 (5).
Paris T and Del Rosario B. 1993. Overview of the Women in Rice Farming Systems Program. Paper presented at the IRRI Rice Research Seminar Series, 7 January.
Parr J F, Padendick R I, Youngberg I G and Meyer R E. 1990. Sustainable agriculture in the United States. In: Edwards C A, Lal R, Madden P, Miller R H and House G (eds) *Sustainable Agricultural Systems*. Soil and Water Conservation Society, Ankeny, Iowa.
Pawson H. 1957. *Robert Blakewell. Pioneer Livestock Breeder*. Crosby Lockwood and Son, London.
Pearson C J and Ison R L. 1990. University education for multiple goal agriculture in Australia. *Agric Systems*.
Perelman M. 1976. Efficiency in agriculture: the economics of energy. In: Merril R (ed) *Radical Agriculture*. Harper and Row, New York.
Peters T. 1987. *Thriving on Chaos: Handbook for a Management Revolution*. Alfred A Knopf, USA (1989 edn, Pan Books, London).
Peters T and Waterman R. 1982. *In Search of Excellence*. Harper and Row, New York.
Phillips D C. 1990. Postpositivistic science. Myths and realities. In: Guba E G (ed) *The Paradigm Dialog*. Sage Publications, Newbury Park.
Pimbert M. 1991. *Participatory Research with Women Farmers*. 30 mins. VHS-PAL Video. ICRISAT Information Series. International Centre for Research in the Semi-Arid Tropics, Hyderabad, India.
Pimbert M. 1993. The making of agricultural biodiversity in Europe. In: Rajan V (ed) *Rebuilding Communities. Experiences and Experiments in Europe*. Resurgence Books.
Pimbert M. and Pretty J N. 1995. *Parks, People and Professionals: Putting 'Participation' into Protected Area Management* UNRISD, Geneva, IIED London and WWF, Geneva.
Pimentel D (ed). 1980. *CRC Handbook of Energy Utilization in Agriculture*. CRC Press, Bocu Raton, Florida.

Pimentel D, Culliney T W, Buttler I W, Reinemann D J and Beckman K S. 1989. Low-input sustainable agriculture using ecological management practices. *Agric. Ecosyst. and Environment* 27: 3–24.
Pingali P. 1991. *Agricultural Growth and the Environment: Conditions for their Compatability in Asia's Humid Tropics.* IRRI Social Science Division Papers 91–12, September 1991.
Platteau J P. 1992. *Local Reform and Structural Adjustment in Sub-Saharan Africa.* FAO, Rome.
Plunkett D L. 1993. Modern crop production technology in Africa: the conditions for sustainability. In: CASIN/SAA/Global 2000. *Africa's Agricultural Development in the 1990s: Can it be Sustained?* CASIN, Geneva, Sasakawa, Africa Foundation, Tokyo and Global 2000, Georgia.
Poffenberger M and Zurbuchen M S. 1980. The economics of village Bali: three perspectives.The Ford Foundation, New Delhi. *Mimeo.*
Poffenberger M. 1990. *Joint Management of Forest Lands. Experiences from South Asia.* The Ford Foundation, New Delhi.
Ponnamperuma F N. 1979. *Soil Problems in the IRRI Farm.* IRRI, Los Baños, Philippines.
Popkewitz T S. 1990. Whose future? Whose past? Notes on critical theory and methodology. In: Guba E G (ed) *The Paradigm Dialog.* Sage Publications, Newbury Park.
Porter D, Allen B and Thompson G. 1991. *Development in Practice: Paved with Good Intentions.* Routledge, London.
Potts G R. 1977. Some effects of increasing the monoculture of cereals. In: Cherrett J M and Sagar G R (eds) *Origins of Pest, Parasite, Disease and Weed Problems.* Blackwell Scientific Publications, Oxford
Pradan N C and Yoder R. 1989. *Improving Irrigation Management Through Farmer to Farmer Training: Examples from Nepal.* International Irrigation Management Institute Working Paper No 12, Kathmandu.
Prairie Horizons Ltd. 1986. Final Report of the Benefit/Cost Team on the Tanzanian Wheat Project. Submitted to the Natural Resources Branch, Canadian International Development Agency, Ottawa.
Pretty J N. 1991. Farmers' extension practice and technology adaptation: agricultural revolution in 17th–19th century Britain. *Agric. and Human Values* VIII (1 and 2), 132–48.
Pretty J N. 1990a. *Rapid Catchment Analysis for Extension Agents.* IIED, London and Ministry of Agriculture, Kenya.
Pretty J N. 1990b. Sustainable agriculture in the Middle Ages: the English Manor. *Agricultural History Review* 38(1), 1–19.
Pretty J N. 1994. Alternative systems of inquiry for sustainable agriculture. *IDS Bulletin* 25(2): 37–40. IDS, University of Sussex.
Pretty J N and Chambers R. 1993a. *Towards a learning paradigm: new professionalism and institutions for sustainable agriculture.* IDS Discussion Paper DP 334. IDS, Brighton
Pretty J N. and Chambers R. 1993b. Towards a learning paradigm: new professionalism and institutions for sustainable agriculture. *IIED Research Series* 1(1): 48–83.
Pretty J N and Howes R. 1993. *Sustainable Agriculture in Britain: Recent Achievements and New Policy Challenges.* IIED, London.
Pretty J N and Sandbrook R. 1991. Operationalising sustainable development at the community level: primary environmental care. Presented to the DAC Working Party on Development Assistance and the Environment. OECD, Paris, October 1991.
Pretty J N and Scoones I. 1994. Institutionalising adaptive planning and local level concerns: looking to the future. In: Nelson N and Wright S (eds). *Power and Participatory Development: Theory and Practice.* IT Publications, London.
Pretty J N and Shah P. 1994. *Soil and Water Conservation in the 20th Century: A History of Coercion and Control.* Rural History Centre Research Series No1. University of Reading, Reading.
Pretty J N, Guijt I, Scoones I. and Thompson J. 1995. *A Trainers' Guide to Participatory Learning and Interaction.* IIED Training Materials Series No 1. IIED, London.
Pretty J N, Guijt I, Scoones I. and Thompson J. 1992. Regenerating agriculture in the agroecology of low-external input and community-based development. In: Holmberg J (ed). *Policies for a Small Planet.* Earthscan Publications Ltd, London.
Pretty J N, Thompson J and Kiara J K. 1994. Agricultural regeneration in Kenya: the catchment approach to soil and water conservation. *Ambio* (in press).
Prigogine I and Stengers I. 1984. *Order out of Chaos: Man's New Dialogue with Nature.* Fontana, London.

Prosterman R L, Temple M N and Hanstad T M. 1990. *Agrarian Reform and Grassroots Development*. Lynne Rienner Publ, Boulder and London.
Quinn A. 1994. 10,000 Pesticide deaths in China. *Pesticides News* 23, 10.
Quinones M A, Foster M A and Sicilima. 1991. The Kilimo/Sasakawa-Global 2000 agricultural project in Tanzania. In: *Africa's Agricultural Development in the 1990s: Can it be Sustained?* CASIN/ SAA/Global 2000. Sasakawa Africa Association, Tokyo.
Rahman M (ed). 1984. *Grass-Roots Participation and Self-Reliance*. Oxford and IBH Publication Co, New Delhi.
Rahnema M. 1992. Participation. In: Sachs W (ed) *The Development Dictionary*. Zed Books Ltd, London.
Rajagopalan V. 1993. Beyond the Green Revolution. Lal Bahadur Shastri Memorial Lecture, New Delhi, 11 February, 1993.
Ramaprasad V and Ramachandran V. 1989. *Celebrating Awareness*. MYRADA, Bangalore and Foster Parents Plan International, New Delhi.
Rands B . 1989. Experiences in soil and water conservation amongst pastoral peoples of northeastern Mali. World Vision International, Bamako.
Rands B. 1992. Experiences in soil conservation work amongst pastoral people in northeastern Mali. In: Tato K and Hurni H (eds) *Soil Conservation for Survival*. Soil and Water Conservation Society, Ankeny, Iowa.
Rao M R and Willey R W. 1980. Evaluation of yield stability in intercropping: studies on sorghum/pigeon-pea. *Experimental Agric.* 16: 105–16.
Ravnborg H. 1992. The CGIAR system in transition. Implications for the poor, sustainability and the national research systems. *Agricultural Administration (Research and Extension) Network Paper 31*. Overseas Development Institute, London.
Reason P and Heron J. 1986. Research with people: The paradigm of cooperative experiential inquiry. *People-Centred Review* 1: 457.
Reddy STS. 1989. Declining groundwater levels in India. *Water Resources Development* 5(3), 183–90.
Redman M. 1992. The economics of organic milk production in the UK. *Ecology and Farming* (IFOAM) 4, 6–8.
Reganold J P, Papendick R I and Parr J F. 1990. Sustainable agriculture. *Scientific American* June 1990: 72–8.
Reichelderfer K. 1990. Environmental protection and agricultural support: are trade-offs necessary? In: Allen K (ed) *Agricultural Policies in a New Decade*. Resources for the Future, Washington DC.
Reij C. 1988. The agroforestry project in Burkina Faso: an analysis of popular participation in soil and water conservation. In: *The Greening of Aid* Conroy C and Litvinoff M (eds). Earthscan Publications Ltd, London, pp 74–7.
Reij C. 1991. *Indigenous Soil and Water Conservation in Africa*. Sustainable Agriculture Programme Gatekeeper Series SA27. IIED, London.
Reij C, Muller P and Begemann L. 1988. *Water Harvesting for Plant Production*. World Bank Technical Paper 91. World Bank, Washington DC.
Reijntjes C, Haverkort B and Waters-Bayer A. 1992. *Farming for the Future: an Introduction to Low-External-Input and Sustainable Agriculture*. The Information Centre for Low-External-Input and Sustainable Agriculture (ILEIA). Macmillan Press Ltd, London.
Repetto R. 1985. *Paying the Price : Pesticide Subsidies in Developing Countries*. World Resources Institute, Washington DC.
Rhoades R and Booth R. 1982. *Farmer Back to Farmer: A Model for Generating Acceptable Agricultural Technology*. Agricultural Administration Network Paper 11. ODI, London.
Rhoades R. 1987. *Farmers and Experimentation*. Agric Admin (R and E) Network Paper 21. ODI, London.
Rhoades R. 1989. *Evolution of Agricultural Research and Development Since 1950: Toward an Integrated Framework*. Sustainable Agriculture Programme Gatekeeper Series SA12. IIED, London.
Rhoades R. 1990. The Coming Revolution in Methods for Rural Development Research. Mimeo. User's Perspective Network International Potato Center, Manila, Philippines.
Rhoades R and Bebbington A. 1988. *Farmers Who Experiment: an Untapped Resource for Agricultural Development*. International Potato Center (CIP), Lima, Peru.
Rhoades R and Booth R. 1982. Farmer-back-to-farmer: a model for generating acceptable agricultural technology. *Agricultural Administration*, 11, 127–37.

Rhône-Poulenc. 1992–4. *Boarded Barns Farm 2nd – 4th Annual Reports*. Rhône Poulenc Agriculture Ltd, Ongar.
Ribaudo M O. 1989. *Water Quality Benefits of the Conservation Reserve Program*. Agricultural Economic Report No 606. Economic Research Service, US Dept. of Agriculture, Washington.
Richards P. 1989. Agriculture as performance. In: Chambers R, Pacey A and Thrupp L A (eds) *Farmer First. Farmer Innovation and Agricultural Research.* IT Publications, London.
Richards P. 1992. Rural development and local knowledge: the case of rice in central Sierra Leone. Discussion paper presented at the *IIED/IDS Beyond Farmer First: Rural People's Knowledge, Agricultural Research and Extension Practice Workshop*, 27–29 October, Institute of Development Studies, University of Sussex, UK. IIED, London.
Risch S J. 1987. Agricultural ecology and insect outbreaks. In: *Insect Outbreaks* Academic Press, New York.
Risch S J, Andow D and Altieri M. 1983. Agroecosystem diversity and pest control: data, tentative conclusions and new research directions. *Environ. Entomol.* 12: 625–9.
Roberts N (ed)s 1989. *Agricultural Extension in Africa*. World Bank, Washington DC.
Robinson D A and Blackman J D. 1990. Soil erosion and flooding. *Land Use Policy* 7, 41–52.
Roche C. 1991. ACORD's experience in local planning in Mali and Burkina Faso. *RRA Notes* 11: 33–41. IIED, London.
Roche C. 1992. *Operationality in Turbulence. The Need for Change*. ACORD, London
Rodale R. 1983. Breaking new ground: the search for a sustainable agriculture. *The Futurist* 1: 15–20.
Rodale R. 1990. Sustainability: an opportunity for leadership. In: Edwards C A, Lal R, Madden P, Miller R H and House G (eds) *Sustainable Agricultural Systems*. Soil and Water Conservation Society, Ankeny, Iowa.
Rogers E M. 1962. *Diffusion of Innovations*. Free Press, New York.
Rogers A. 1985. *Teaching Adults*. Open University Press, Milton Keynes.
Rohn A R. 1963. Prehistoric soil and water conservation on Chapin Mesa, southwestern Colorado. *American Antiquity* 28, 441–55.
Rola A. 1989. *Pesticides, Health Risks and Farm Productivity: A Philippine Experience*. Agric. Policy Research Program Monograph No 89–01. University of the Philippines at Los Baños.
Rola A. and Pingali P. 1993. Pesticides, rice productivity and helath impacts in the Philippines. In Faeth P (ed) *Agricultural Policy and Sustainability*. World Resources Institute, Washington DC.
Röling N. 1988. *Extension Science: Information Systems in Agricultural Development.* Cambridge University Press, Cambridge.
Röling N. 1992. Facilitating sustainable agriculture: turning policy models upside down. Paper presented at *IIED/IDS Beyond Farmer First: Rural People's Knowledge, Agricutlural Research and Extension Practice Conference*. Brighton, October 1992.
Röling N. 1994. Platforms for decision making about ecosystems. In: Fresco L (ed) *The Future of the Land*. John Wiley and Sons, Chichester.
Röling N and Engel P. 1989. IKS and knowledge management: utilizing indigenous knowledge in institutional knowledge systems. In: Warren D et al (eds) *Indigenous Knowledge Systems: Implications for Agriculture and International Development*. Studies in Technology and Social Change 11. Technology and Social Change Program, Iowa State University, Ames.
Röling N and Jiggins J. 1993. Policy paradigm for sustainable farming. *European J. of Agric. Education and Extension* (forthcoming).
Röling N and van der Fliert E. 1994. Transforming the extension for sustainable agriculture: the case of integrated pest management in rice in Indonesia. *Agric. and Human Values* (forthcoming).
Rorty R. 1989. *Contingency, Irony and Solidarity*. Cambridge University Press, Cambridge.
Rose J Sir. 1993. The farmer and the market: a reassessment. Paper presented to the *Centre for Agricultural Strategy/Small Farmers Association Symposium* 30–31 March, University of Reading.
Rossert P and Benjamin M (eds). 1993. *Two Steps Backward, One Step Forward. Cuba's Nationwide Experiment with Organic Agriculture.* Global Exchange, San Francisco.
Rossiter M W. 1975. *The Emergence of Agricultural Science*. Yale University Press, New Haven and London.

RRA Notes. 1988–95. Issues 1–present. Sustainable Agriculture Programme, IIED, London.
RRA Notes. 1992. Special Issue on Wealth Ranking. RRA Notes No 15. IIED, London.
Ruddell E D. 1993. *Engaging Peasants to Conduct Site-Specific Scientific Field Trials*. World Neighbors Santiago, Chile.
Russell D B. Ison R L, Gamble D R and Williams R K. 1989. *A Critical Review of Rural Extension Theory and Practice*. Faculty of Agriculture and Rural Development, University of Sydney, Australia.
Russell D B. and Ison R L. 1991. The research-development relationship in rangelands: an opportunity for contextual science. Plenary paper for *4th International Rangelands Congress*. Montpellier, France, 22–26 April 1991.
Russell E J. 1946. *British Agricultural Research: Rothamsted*. The British Council, Longman, Green & Co,London.
Russell E J. 1966. *A History of Agricultural Science in Great Britain, 1620–1954*. George Allen and Unwin, London.
SAA/Global 2000/CASIN. 1991. *Africa's Agricultural Development in the 1990s: Can it be sustained?* Sasakawa Africa Association, Tokyo.
SAA/Global 2000/CASIN. 1993. *Developing African Agriculture: New Initiatives for Institutional Cooperation*. Sasakawa Africa Association, Tokyo.
Salazar R. 1992. Community plant genetic resources management: experiences in southeast Asia. In: Cooper D, Vellve R and Hobbelink H (eds) *Growing Diversity: Genetic Resources and Local Food Security*. IT Publications, London.
Sampson H C. 1930. Soil erosion in Tropical Africa. *Rhodesian Agric. Journ.* 33, 197–205.
San Valentin G O. 1991. Utilization of Azolla in rice-based farming systems in the Philippines. In: *Asian Experiences in Integrated Plant Nutrition: Report of the Expert Consultation of the Asia Network of Bio and Organic Fertilizers*. Regional Office for Asia and the Pacific (RAPA). FAO, Bangkok.
Sandford S. 1983. *Management of Pastoral Development in the Third World*. John Wiley and Sons, London.
Sanghi N K. 1987. Participation of farmers as co-research workers: some case studies in dryland agriculture. Paper presented to *IDS Workshop Farmers and Agricultural Research: Complementary Methods*. IDS, Sussex.
Sardamon K. 1991. *Filling the Rice Bowl: Women in Paddy Cultivation*. Sangam Books, Hyderabad.
Sawit M M and Manwan I. 1991. The beginnings of the new Supra Insus rice intensification program:the case of the north east of West Java and south Sulawesi. *Bulletin of Indonesian Econ. Studies* 27(1), 81–103.
Saxena R C. 1987. Antifeedants in tropical pest management. *Insect Sci. Applic.* 8, 731–6.
Schiff M and Valdes A. 1992. *The Plundering of Agriculture in Developing Countries*. World Bank, Washington DC.
Schrimpf B and Dziekan I. 1989. Working with farmers on natural crop protection. *ILEIA Newsletter*. 1989(3), 23–24.
Schwabe C W and Kuojok J M. 1981. Practices and beliefs of the traditional Dinka herder. *Human Organisation* 40(3), 231–8.
Scoones I and Toulmin C. 1993. Socio-economic dimensions of nutrient cycling in agropastoral systems in dryland Africa. Paper for *ILCA Nutrient Cycling Conference*. August 1993. ILCA, Addis Ababa.
Scoones I and Thompson J. 1994. *Beyond Farmer First*. IT Publications, London.
Scoones I, Melnyk M and Pretty J N. 1992. *The Hidden Harvest: Wild Foods and Agricultural Systems. An Annotated Bibliography*. IIED, London with WWF, Geneva and SIDA, Stockholm.
Scoones I. 1991. Weltands in drylands: key resources for agricultural and pastoral production in Africa. *Ambio* XX (8), 366–71.
SDC 1992. *Mirror, Mirror. A Guide to Self-Evaluation*. Swiss Development Cooperation. Berne.
Sen C K. 1993. Nepal: group extension. *Rural Extension Bulletin* 3, 17–23. University of Reading, Reading.
Shah P. 1994a. Village-managed extension systems in India: implications for policy and practice. In: Scoones I and Thompson J (eds) *Beyond Farmer First*. IT, Publications London.
Shah P. 1994b. Participatory Watershed Management in India: the experience of the Aga Khan Rural Support Programme. In: Scoones I and Thompson J (eds) *Beyond Farmer First*. IT Publications London.

Shah P, Bharadwaj G and Ambastha R. 1991. Participatory impact monitoring of a soil and water conservation programme by farmers, extension volunteers, and AKRSP. In: Mascarenhas J et al. (eds) Participatory Rural Appraisal. *RRA Notes* 13: 127–31. IIED, London.

Shah T. 1990. *Sustainable Development of Groundwater Resources: Lessons from Amrapur and Husseinabad Villages, India*. ODI Irrigation Management Network Paper 90/3d, ODI, London.

Shapiro M H. 1988. Judicial selection and the design of clumsy institutions. *Southern California Land Review* 61(6), 1555–69.

Shaxson T F and Schlolo D M. 1993. *Production Through Conservation: the PTC II Programme*. Report of the Informal Evaluation Mission. Swedforest/SIDA/Govt of Lesotho, Maseru.

Shaxson T F, Hudson N W, Sanders D W, Roose E and Moldenhauer W C. 1989. *Land Husbandry. A Framework for Soil and Water Conservation*. Soil and Water Conservation Society, Ankeny, Iowa.

Sheldrake R. 1988. *The Presence of the Past: Morphic Resonance and the Habits of Nature*. Collins, London.

Sherief A K. 1991. Kerala, India: group farming. *AERDD Bulletin* (University of Reading) 32, 14–17.

Shirley C. 1991. Almond brother still gambling. *The New Farm* July–Aug, 32–35.

Showers K B and Malahleha G. 1990. Pilot study for the development of methodology to be used an historical environmental impact assessment of colonial conservation schemes. Paper presented at *Workshop on Conservation in Africa: Indigenous Knowledge and Conservation Strategies*. Harare, Zimbabwe.

Showers K B. 1989. Soil erosion in the Kingdom of Lesotho: origins and colonial response. 1830s–1950s. *Journ. Southern African Studies* 15, 263–86.

SIDA. 1984. *Soil Conservation in Borkana Catchment. Evaluation Report*. Final Report, Swedish International Development Authority, Stockholm.

Siedenberg K. 1992. Philippine pesticide bans show NGO strength. *Global Pesticide Campaigner* 2(3) 1, 6–7.

Sikana P. 1993. *Indigenous Soil Characterisation and Farmer Participation in Northern Zambia: Implications for Research and Extension Delivery*. Rural People's Knowledge, Agricultural Research and Extension Practice, Research Series Vol 1, No2. IIED, London.

Singh S. 1990. People's participation in forest management and the role of NGOs and voluntary agencies. National Wastelands Development Board, Ministry of Environment and Forests, New Delhi.

Singh K. 1992. *Agricultural Policy in India: Need for a Fresh Look*. Working Paper 42. Institute of Rural Management Anand, India.

Singh A J and Miglani S S. 1976. An economic analysis of energy requirements in Punjab agriculture. *Indian J. of Agric. Econ.* July–Sept.

Singh L R and Singh B. 1976. Level and pattern of energy consumption in an agriculturally advanced area of Uttar Pradesh. *Indian J. of Agric. Econ.* July–Sept 197: 160–6.

Siripatra D C and Lianchamroon W. 1992. An integrated NGO approach in Thailand. In: Cooper D, Vellve R and Hobbelink H (eds) *Growing Diversity: Genetic Resources and Local Food Security*. IT Publications, London.

Small Farm Viability Project. 1977. *The Family Farm in California: Report on the Small Farm Viability Project*. Employment Development, Governor's Office of Planning and Research. Dept of Food and Agriculture, Sacremento, California.

Smil V, Nachman P and Long T V. 1982. *Energy Analysis and Agriculture. An Application to US Corn Production*. Westview Press, Boulder, Colorado.

Smith J K 1990. Alternative research paradigms and the problem of criteria. In: Guba E (ed) *The Paradigm Dialog*. Sage Publications, Newbury Park.

Smith R F and van den Bosch R. 1967. Integrated control. In: Kilgore W W and Doutt R C. (eds) *Pest Control – Biological, Physical, and Selected Chemical Methods*. Academic Press, New York.

Smolik J D and Dobbs T L. 1991. Crop yields and economic returns accompanying the transition to alternative farming systems. *J. Prod. Agric.* 4, 153–61.

Soetomo D. 1992. Growing community seed banks in Indonesia. In: Cooper D, Vellve R and Hobbelink H (eds) *Growing Diversity: Genetic Resources and Local Food Security*. IT Publications, London.

Soetrisno L. 1982. Further agricultural intensification in Indonesia: who gains and who loses? Paper prepared for working group meeting on agricultural intensification in Indonesia, Puncak. 25–7 June.

Sommer M. 1993. *Whose Values Matter? Experiences and Lessons from the Self-Evaluation in PIDOW Project*. SDC, Bangalore.

Sowbaghya et al. 1983. Chlorinated insecticide residues in certain food samples, *Indian J. Med. Res.* 78, 403–6.

Sperling L, Loevinsohn M E and Ntabomvura B. 1993. Rethinking the farmers' role in plant breeding: local bean experts and on-station selection in Rwanda. CIAT, Butare, Rwanda.

SPWD. 1992. *Joint Forest Management: Concept and Opportunities*. Society for Promotion of Wastelands Development, New Delhi.

Sriskandarajah N, Bawden R J and Packham R G. 1991. Systems agriculture: a paradigm for sustainability. *Association for Farming Systems Research-Extension Newsletter* 2(2), 1–5.

Stakman E C, Bradfield R and Mengelsdorf P. 1967. *Campaigns Against Hunger*. Belknap Press, Cambridge, Mass.

Stanhill G. 1979. A comparative study of the Egyptian agroecosystem. *Agro-Ecosystems* 5, 213–30.

Steier F (ed). 1991. *Research and Reflexivity*. Sage Publications, London.

Steinmann R. 1983. *Der Biologische Landbau-ein Betriebswirtschaftlicher Vergleich*. Schriftenriehe der Eidg. Forschungsanstalt für Betrriebwirtschaft und Landtechnik FAT, Tänikon.

Stock T. 1994. *The impact of integrated pest management farmer field schools in the Cordillera Region of the Philippines*. MSc thesis, University of Reading.

Stocking M. 1985. Soil conservation policy in colonial Africa *Agric History* 59 148–61.

Stocking M. 1993. Soil erosion in developing countries:where geomorphology fears to tread. *Discussion Paper* No 241. School of Development Studies, Univ. of East Anglia, Norwich.

Stoll G. 1987. *Natural Crop Protection*. Agroecol Josef Margraf, Langen. Germany, 2nd edn.

Stone R. 1992. Researchers score victory over pesticides and-pests-in Asia. *Science* 256, 1272

Stout B A. 1979. *Energy for World Agriculture*. FAO, Rome.

Svendsen M. 1993. The impact of financial autonomy on irrigation system performance in the Philippines. *World Development* 21(6), 989–1006.

Swanson B E,, Farmer B J and Bahal R 1990. *The Current Status of Agricultural Extension Worldwide*. Report of the Global Consultation on Agricultural Extension. FAO, Rome.

SWCB. 1994. *The Impact of the Catchment Approach to Soil and Water Conservation: A Study of Six Catchments in Western, Rift Valley and Central Provinces, Kenya*. Ministry of Agriculture, Livestock Development and Marketing, Nairobi.

TAC. 1988. *Sustainable Agricultural Production: Implications for International Agricultural Research*. TAC Secretariat, FAO, Rome.

Tacio H D. 1991. Soil erosion control: the experience of the Mindinao Baptist Rural Life Center. *Contour* III(1), 13–15.

Tacio H D. 1992. Contour farming and livestock raising: a likely combination. *Contour* IV(1) 12–15.

Tamang D. 1993. *Indigenous Soil Fertility Management in the Hills of Nepa*l. Sustainable Agriculture Programme Gatekeeper Series SA41. IIED, London.

Tato K and Hurni H. 1992. *Soil Conservation for Survival*. Soil and Water Conservation Society, Ankeny, Iowa.

Thai V C and Loan L D. 1991. Research on recovery of the fertility of degraded red-brown ferrasols on basalt by green manure plants. In: *Asian Experiences in Integrated Plant Nutrition: Report of the Expert Consultation of the Asia Network of Bio and Organic Fertilizers*. Regional Office for Asia and the Pacific (RAPA), FAO, Bangkok.

Thatcher L E 1925. The soybean in Ohio. *Ohio Agric. Exp. Station Bull.* 384.

Thirsk J. 1985. Agricultural innovations and their diffusion. In Thirsk J (ed) *The Agrarian History of England and Wales. Volume V. 1640–1750. II. Agrarian Change*. Cambridge University Press, Cambridge.

Thomen A. 1992. Dominican Republic bans dirty dozen – industry fights back. *Global Pesticide* Campaigner 2(1), 3.

Thompson M, Warburton M and Hatley T. 1986. *Uncertainty on a Himalyan Scale*. Ethnographica, London.

Thompson M and Trisoglio A. 1993. Managing the Unmanageable. Paper presented at 2nd *Environmental Management of Enclosed Coastal Seas Conference*, Baltimore Maryland. 10–13 November 1993.

Thrupp L A. 1990. Entrapment and escape from fruitless insecticide use: lessons from the banana sector of Costa Rica. *Intern. J. Environmental Studies* 36, 173–89.

Tiffen M, Mortimore M and Gichuki F. 1993. *More People, Less Erosion. Environmental Recovery in Kenya*. John Wiley and Sons, Chichester.

Tjernström R. 1992. Yields from terraced and non-terraced fields in the Machakos District of Kenya. In: *Soil Conservation for Survival* Tato K and Hurni H (eds). Soil Conservation Society, Ankeny, Iowa, pp 251–65.

TNAU/IIED. 1993. *Participatory Rural Appraisal (PRA) for Agricultural Research at Aruppukottai and Paiyur, Tamil Nadu*. Tamil Nadu Agricultural University, Coimbatore and IIED, London.

Toulmin C, Scoones I and Bishop J. 1992. The Future of Africa's Drylands: Is Local Resource Management the Answer? In: Holmberg J (ed) *Policies for a Small Planet* Earthscan Publications Ltd, London.

Treacy J M. 1989. Agricultural terraces in Peru's Colca Valley: promises and problems of an ancient technology. In: Browder (ed) *Fragile Lands of Latin America : Strategies for Sustainable Development*. Westview Press, Boulder, Colorado.

Trenbath B R 1974. Biomass productivity of mixtures. *Adv. Agron*. 26: 177–250.

Trenbath B R. 1976. Plant interactions in mixed crop communities. In: Papendick R I, Sanchez P A. and Triplett G B. (eds) *Multiple Cropping*. American Society of Agronomy, Madison, WI.

Trimble S W. 1985. Perspectives on the history of soil erosion control in the eastern United States. *Agric. History* 59, 162–80.

Tripp R. 1991. The limitations of on-farm research. In: *Planned Change in Farming Systems: Progress in On-Farm Research*. John Wiley and Sons, Chichester.

Tsoukas H. 1992. Panoptic reason and the search for totality: a critical assessment of the critical systems perspective. *Human Relations* 45(7), 637–57.

UNDP. 1992. *The Benefits of Diversity. An Incentive Toward Sustainable Agriculture*. United Nations Development Program, New York.

UNEP. 1983. Rainwater Harvesting for Agriculture. UNEP, Nairobi.

UoN/SIDA. 1989. *Soil and Water Conservation in Kenya*. Proceedings of the 3rd National Workshop, Kabete, Nairobi, University of Nairobi and SIDA, Nairobi.

UoS/GC. 1992. *Helping Nature to Control Pests*. University of Southampton and Game Conservancy. Rhône-Poulenc, Ongar.

Upaswansa G K. 1989. Ancient methods for modern dilemmas. *ILEIA Newsletter* 3, 9–11.

Uphoff N. 1990. Paraprojects: a new role of international development assistance. *World Development* 18, 1401–11.

Uphoff N. 1992a. *Learning from Gal Oya: Possibilities for Participatory Development and Post-Newtonian Science*. Cornell University Press, Ithaca.

Uphoff N. 1992b. Approaches and methods for monitoring and evaluation of popular participation in World Bank-assisted projects. Paper for World Bank Workshop on Popular Participation, Washington DC 26–27 February, 1992.

Uphoff N. 1992c. *Local Institutions and Participation for Sustainable Development*. Gatekeeper Series SA31. IIED, London.

Uphoff N. 1994. Local organisations for supporting people-based agricultural research and extension: lessons from Gal Oya, Sri Lanka. In: Scoones I and Thompson J (eds) *Beyond Farmer First*. IT Publications, London.

UPWARD. 1990. Proceedings of the Inaugural Planning Workshop on the User's Perspective With Agricultural Research and Development. Los Banos, Philippines.

USACE. 1990. *Public Involvement, Conflict Management, and Dispute Resolution in Water Resources and Environment Decision Making*. Delli Priscoli J., (ed). US Army Corps of Engineers Working Paper 42, Fort Belvoir, Virginia.

USAID. 1987. *Women in Development: A I D's Experience, 1973–1985*. AID Program Evaluation Report No 18. Agency for International Development, Washington DC.

Utting P. 1993. *Trees, People and Power. Social Dimensions of Deforestation and Forest Protection in Central America*. Earthscan Publications Ltd, London.

van der Fliert E. 1993. *Integrated Pest Management: Famer Field Schools Generate Sustainable Practices*. Wageningen Agricultural University Paper 93–3. WAU, The Netherlands.

van der Werf E and de Jager A.1992. *Ecological Agriculture in South India: An Agro-Economic Comparison and Study of Transition*. Landbouw-Economisch Institut, The Hague and ETC-Foundation, Leusden.

Vandermeer J. 1989. *The Ecology of Intercropping*. Cambridge University Press, Cambridge.
Vellve R. 1992. *Saving the Seed: Genetic Diversity and European Agriculture*. Earthscan Publications Ltd, London.
Vereijken P. 1990. Research on integrated arable farming and organic mixed farming in the Netherlands. In: Edwards C A, Lal R, Madden P, Miller R H and House G (eds) *Sustainable Agricultural Systems*. Soil and Water Conservation Society, Ankeny, Iowa.
Vermillion D L. 1989. Second approximations: unplanned farmer contributions to irrigation design. *Irrigation Management Network* 89/2c. ODI, London.
Vickers G. 1981. Some implications of systems thinking. In: *Systems Behaviour* Open Systems Group, ed Harper and Row, London with the Open University Press. 3rd edn.
Vine A and Bateman D I. 1983. *Organic Farming Systems in England and Wales: Practice, Performance and Implications*. Department of Agricultural Economics, University College Wales, Aberystwyth.
Vipond J E, Swift G, McClelland T H, Fitzsimons J, Milne J A and Hunter E A. 1992. A comparison of diploid and tetraploid ryegrass and tetraploid ryegrass/white clover swards under continuous sheep stocking at controlled sward heights. 2. Animal production. The Scottish Agricultural College, Edinburgh.
Vlek PLG. 1990. The role of fertilizers in sustaining agriculture in sub-Saharan Africa. International Fertilizer Development Center, Lomé, Togo.
Volke Haller and Sepulveda-González. 1987. *Agricultura de Subsistencia y Desarrollo Rural*. Editorail Trillas, Mexico, D F.
Von Hildebrand A. 1993. Integrated pest management in rice: The case of the paddy fields in the region of Lake Alaotra. Paper presented at *East/Central/Southern Africa Integrated Pest Management Implementation Workshop*. Harare, Zimbabwe. 19–24 April 1993.
Vorley W. 1993. Sustainable agriculture and pesticide use. Leopold Center, Iowa State University. *Mimto.*
VROM (Ministry of Housing, Physical Planning and Environment). 1989. *To Choose or To Lose: National Environmental Policy Plan*. VROM, The Hague, Netherlands.
VROM. (Ministry of Housing, Physical Planning and Environment). 1990. *National Environmental Policy Plan Plus*. VROM, The Hague, Netherlands.
Waage J K and Greathead D J. 1988. Biological control: challenges and opportunities. *Phil. Trans. R. Soc. Lond.* B 318, 111–28.
Waldrop M M. 1992. *Complexity and the Emerging Science at the Edge of Order and Chaos*. Simon and Shuster, New York.
Wale S. 1993. Reducing fungicide use on spring barley with confidence. In: *HGCA Proceedings Of The Cereals R&D Conference*. Robinson College, Cambridge. 5–6 January 1993.
Walters R F. 1971. *Shifting Cultivation in Latin America*. FAO, Rome.
Wanjohi B. 1987. Women's groups, gathered plants and their agroforestry potentials in the Kathama Area. In: Wachiira K K *Women's Use of Off-Farm and Boundary Lands: Agroforestry Potentials*, pp 61–104. Final Report, ICRAF, Nairobi, Kenya.
Ward N. 1993. Environmental concern and the decline of the dynastic family farm. Paper presented to the *Centre for Agricultural Strategy/Small Farmers Association Symposium*. 30–31 March. University of Reading, Reading.
Wardman A and Salas L G. 1991. The implementation of anti-erosion techniques in the Sahel: A case study for Kenya, Burkina Faso. *J. Developing Areas* 26, 65–80.
Warren D. 1991. The role of indigenous knowledge in facilitating a participatory approach to agricultural extension. Paper presented at the *International Workshop on Agricultural Knowledge Systems and the Role of Extension*, Bad Boll, Germany. 21–24 May.
Watanabe I, Espinas C R, Berja N S and Alimango B V. 1977. *Utilisation of the Azolla-Anabaena Complex as a Nitrogen Fertilizer for Rice*. IRRI Research Paper Series 11. IRRI, Los Banos.
WCED. 1987. *Our Common Future*. Oxford University Press, Oxford and New York.
Wenner C G. 1992. *The Revival of Soil Conservation in Kenya*. Carl Gosta Wenner's personal notes, 1974–81. Edited by A. Eriksson. RSCU/SIDA, Nairobi.
West P C and Brechin S R. 1992. *Resident People and National Parks*. University of Arizona Press, Tucson.
Whitehead A. 1985. Effects of technological change on rural women: a review of analysis and concepts. In: Ahmed I (ed) *Technology and Rural Women: Conceptual and Empirical Issues*. George Allen and Unwin, London.
WHO. 1990. *Public Health Impact of Pesticides Used in Agriculture*. WHO, Geneva.

WHO. 1992. *Our Planet, Our Health*. WHO, Geneva.

Wibberley J. 1991. Farmer-dominant study groups. *AERDD Bulletin (University of Reading)* 32, 8–13.

Wibisana R. 1987. Letter from a Javanese farmer to *Kompas* newspaper. Jakarta, Indonesia.

Wilken G C. 1987. *Good Farmers. Traditional Agricultural Resource Management in Mexico and Central America*. University of California Press, Berkeley.

Willey R W. 1979. Intercropping – its importance and its research needs, Part II: agronomic relationships. *Field Crop Abstracts* 32: 73–85.

Williams M C and Antholt C H. 1992. *Managing for Innovations: Lessons for Agricultural Research*. Asia Technical Dept, World Bank, Washington DC.

Wilson K B. 1989. Indigenous conservation in Zimbabwe: soil erosion, land-use planning and rural life. Paper presented to *Conservation and Rural People*. African Studies Association of UK Conference, Cambridge. September 1988.

Winarto Y. 1992. Farmers' agroecological knowledge and integrated pest management in North West Java. Case study presented at the *IIED/IDS Beyond Farmer First: Rural People's Knowledge, Agricultural Research and Extension Practice Workshop*, 27–29 October, Institute of Development Studies, University of Sussex, UK. IIED, London.

Winarto Y. 1993. Farmers' Agroecological Knowledge Construction: The Case of Integrated Pest Management Among Rice Farmers on the North Coast of West Java. *Rural people's Knowledge, Agricultural Research and Extension Practice*. Research Series, Vol 1, No 3. IIED, London.

Winarto Y. 1994. State intervention and farmer creativity: integrated pest management among rice farmers in Subang, West Java. Paper presented at 1994 Asian Studies Association of Australia Biennial Conference, Murdoch University, Perth. 13–16 July.

Witt J M. 1980. A discussion of the suspension of 2,4,5-T and the EPA Alsea II study, Special Report, unpublished. Oregon State University, Corvallis.

Wolfe M S and Barratt J A. 1986. Responses of plant pathogens to fungicides. In: National Research Council. *Pesticide Resistance: Strategies and Tactics for Management*. National Academy Press, Washington DC.

Wolfe M S. 1981. The use of spring barley cultivar mixtures as a technique for the control of powdery mildew. *Proc 1981 Brit. Crop Protection Conf* 1, 233–39.

Woodhill J. 1992. *Landcare – Who Cares? Current Issues and Future Directions for Landcare in NSW*. Centre for Rural Development, University of Western Sydney-Hawkesbury, Australia.

Woodhill J. 1993. Science and the facilitation of social learning: a systems perspective. Paper for *37th Annual Meeting of The International Society for the Systems Sciences*. University of Western Sydney. July 1993.

World Bank. 1994a. *World Development Report*. Oxford University Press, Oxford.

World Bank. 1993. *Agricultural Sector Review*. Agriculture and Natural Resources Department, Washington DC.

World Bank. 1994b. *The World Bank and Participation*. Report of the Learning Group on Participatory Development. April 1994. World Bank, Washington, DC.

World Bank. 1994c. *Agricultural Extension: Lessons from Completed Projects Operations*. Evaluation Dept, World Bank, Washington DC.

World Conservation Monitoring Centre. 1992. *Global Biodiversity: Status of the Earth's Living Resources*. Chapman and Hall, London.

Worster D. 1979. *Dust Bowl. The Southern Plains in the 1930s*. Oxford University Press, Oxford.

WPPR. 1994. *Report of Working Party on Pesticides Residues*. MAFF, London.

Wratten S. 1992. Farmers weed out the cereal killers. *New Scientist* 22 August, 31–5.

WRI. 1994. *World Resources 1994–95*. World Resources Institute, Washington. Oxford University Press, Oxford.

Wynne B. 1992. Uncertainty and environmental learning. Reconceiving science and policy in the preventive paradigm. *Global Environmental Change* June: 111–27.

Wynne B and Mayer S. 1993. How science fails the environment *New Scientist* 5 June: 33–5.

Yixian G. 1991. Improving China's rice cropping systems. *Shell Agriculture* 10, 28–30.

Yoder R. 1991. Peer training as a way to motivate institutional change in farmer-managed irrigation systems. Paper presented for Workshops on Democracy and Governance. USAID, Washington DC, IIMI, Nepal.

Young A. 1767. *The Farmer's Letters to the People of England*. London. 1st edn.

Young A. 1770. *Experimental Agriculture*. 2 vols. London.
Young, A. 1989. *Agroforestry for Soil Conservation*. CAB International, Wallingford.
Young E C. 1974. The epizootiology of two pathogens of the coconut palm rhinoceros beetle. *J. Invert. Pathol.* 24 82–92.
Young R. 1983. Canadian Development Assistance to Tanzania. *The North South Institute*. Ottawa.
Younie D. 1992. Potential output from forage legumes in organic systems. Paper presented at the CEC Workshop on Organic Farming: *Potential and Limits of Organic Farming*. Louvain-la-Neuve, Belgium. September 1992.
Yudelman M. 1993. Demand and supply of foodstuffs up to 2050 with special reference to irrigation. WWF, Washington DC.
Yuryevic A. 1994. Community-based sustainable development in Latin America: the experience of CLADES. Paper presented at *IIED In Local Hands: Community-Based Sustainable Development Symposium*. 4–8 July, Brighton.
Zadek S and Mayo E. 1994. The rise and fall of the development non-governmentals. A matter of choice? Paper presented at New Economics Foundation, Development Alternatives seminar on *Development and Power*. London. 24 March 1994.
Zehnder G W and Warthen J D. 1988. Neem and the Colorado Potato Beetle. *International Ag-Sieve* 1(3), 4.
Zhaohua Z. 1988. A new farming system – crop/Paulownia intercropping. In: *Multipurpose Tree Species for Small Farm Development*. IDRC/Winrock.
Zhaogiamg L and Ning W . 1992. A local resource-centred approach to rural transformation: agro-based cottage industries in Western Sichuan, China. In: Jodha N S, Bonskotu M and Partap T (eds) *Sustainable Mountain Development. Volume 1*. Oxford and IBH, New Delhi.
Zhu C S and Luo S M. 1992. Red deserts turn to green oceans. *ILEIA Newsletter* 8(4), 25–26.
Zijp W. 1993. Being a good communicator doesn't solve all of extension's problems. *Development Communication Report* No 80, 20.

INDEX